MW00388575

© 2019 by Dale Hoffman, Bellevue College.
Additional content by Jeff Eldridge, Edmonds Community College.

Author web page: http://scidiv.bellevuecollege.edu/dh/
Report typographical errors to: jeldridg@edcc.edu

A free, color PDF version is available online at: http://contemporarycalculus.com

This text is licensed under a Creative Commons Attribution — Share Alike 3.0 United States License. You are free:

- to **Share** — to copy, distribute, display and perform the work
- to **Remix** — to make derivative works

under the following conditions:

- **Attribution:** You must attribute the work in the manner specified by the author (but not in any way that suggests that they endorse you or your work)
- **Share Alike:** If you alter, transform or build upon this work, you must distribute the resulting work only under the same, similar or compatible license.

First beta edition of this version printed: April 1, 2019; page numbers not finalized

ISBN-13: 978-1092279581
ISBN-10: 1092279581

DALE HOFFMAN

CONTEMPORARY CALCULUS

Contents

9
Sequences and Series

This chapter introduces two special topics in calculus: sequences and series. It provides a bridge between the concepts from Section 8.7 (where we used "easy" polynomials to approximate values of "complicated" functions like $\sin(x)$ and e^x) and Chapter 10 (where we will let the degrees of the approximating polynomials become infinite).

From our work in Section 8.7, we know, for example, that:

$$e^x = \exp(x) \approx 1 + x + \frac{1}{2}x^2 + \frac{1}{2 \cdot 3}x^3 + \frac{1}{2 \cdot 3 \cdot 4}x^4$$

so that:

$$e = e^1 \approx 1 + 1 + \frac{1}{2} + \frac{1}{6} + \frac{1}{24} = \frac{65}{24} \approx 2.7083$$

This is a "pretty good" approximation of e, but you can do better by using a fifth-degree polynomial approximation:

$$e = e^1 \approx 1 + 1 + \frac{1}{2} + \frac{1}{6} + \frac{1}{24} + \frac{1}{120} = \frac{163}{60} \approx 2.7167$$

You can achieve better approximations by adding more terms to your polynomial approximation. But what if you never stopped adding? Does it make sense to add up an infinite number of numbers? And does this infinite sum add up to the finite value you want, namely e?

In this chapter we will examine what it means to add up an infinite number of numbers and learn how to determine if this sum is finite or infinite. Before examining sums, however, we need to lay a foundation: Section 9.1 investigates lists of numbers, called **sequences** and examines some specific sequences we will need later; Section 9.2 introduces the concept of the convergence of a sequence. The rest of the chapter focuses on what it means to add up an infinite number of numbers—an infinite **series**—and how we can determine whether the resulting sum is a finite number. Chapter 10 will generalize the idea of an infinite series of numbers to infinite series that contain a variable. We will use these series that contain powers of a variable—called **power series**—to represent and approximate functions such as $\sin(x)$ and e^x, extending the concepts of Section 8.7 to the infinite realm.

9.1 *Sequences*

Sequences play important roles in several areas of theoretical and applied mathematics. As you continue your study of mathematics, you will encounter them again and again. In this course, however, their primary role is as a foundation for our study of power series ("big polynomials"). In order to understand how and where it is valid to represent a function such as sine as a power series, we need to examine what it means to add together an infinite number of values. And in order to understand this infinite addition we need to analyze lists of numbers (called **sequences**) and determine whether or not the numbers in the list are converging to a single value. This section examines sequences and how to represent sequences graphically.

Example 1. You deposit \$100 in an account that pays 8% interest at the end of each year. How much money will be in the account at the end of 1 year? 2 years? 3 years? n years?

Solution. After one year, the total amount in the account is the principal (the amount you originally deposited) plus the interest:

$$100 + (0.08)100 = 100(1 + 0.08) = 100(1.08) = \$108$$

or 108% of the amount at the beginning of the first year.

At the end of the second year, the amount is 108% of the amount at the start of the second year:

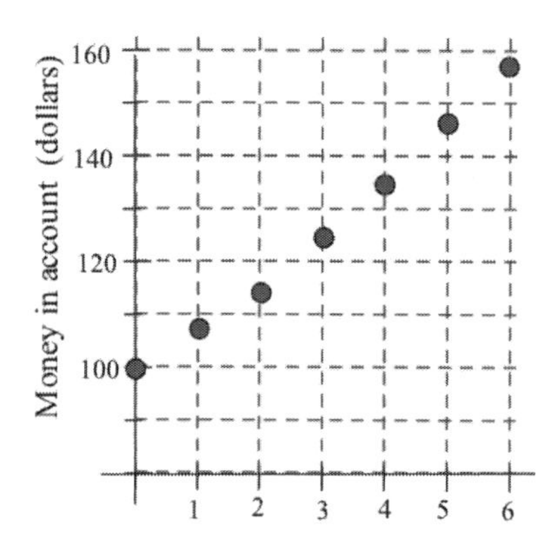

$$(1.08)\left[(1.08)100\right] = (1.08)^2(100) = \$116.64$$

At the end of the third year, the amount is 108% of the amount at the start of the third year:

$$(1.08)\left[(1.08)^2 100\right] = (1.08)^3(100) = \$125.97$$

See the margin for a graph of these results. In general, at the end of the n-th year, the amount in the account is $(1.08)^n(100)$ dollars. ◀

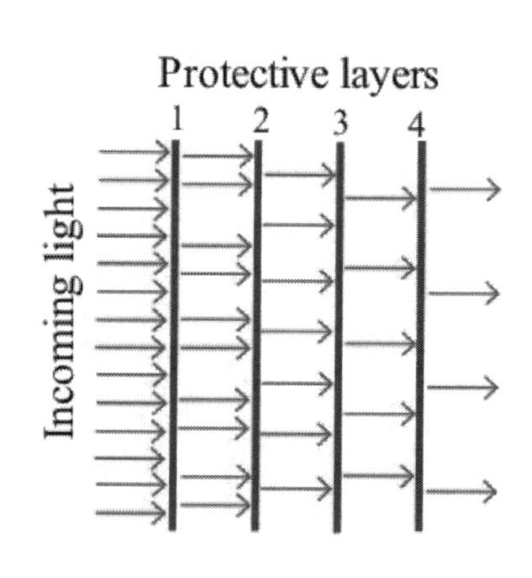

Practice 1. A layer of protective film transmits two-thirds of the light that reaches that layer (see margin). How much of the incoming light is transmitted through 1 layer? 2 layers? 3 layers? n layers?

The Example and Practice problems above each asked for a list of numbers in a definite order: a first number, then a second number, and so on. Such a list of numbers in a definite order is called a **sequence**. An **infinite sequence** is a sequence that keeps going and has no last number. Often the pattern of a sequence becomes clear from the first few numbers, but in order to precisely specify a sequence, we usually state a rule for finding the value of the n-th term, a_n ("a sub n").

Example 2. List the next two numbers in each sequence and give a rule for calculating the n-th number, a_n:

(a) $1, 4, 9, 16, \ldots$ (b) $-1, 1, -1, 1, \ldots$ (c) $\frac{1}{2}, \frac{1}{4}, \frac{1}{8}, \frac{1}{16}, \ldots$

Solution. (a) $a_5 = 25$, $a_6 = 36$ and $a_n = n^2$ (b) $a_5 = -1$, $a_6 = 1$ and $a_n = (-1)^n$ (c) $a_5 = \frac{1}{32}$, $a_6 = \frac{1}{64}$ and $a_n = \frac{1}{2^n} = \left(\frac{1}{2}\right)^n$. ◀

Practice 2. List the next two numbers in each sequence and give a rule for calculating the n-th number, a_n:

(a) $1, \frac{1}{2}, \frac{1}{3}, \frac{1}{4}, \ldots$ (b) $-\frac{1}{2}, \frac{1}{4}, -\frac{1}{8}, \frac{1}{16}, \ldots$ (c) $2, 2, 2, 2, \ldots$

Definition and Notation

Because a sequence gives a single value for each integer n, a sequence is a function whose domain is restricted to the integers.

> **Definition**
> A **sequence** is a function with a domain consisting of all integers greater than or equal to a starting integer.

Most sequences we will encounter have a starting integer of 1, but sometimes it is convenient to start with 0 or another integer value.

In many computer languages, arrays (multi-dimensional lists) employ 0-based indexing.

> **Notation**
> a_n represents a single number called the n-th term.
> $\{a_n\}$ represents the set of all terms in the sequence.
> $\{\text{rule}\}$ represents the sequence generated by the rule.
> $\{a_n\}_{n=3}$ represents a sequence that starts with $n = 3$.

Because sequences are functions, we can add, subtract, multiply and divide them, and combine them with other functions to form new sequences. We can also graph sequences: these graphs can sometimes help us describe and understand their behavior.

Example 3. For the sequences given by $a_n = 3 - \frac{1}{n}$ and $b_n = \frac{1}{2^n}$, graph the points (n, a_n) and (n, b_n) for $n = 1$ to 5. Compute the first five terms of $c_n = a_n + b_n$ and graph the points (n, c_n).

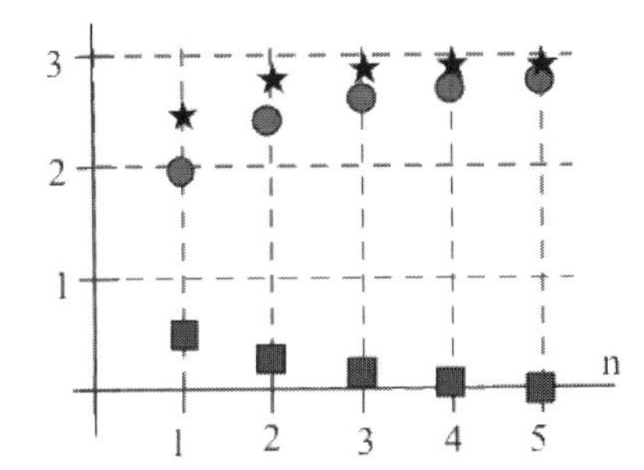

Solution. $c_1 = \left(3 - \frac{1}{1}\right) + \frac{1}{2^1} = 2.5$, $c_2 = 2.75$, $c_3 \approx 2.792$, $c_4 = 2.8125$ and $c_5 = 2.83125$. The graphs appear in the margin. ◀

Practice 3. For a_n and b_n in the previous example, compute the first five terms of $c_n = a_n - b_n$ and $d_n = (-1)^n b_n$, and graph the points (n, c_n) and (n, d_n).

Recursive Sequences

A **recursive sequence** is a sequence defined by a rule that gives the value of each new term in the sequence as a combination of one or more of the previous terms. We already encountered a recursive sequence when we studied Newton's Method for approximating roots of a function (Section 2.7).

Example 4. Let $f(x) = x^2 - 4$. Take $x_1 = 3$ and apply Newton's method to calculate x_2 and x_3. Give a rule for x_n.

Solution. $f(x) = x^2 - 4 \Rightarrow f'(x) = 2x$; applying Newton's method:

$$x_2 = x_1 - \frac{f(x_1)}{f'(x_1)} = 3 - \frac{f(3)}{f'(3)} = 3 - \frac{5}{6} = \frac{13}{6} \approx 2.1667$$

$$x_3 = x_2 - \frac{f(x_2)}{f'(x_2)} = \frac{13}{6} - \frac{f\left(\frac{13}{6}\right)}{f'\left(\frac{13}{6}\right)} = \frac{13}{6} - \frac{25}{156} = \frac{313}{156} \approx 2.0064$$

In general:

$$x_n = x_{n-1} - \frac{f(x_{n-1})}{f'(x_{n-1})} = x_{n-1} - \frac{(x_{n-1})^2 - 4}{2x_{n-1}}$$

The terms $x_1, x_2, x_3, \ldots$ appear to be approaching the value 2, one solution of $x^2 - 4 = 0$. The sequence $\{x_n\}$ is a recursive sequence because each x_n is defined as a function of the previous term x_{n-1}. ◀

Successive iterations of a function also generate a recursive sequence.

Practice 4. Let $f(x) = 2x - 1$ and define $a_n = f(f(f(\cdots f(a_0)\cdots)))$, where the function is applied n times. Put $a_0 = 3$ and compute a_1, a_2 and a_3. Note that a_n can be defined recursively as $a_n = f(a_{n-1})$.

Example 5. Let $a_k = \frac{1}{2^k}$ and define a second sequence $\{s_n\}$ by the rule that s_n is the sum of the first n terms of $\{a_k\}$. Compute the values of s_n for $n = 1$ to 5.

Solution. Computing the first three terms of $\{s_n\}$:

$$s_1 = a_1 = \frac{1}{2}$$

$$s_2 = a_1 + a_2 = \frac{1}{2} + \frac{1}{4} = \frac{3}{4}$$

$$s_3 = a_1 + a_2 + a_3 = \frac{1}{2} + \frac{1}{4} + \frac{1}{8} = \frac{7}{8}$$

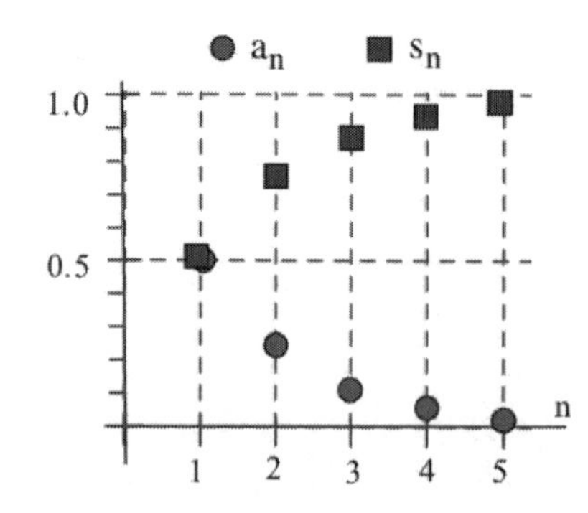

Similarly, $s_4 = \frac{15}{16}$ and $s_5 = \frac{31}{32}$. You should notice two patterns in these sums (and in the associated graph—see margin). First, it appears that:

$$s_n = \frac{2^n - 1}{2^n} = 1 - \frac{1}{2^n}$$

(so we have an example of a sequence defined recursively that can also be written explicitly as a function of n).

Second, you can simplify the addition process: each term s_n is the sum of the previous term s_{n-1} and the a_n term: $s_n = s_{n-1} + a_n$. (For example, check that $s_3 = s_2 + a_3$.) We will meet this second pattern again in Section 9.3. ◀

Practice 5. Let $b_0 = 0$ and, for $n > 0$, define $b_n = b_{n-1} + \frac{1}{3^n}$. Compute b_n for $n = 1$ to 4

9.1 Problems

Problems 1–6 provide the first four numbers a_1, a_2, a_3 and a_4 in a list. (a) Write the next two numbers in the list. (b) Write a formula for the fifth number, a_5, as an expression involving 5. (c) Write a formula for the n-th number, a_n, in terms of n.

1. $2, 4, 8, 16, \_\_, \_\_$
2. $3, 9, 27, 81, \_\_, \_\_$
3. $-1, 1, -1, 1, \_\_, \_\_$
4. $1, \frac{1}{2}, \frac{1}{3}, \frac{1}{4}, \_\_, \_\_$
5. $1, 2, 6, 24, \_\_, \_\_$
6. $1, 4, 9, 16, \_\_, \_\_$

For 7–11, evaluate each of the given expressions and write the next two numbers in the list.

7. $1, 1+\frac{1}{2}, 1+\frac{1}{2}+\frac{1}{3}, 1+\frac{1}{2}+\frac{1}{3}+\frac{1}{4}, \_\_, \_\_$
8. $1, 1+\frac{1}{2}, 1+\frac{1}{2}+\frac{1}{4}, 1+\frac{1}{2}+\frac{1}{4}+\frac{1}{6}, \_\_, \_\_$
9. $1, 1-\frac{1}{2}, 1-\frac{1}{2}+\frac{1}{4}, 1-\frac{1}{2}+\frac{1}{4}-\frac{1}{8}, \_\_, \_\_$
10. $1, 1+2, 1+2+4, 1+2+4+8, \_\_, \_\_$
11. $1, 1-1, 1-1+1, 1-1+1-1, \_\_, \_\_$

For Problems 12–15, (a) fill in the next two entries in the table for the specified function and (b) plot the values of that function given in the table. (These functions are defined only for integer values of x.)

12. $f(x)$
13. $g(x)$
14. $s(x)$
15. $t(x)$

x	1	2	3	4	5	6
$f(x)$	2	4	8	16		
$g(x)$	−1	1	−1	1		

x	$s(x)$	$t(x)$
1	$1+\frac{1}{2}$	$1-\frac{1}{2}$
2	$1+\frac{1}{2}+\frac{1}{3}$	$1-\frac{1}{2}+\frac{1}{4}$
3	$1+\frac{1}{2}+\frac{1}{3}+\frac{1}{4}$	$1-\frac{1}{2}+\frac{1}{4}-\frac{1}{8}$
4	$1+\frac{1}{2}+\frac{1}{3}+\frac{1}{4}+\frac{1}{5}$	$1-\frac{1}{2}+\frac{1}{4}-\frac{1}{8}+\frac{1}{16}$
5		
6		

In Problems 16–21, find a rule that describes the n-th term in the sequence.

16. $1, \frac{1}{8}, \frac{1}{27}, \frac{1}{64}, \frac{1}{125}, \ldots$

17. $1, \frac{1}{4}, \frac{1}{9}, \frac{1}{16}, \frac{1}{25}, \ldots$

18. $-1, \frac{1}{3}, -\frac{1}{9}, \frac{1}{27}, \ldots$

19. $0, \frac{1}{2}, \frac{2}{3}, \frac{3}{4}, \frac{4}{5}, \ldots$

20. $7, 7, 7, 7, 7, \ldots$

21. $\frac{1}{2}, \frac{2}{4}, \frac{3}{8}, \frac{4}{16}, \frac{5}{32}, \ldots$

In 22–33, compute the first six terms (starting with $n = 1$) of each sequence and graph these terms.

22. $\left\{3 + \frac{1}{n^2}\right\}$

23. $\left\{1 - \frac{2}{n}\right\}$

24. $\left\{\frac{\ln(n)}{n}\right\}$

25. $\left\{\frac{n}{2n-1}\right\}$

26. $\{4 + (-1)^n\}$

27. $\left\{3 + \frac{(-1)^n}{n}\right\}$

28. $\left\{\cos\left(\frac{n\pi}{2}\right)\right\}$

29. $\left\{(-1)^n \frac{n-1}{n}\right\}$

30. $\left\{\frac{n+1}{n!}\right\}$

31. $\left\{\frac{1}{n!}\right\}$

32. $\left\{\left(1 + \frac{1}{n}\right)^n\right\}$

33. $\left\{\frac{2^n}{n!}\right\}$

In Problems 34–39, compute the first 10 terms (starting with $n = 1$) of each sequence.

34. $a_1 = 3$ and $a_{n+1} = \frac{1}{a_n}$

35. $b_1 = 2$ and $b_{n+1} = -b_n$

36. $a_1 = 2$, $a_2 = 3$ and (for $n \geq 3$), $a_n = a_{n-1} - a_{n-2}$

37. $\left\{\sin\left(\frac{2\pi n}{3}\right)\right\}$

38. $c_n =$ the sum of the first n prime numbers (2 is the first prime)

39. $d_n =$ the sum of the first n positive integers

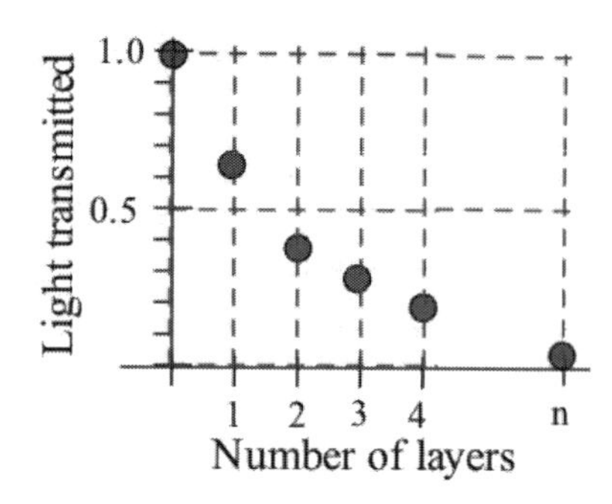

9.1 *Practice Answers*

1. One layer transmits $\frac{2}{3}$ of the original amount of light. Two layers transmit $\left(\frac{2}{3}\right)\left(\frac{2}{3}\right) = \left(\frac{2}{3}\right)^2 = \frac{4}{9}$ of the original amount of light. Three layers transmit $\left(\frac{2}{3}\right)^3 = \frac{8}{27}$ of the original light, and, in general, n layers transmit $\left(\frac{2}{3}\right)^n$ of the original light.

2. (a) $\frac{1}{5}, \frac{1}{6}, \ldots, \frac{1}{n}, \ldots$ (b) $-\frac{1}{32}, \frac{1}{64}, \ldots, \left(-\frac{1}{2}\right)^n, \ldots$ (c) $2, 2, \ldots, 2, \ldots$

3.
$$c_1 = \left(3 - \frac{1}{1}\right) - \frac{1}{2} = \frac{3}{2} = 1.5 \qquad d_1 = (-1)^1\left(\frac{1}{2}\right) = -\frac{1}{2}$$
$$c_2 = \left(3 - \frac{1}{2}\right) - \frac{1}{2^2} = \frac{9}{4} = 2.25 \qquad d_2 = (-1)^2\left(\frac{1}{2^2}\right) = \frac{1}{4}$$
$$c_3 = \left(3 - \frac{1}{3}\right) - \frac{1}{2^3} = \frac{61}{24} \approx 2.542 \qquad d_3 = (-1)^3\left(\frac{1}{2^3}\right) = -\frac{1}{8}$$
$$c_4 = \left(3 - \frac{1}{4}\right) - \frac{1}{2^4} = \frac{43}{16} = 2.6875 \qquad d_4 = \frac{1}{16}, d_5 = -\frac{1}{32}$$
$$c_5 = \left(3 - \frac{1}{5}\right) - \frac{1}{25} = \frac{443}{160} \approx 2.769$$

4. $a_0 = 3$, $a_1 = f(a_0) = 2(3) - 1 = 5$, $a_2 = f(a_1) = 2(5) - 1 = 9$, $a_3 = f(a_2) = 2(9) - 1 = 17$

5. $b_1 = b_0 + \frac{1}{3} = 0 + \frac{1}{3} = \frac{1}{3}$, $b_2 = b_1 + \frac{1}{9} = \frac{4}{9}$, $b_3 = b_2 + \frac{1}{27} = \frac{13}{27}$, $b_4 = b_3 + \frac{1}{81} = \frac{40}{81}$

9.2 *Limits of Sequences*

Because sequences are discrete functions defined only for integers, many calculus concepts that hold for continuous functions do not apply to sequences, with one important exception: the limit as n approaches infinity. Do the values a_n eventually approach (or equal) some number?

We say that the **limit** of a sequence $\{a_n\}$ is L if the terms a_n are all arbitrarily close to L for sufficiently large values of n. The terms at the beginning of the sequence can take on any values, but for large values of n, the a_n terms must all be close to L. The following definition makes these concepts of "close" and "large" more precise.

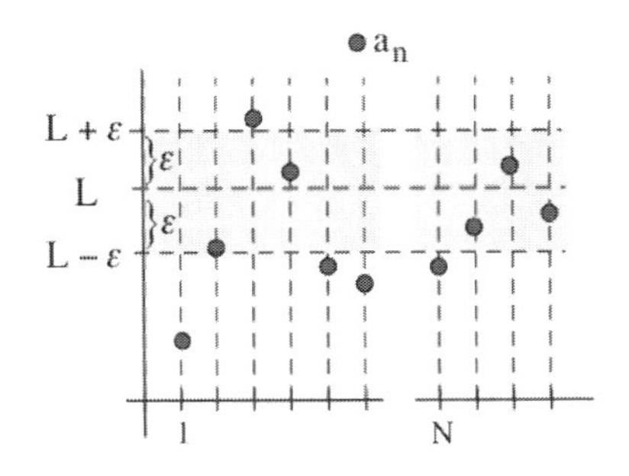

> **Definition**
> $\lim_{n\to\infty} a_n = L$ means that for any $\epsilon > 0$ ("epsilon > 0") there is an index N (typically depending on ϵ) so that a_n is within ϵ of L whenever n is larger than N:
>
> $$n > N \Rightarrow |a_n - L| < \epsilon$$

If a sequence has a finite limit L, we say that the sequence "**converges** to L." If a sequence does not have a finite limit, we say the sequence "**diverges**." Typically a sequence diverges because its terms grow infinitely large (positively or negatively) or because the terms oscillate and do not approach a single value.

Example 1. For $a_n = 3 + \frac{1}{n^2}$ show that $\lim_{n\to\infty} a_n = 3$.

Solution. Given $\epsilon > 0$, we need to find a number N so that $|a_n - L|$ is less than ϵ whenever n is larger than N:

$$n > N \quad \Rightarrow \quad \left|\left(3 + \frac{1}{n^2}\right) - 3\right| < \epsilon$$

To determine how large N must be, solve this inequality for n:

$$\left|\frac{1}{n^2}\right| < \epsilon \quad \Rightarrow \quad \frac{1}{\epsilon} < n^2 \quad \Rightarrow \quad n > \frac{1}{\sqrt{\epsilon}}$$

Taking $N = \left\lfloor \frac{1}{\sqrt{\epsilon}} \right\rfloor + 1$ (the next integer after $\frac{1}{\sqrt{\epsilon}}$):

$$n > N \quad \Rightarrow \quad n > \left\lfloor \frac{1}{\sqrt{\epsilon}} \right\rfloor + 1 \geq \frac{1}{\sqrt{\epsilon}} \quad \Rightarrow \quad \frac{1}{n^2} < \epsilon$$

so $\left|\left(3 + \frac{1}{n^2}\right) - 3\right| < \epsilon$, as desired. ◀

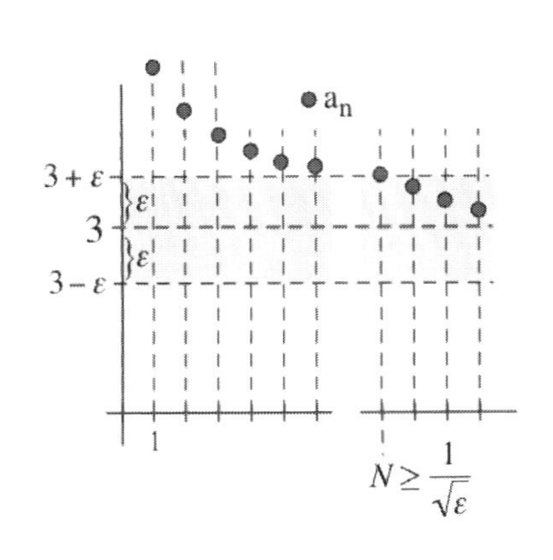

Practice 1. Show that $\lim_{n\to\infty} \frac{n+1}{n} = 1$.

The limit of a sequence, as n approaches infinity, depends only on the behavior of the terms of the sequence corresponding to "large" values of n (the "tail end" of the sequence) and not on the values of the first few (or few thousand, or few billion) terms. Consequently, we can insert or delete any *finite* number of terms without changing the convergence behavior of the sequence.

Because sequences are functions, limits of sequences possess certain properties of limits of more general functions.

Uniqueness Theorem

If a sequence converges to a limit
then the limit is unique:
a sequence cannot converge to two different values.

Proof. We will use the "proof by contradiction" technique.

Suppose that a sequence $\{a_n\}$ converges to two different limits. Call these limits L_1 and L_2. Let $d = |L_1 - L_2| > 0$ be the distance between L_1 and L_2. (We know $d > 0$ because $L_1 \neq L_2$.)

We chose $\epsilon = \frac{d}{3}$ because it "works" for our purpose by leading to a contradiction. Many other choices (any ϵ less than $\frac{d}{2}$) also lead to the contradiction. Because the definition of limit says "for any $\epsilon > 0$," we picked one we wanted.

Because $\{a_n\}$ converges to L_1, for any $\epsilon > 0$ there is an N_1 so that $n > N_1 \Rightarrow |a_n - L_1| < \epsilon$. If $\epsilon = \dfrac{d}{3}$, then there is an N_1 so that $n > N_1 \Rightarrow |a_n - L_1| < \frac{d}{3}$. Similarly, there is an N_2 so that $n > N_2 \Rightarrow |a_n - L_2| < \frac{d}{3}$.

Now let N be the larger of N_1 and N_2. When $n > N$, both $|a_n - L_1| < \dfrac{d}{3}$ and $|a_n - L_2| < \dfrac{d}{3}$ hold, so:

$$d = |L_1 - L_2| = |(a_n - L_2) - (a_n - L_1)|$$

Using the triangle inequality:

$$d \leq |a_n - L_2| + |a_n - L_1| < \frac{d}{3} + \frac{d}{3} = \frac{2}{3}d$$

We have shown that $0 < d < \frac{2}{3}d$, which is impossible, so we can conclude that our original assumption, that $L_1 \neq L_2$, must be false. □

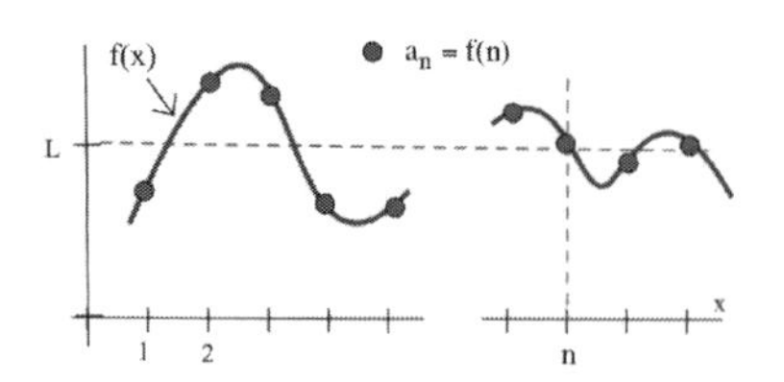

Sometimes it is useful to replace a sequence $\{a_n\}$, a function whose domain consists of integers, with a function f whose domain consists of the real numbers and satisfying $a_n = f(n)$. If $f(x)$ has a limit as "$x \to \infty$" then $f(n)$ has the same limit as "$n \to \infty$." This replacement of x with n allows us to use earlier results about limits of functions, particularly L'Hôpital's Rule, to compute limits of sequences.

Theorem:

If $a_n = f(n)$ and $\lim\limits_{x\to\infty} f(x) = L$
then $\{a_n\}$ converges to L: $\lim\limits_{n\to\infty} a_n = L$

Example 2. Compute $\lim_{n\to\infty}\left(1+\frac{2}{n}\right)^n$.

Solution. With $a_n = \left(1+\frac{2}{n}\right)^n$, we can define $f(x) = \left(1+\frac{2}{x}\right)^x$ so that $a_n = f(n)$. Using L'Hôpital's Rule (as in Example 7 from Section 3.7), we can show that $\lim_{x\to\infty}\left(1+\frac{2}{x}\right)^x = e^2$ and conclude that $\lim_{n\to\infty}\left(1+\frac{2}{n}\right)^n = e^2 \approx 7.389$. ◀

Practice 2. Compute $\lim_{n\to\infty}\frac{\ln(n)}{n}$.

A **subsequence** is an infinite subset of terms from a sequence that occur in the same order as they appear in the original sequence. The sequence of even integers $\{2,4,6,\ldots\}$ is a subsequence of the sequence of all positive integers $\{1,2,3,4,\ldots\}$. The sequence of reciprocals of primes $\left\{\frac{1}{2},\frac{1}{3},\frac{1}{5},\frac{1}{7},\ldots\right\}$ is a subsequence of the sequence of the reciprocals of all positive integers $\left\{1,\frac{1}{2},\frac{1}{3},\frac{1}{4},\frac{1}{5},\ldots\right\}$. Subsequences inherit certain properties from their original sequences.

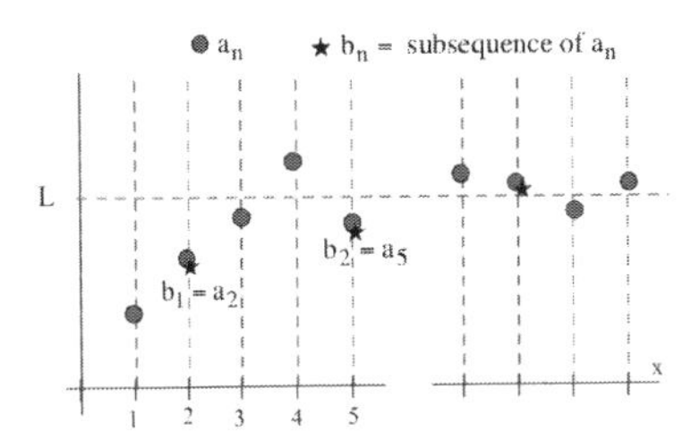

Subsequence Theorem
Every subsequence of a convergent sequence converges to the same limit as the original sequence:

If $\lim_{n\to\infty} a_n = L$ and $\{b_n\}$ is a subsequence of $\{a_n\}$
then $\lim_{n\to\infty} b_n = L$.

The proof is left to you. Use proof by contradiction and start by assuming that $\lim_{n\to\infty} b_n \neq L$. Then for some $\epsilon > 0$, infinitely many terms of $\{b_n\}$ satisfy $|b_n - L| > \epsilon$, hence infinitely many terms of $\{a_n\}$ satisfy $|a_n - L| > \epsilon$.

If the sequence $\{a_n\}$ does *not* converge, then the subsequence $\{b_n\}$ **may or may not** converge.

Corollary

If two subsequences of $\{a_n\}$ converge to two different limits
then $\{a_n\}$ diverges.

The proof is left to you. Use proof by contradiction and start by assuming that $\lim_{n\to\infty} a_n \neq L$. By the Subsequence Theorem, any two subsequences must also converge to L.

Example 3. Show that the sequence $\left\{(-1)^n\frac{n}{n+1}\right\}$ diverges.

Solution. For the even-indexed terms, $a_n = (-1)^n\frac{n}{n+1} = \frac{n}{n+1}$, so the subsequence of even-indexed terms converges to 1. For the odd-indexed terms, $a_n = (-1)^n\frac{n}{n+1} = \frac{-n}{n+1}$, so the subsequence of odd-indexed terms converges to -1. Because two subsequences converge to different values, the original sequence diverges. ◀

Practice 3. Show that the sequence $\left\{ \sin\left(\frac{n\pi}{2}\right) \right\}$ diverges.

Bounded and Monotonic Sequences

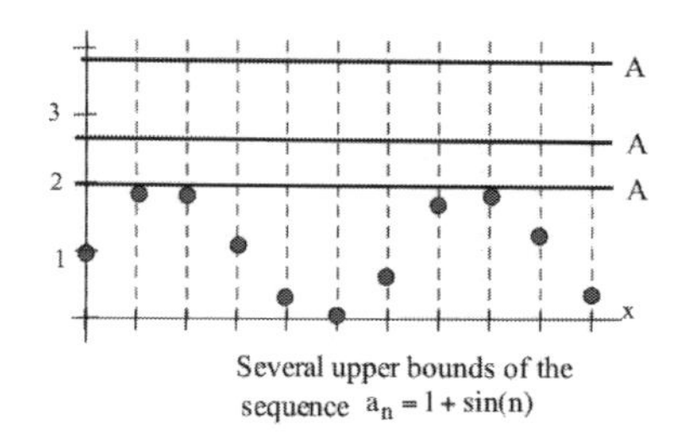

Several upper bounds of the sequence $a_n = 1 + \sin(n)$

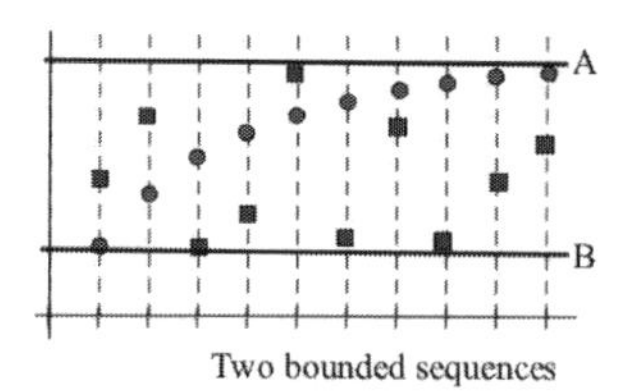

Two bounded sequences

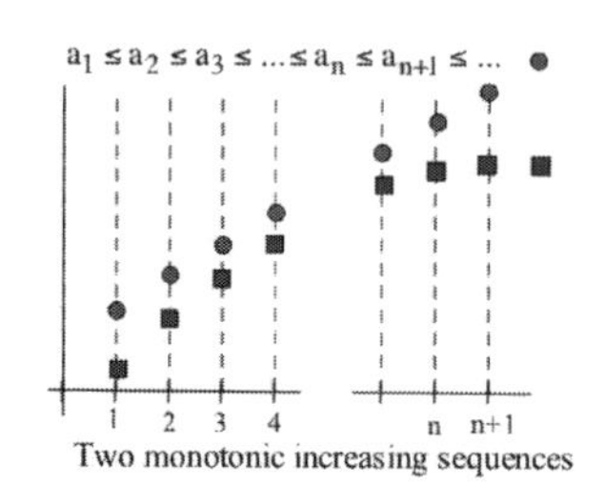

Two monotonic increasing sequences

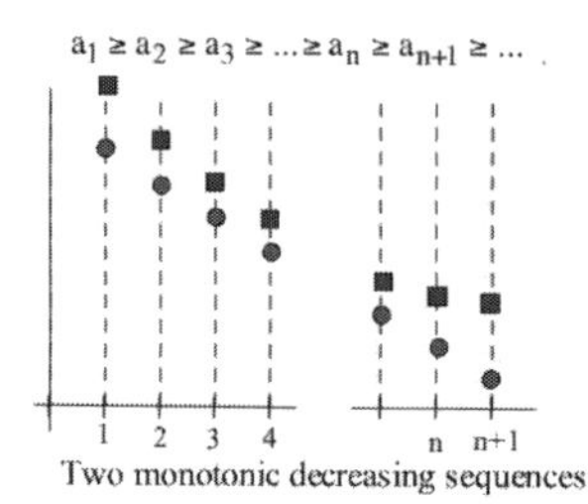

Two monotonic decreasing sequences

A sequence $\{a_n\}$ is **bounded above** if there is a value A so that $a_n \le A$ for all values of n: we call A an **upper bound** of $\{a_n\}$. Similarly, $\{a_n\}$ is **bounded below** if there is a value B so that $a_n \ge B$ for all values of n: we call B a **lower bound** of the sequence. A sequence is **bounded** if it has an upper bound and a lower bound. All of the terms of a bounded sequence are between (or equal to) the upper and lower bounds.

A **monotonically increasing** sequence is a sequence for which each term is at least as big as the previous term:

$$a_1 \le a_2 \le a_3 \le \ldots$$

A **monotonically decreasing** sequence is one for which each term is no bigger than the previous term:

$$a_1 \ge a_2 \ge a_3 \ge \ldots$$

A monotonic sequence does not oscillate: if one term is larger than a previous term and another term is smaller than a previous term, then the sequence is neither monotonic increasing nor monotonic decreasing.

There are three common ways to demonstrate a sequence is monotonically increasing. You can do this by showing that:

- $a_{n+1} \ge a_n$ for all n
- $a_n > 0$ and that $\frac{a_{n+1}}{a_n} \ge 1$ for all n
- $a_n = f(n)$ for integer values n and some differentiable function f for which $f'(x) \ge 0$ for all $x > 0$

Practice 4. List three ways to demonstrate that a sequence is monotonically decreasing.

Example 4. Show that $\{a_n\} = \left\{ \frac{2^n}{n!} \right\}$ is monotonically decreasing.

Solution. We will use the second technique from Practice 4 (we can't use the third technique—why not?) and write:

$$a_n = \frac{2^n}{n!} = \frac{2^n}{1 \cdot 2 \cdot 3 \cdots n}$$

$$a_{n+1} = \frac{2^{n+1}}{(n+1)!} = \frac{2^n \cdot 2}{1 \cdot 2 \cdot 3 \cdots n \cdot (n+1)}$$

so that:

$$\frac{a_{n+1}}{a_n} = \frac{2^n \cdot 2}{1 \cdot 2 \cdot 3 \cdots n \cdot (n+1)} \cdot \frac{1 \cdot 2 \cdot 3 \cdots n}{2^n} = \frac{2}{n+1}$$

When $n \ge 1$, then $n+1 \ge 2$ so that $\frac{2}{n+1} \le 1$. ◀

Practice 5. Show that $\left\{\left(\frac{2}{3}\right)^n\right\}$ is monotonically decreasing.

Because the behavior of a monotonic sequence is so "regular," with a bit more information we can determine whether or not a monotonic sequence has a finite limit: we just need to show that it is bounded.

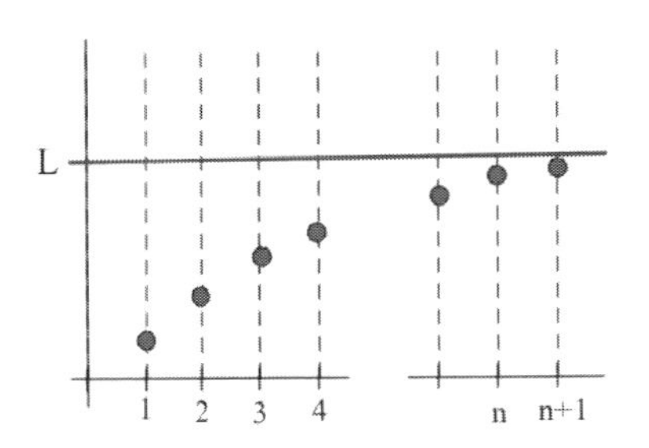

The idea for a monotonically decreasing sequence is similar.

> **Monotone Convergence Theorem**
>
> If a monotonic sequence is bounded
> then the sequence converges.

Idea for a proof. If a monotonically increasing sequence $\{a_n\}$ is bounded above, then $\{a_n\}$ has an infinite number of upper bounds, each of them larger than every a_n. Let L be the smallest of these upper bounds (the **least upper bound**) of $\{a_n\}$. Then there is a value a_N as close as we want to L (otherwise there would be an upper bound smaller than L), so given $\epsilon > 0$, find an N so that $|a_N - L| < \epsilon$. Because $\{a_n\}$ is increasing, all of the later values in the sequence (with $n \geq N$), must satisfy $a_n \geq a_N$; because L is an upper bound for $\{a_n\}$, it must also be true that $\{a_n\} \leq L$ for $n > N$. So if $n > N$, then $a_N \leq a_n \leq L$, hence:

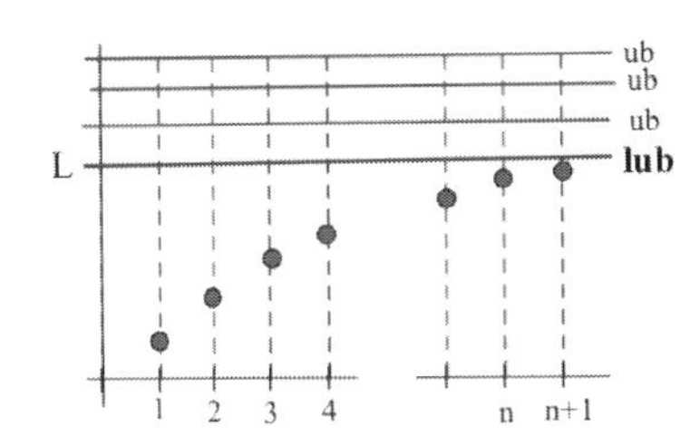

$$a_N - L \leq a_n - L \leq 0 \quad \Rightarrow \quad |a_n - L| \leq |a_N - L| < \epsilon$$

Cauchy and other mathematicians accepted this theorem on intuitive and geometric grounds similar to the "idea for a proof" given above, but later mathematicians felt more rigor was needed. Yet even the mathematician Dedekind, who supplied much of that rigor, recognized the usefulness of geometric intuition.

"Even now such resort to geometric intuition in a first presentation of differential calculus, I regard as exceedingly useful, from a didactic standpoint, and indeed indispensable if one does not wish to lose too much time." (Dedekind, *Essays on the Theory of Numbers*, 1901)

9.2 *Problems*

In Problems 1–4, state whether each sequence appears to be converging or diverging. If you think the sequence is converging, mark its limit as a value on the vertical axis. (Important: The behavior of a sequence can change drastically after the first terms, which have no influence on whether or not the sequence converges, but sometimes we need to reach a tentative conclusion based on the few sequence values we know.)

1.

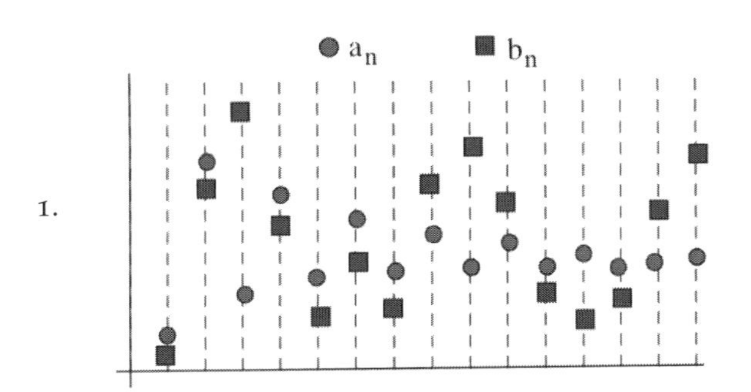

2.

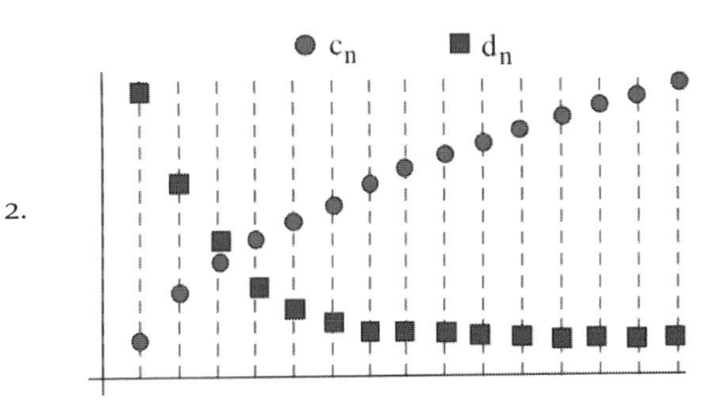

3.

4.

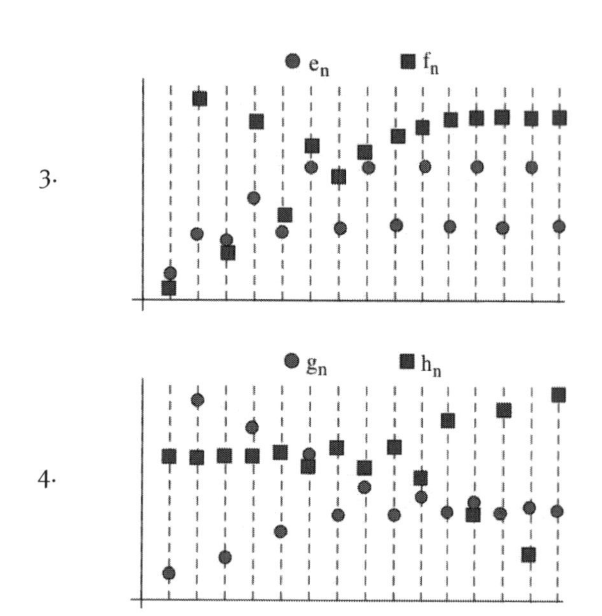

In 5–19, state whether each sequence converges or diverges. If the sequence converges, find its limit.

5. $\left\{1-\frac{2}{n}\right\}$

6. $\left\{n^2\right\}$

7. $\left\{\frac{n^2}{n+1}\right\}$

8. $\left\{3+\frac{1}{n^2}\right\}$

9. $\left\{\frac{n}{2n-1}\right\}$

10. $\left\{\frac{\ln(n)}{n}\right\}$

11. $\left\{\ln\left(3+\frac{7}{n}\right)\right\}$

12. $\left\{3+\frac{(-1)^{n+1}}{n}\right\}$

13. $\{4+(-1)^n\}$

14. $\left\{(-1)^n\frac{n-1}{n}\right\}$

15. $\left\{\frac{1}{n!}\right\}$

16. $\left\{\left(1+\frac{3}{n}\right)^n\right\}$

17. $\left\{\left(1-\frac{1}{n}\right)^n\right\}$

18. $\left\{\frac{\sqrt{n}-3}{\sqrt{n}+3}\right\}$

19. $\left\{\frac{(n+2)(n+5)}{n^2}\right\}$

In 20–23, prove that the sequence converges to the given limit by showing that, for any $\epsilon > 0$, you can find an N that satisfies the definition of convergence.

20. $\lim_{n\to\infty}\left[2-\frac{3}{n}\right]=2$

21. $\lim_{n\to\infty}\frac{3}{n^2}=0$

22. $\lim_{n\to\infty}\frac{7}{n+1}=0$

23. $\lim_{n\to\infty}\frac{3n-1}{n}=3$

In Problems 24–29, use subsequences to help determine whether the given sequence converges or diverges. If the sequence converges, find its limit.

24. $\{(-1)^n\cdot 3\}$

25. $\left\{\frac{1}{n\text{-th prime}}\right\}$

26. $\left\{(-1)^n\cdot\frac{n+1}{n}\right\}$

27. $\left\{(-2)^n\cdot\left(\frac{1}{2}\right)^n\right\}$

28. $\left\{\left(1+\frac{1}{3n}\right)^{3n}\right\}$

29. $\left\{\left(1+\frac{5}{n^2}\right)^{n^2}\right\}$

In problems 30–35, calculate $a_{n+1}-a_n$ and use that value to determine whether $\{a_n\}$ is monotonic increasing, monotonic decreasing, or neither.

30. $\left\{\frac{3}{n}\right\}$

31. $\left\{7-\frac{2}{n}\right\}$

32. $\left\{\frac{n-1}{2n}\right\}$

33. $\{2^n\}$

34. $\left\{1-\frac{1}{2^n}\right\}$

35. $\left\{5+\frac{7}{3^n}\right\}$

In Problems 36–41, calculate $\frac{a_{n+1}}{a_n}$ and use that value to determine whether each sequence is monotonic increasing, monotonic decreasing, or neither.

36. $\left\{\frac{n}{n+1}\right\}$

37. $\left\{\frac{n+1}{n!}\right\}$

38. $\left\{\frac{n^2}{n!}\right\}$

39. $\left\{\left(\frac{5}{4}\right)^n\right\}$

40. $\left\{\frac{n^2}{2^n}\right\}$

41. $\left\{\frac{n}{e^n}\right\}$

In Problems 42–46, use derivatives to determine whether each sequence is monotonic increasing, monotonic decreasing, or neither.

42. $\left\{\frac{n+1}{n}\right\}$

43. $\left\{5-\frac{3}{n}\right\}$

44. $\{n\cdot e^{-n}\}$

45. $\left\{\cos\left(\frac{1}{n}\right)\right\}$

46. $\left\{\left(1+\frac{1}{n}\right)^3\right\}$

In 47–51, show that each sequence is monotonic.

47. $\left\{\frac{n+3}{n!}\right\}$

48. $\left\{\frac{n}{n+1}\right\}$

49. $\left\{1-\frac{1}{2^n}\right\}$

50. $\left\{\sin\left(\frac{1}{n}\right)\right\}$

51. $\left\{\frac{n+1}{e^n}\right\}$

52. The **Fibonacci sequence**, named after Leonardo Fibonacci (1170–1250), who used it to model a population of rabbits, is obtained by setting the first two terms equal to 1 and then defining each new term as the sum of the two previous terms: $a_n = a_{n-1} + a_{n-2}$ for $n \geq 3$. (a) Write the first seven terms of this sequence. (b) Compute the successive ratios of the terms, $\frac{a_n}{a_{n-1}}$. (These ratios approach the "golden ratio," $\frac{1+\sqrt{5}}{2} \approx 1.618$.)

53. To approximate the square root of a positive number N, **Heron's method** puts $a_1 = N$ and $a_{n+1} = \frac{1}{2}\left(a_n + \frac{N}{a_n}\right)$. Then $\{a_n\}$ converges to $\sqrt{N}$. Compute a_1 through a_4 for $N = 4$, 9 and 5. (Heron's method is equivalent to Newton's method applied to the function $f(x) = x^2 - N$.)

54. Using an initial or "seed" value h_0, define the **hailstone sequence** by the rule:

$$h_n = \begin{cases} 3 \cdot h_{n-1} + 1 & \text{if } h_{n-1} \text{ is odd} \\ \frac{1}{2} \cdot h_{n-1} & \text{if } h_{n-1} \text{ is even} \end{cases}$$

Define the **length** of the sequence to be the first value of n so that $h_n = 1$. If the seed value is $h_0 = 3$, then $h_1 = 3(3) + 1 = 10$, $h_2 = \frac{1}{2} \cdot 10 = 5$, $h_3 = 3(5) + 1 = 16$, $h_4 = \frac{1}{2} \cdot 16 = 8$, $h_5 = 4$, $h_6 = 2$ and $h_7 = 1$, so the length is 7.

(a) Find the length of the hailstone sequence for each seed value from 2 through 10.

(b) Find the length of the hailstone sequence for a seed value of the form $h_0 = 2^n$. .

(c) Is the length of the hailstone sequence finite for every seed value? (This is an open question — no one has yet been able to answer it.)

Called the hailstone sequence because (for some seed values) the terms of the sequence rise and drop like the path of a hailstone as it forms, this sequence has been attributed to Lothar Collatz. The open question in part (c) is also called Ulam's conjecture, the Syracuse problem, Kakutani's problem and Hasse's algorithm.

55. Suppose that individuals with the gene combination "aa" do not reproduce and those with the combinations "aA" and "AA" do reproduce. When the initial proportion of individuals with "aa" is $a_0 = p$ (typically a small number), then the proportion of individuals with "aa" in the k-th generation is $a_k = \frac{p}{kp+1}$. Use this formula for a_k to answer the following questions.

(a) If 2% of a population initially have the combination "aa" and these individuals do not reproduce, then how many generations will it take for the proportion of individuals with "aa" to drop below 1%?

(b) In general, find the number of generations until the proportion of individuals with "aa" is half of the initial proportion.

"Negative eugenics" was a strategy proposed during the early 20th century in which individuals with an undesirable trait were discouraged or forcibly prevented from reproducing. Mathematics shows that this is not an effective strategy for reducing the occurrence of traits carried by recessive genes (the above example) that are uncommon (p small) in a species that reproduces slowly (people).

56. The fractional part of a number is the difference between the number and its integer part: $x - \lfloor x \rfloor$. The sequence of fractional parts of multiples of a number x is the sequence with terms $a_n = n \cdot x - \lfloor n \cdot x \rfloor$. The behavior of the sequence of fractional parts of the multiples of a number is one way in which rational numbers differ from irrational numbers.

(a) Let $a_n = n \cdot x - \lfloor n \cdot x \rfloor$ be the fractional part of the n-th multiple of x. Calculate a_1 through a_6 for $x = \frac{1}{3}$. These are the fractional parts of the first six multiples of $\frac{1}{3}$.

(b) Calculate the fractional parts of the first nine multiples of $\frac{3}{4}$ and $\frac{2}{5}$.

(c) Calculate the fractional parts of the first five multiples of π.

(d) Let $a_n = n \cdot \pi - \lfloor n \cdot \pi \rfloor$ be the fractional part of the n-th multiple of π. Can two different multiples of π have the same fractional part? (Suggestion: Assume the answer is yes and obtain a contradiction.)

An Alternative Way to Visualize Sequences and Convergence

A sequence is a function, and we have graphed sequences in the xy-plane in the same way we graphed other functions: because $a_n = f(n)$, we plotted the point (n, a_n). If the sequence $\{a_n\}$ converges to L, then the points (n, a_n) eventually (for big values of n) all lie close to—or on—the horizontal line $y = L$.

We can also graph a sequence $\{a_n\}$ in one dimension, using the x-axis alone. For each value of n, plot the point $x = a_n$. Then the graph of $\{a_n\}$ consists of a collection of points on the x-axis. The margin figures show the one-dimensional graphs of $a_n = \frac{1}{n}$, $b_n = 2 + \frac{(-1)^n}{n}$ and $c_n = (-1)^n$.

$a_n = 1/n$

$a_5 a_4 a_3$ a_2 a_1

0 1

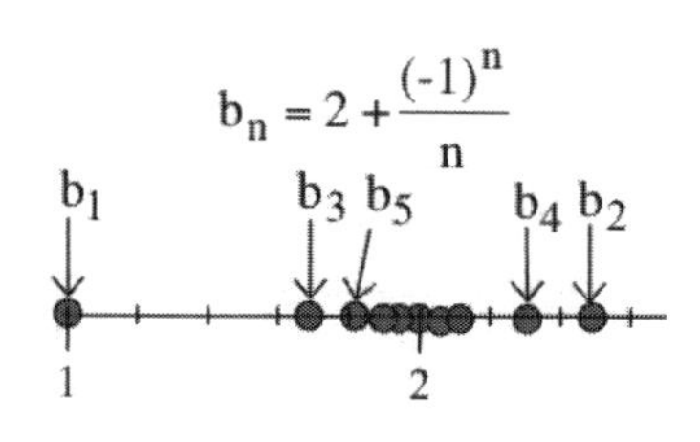

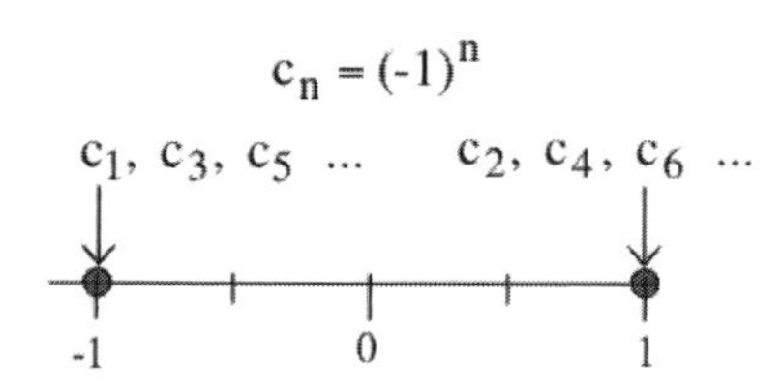

If $\{a_n\}$ converges to L, then the points $x = a_n$ eventually (for big values of n) all lie close to—or on—the point $x = L$. If we build a narrow box, with width $2\epsilon > 0$, and center the box at the point $x = L$, then all of the points a_n will sit inside the box once n becomes larger than some value N.

In more advanced courses, these points $x = L$ about which the terms in the sequence eventually "cluster" are called **cluster points** or **accumulation points**. A convergent sequence has exactly one accumulation point; a divergent sequence with more than one convergent subsequence has multiple accumulation points.

57. Suppose that the sequence $\{a_n\}$ converges to 3 and that you place a single grain of sand at each point $x = a_n$ on the x-axis. Describe the likely result (a) after you have placed a few grains and (b) after you have placed a lot (thousands or millions) of grains.

58. Suppose the sequence $\{a_n\}$ converges to 3, the sequence $\{b_n\}$ converges to 1, and that you place a single grain of sand at each point $(x, y) = (a_n, b_n)$ on the xy-plane. Describe the likely result (a) after you have placed a few grains and (b) after you have placed a lot (thousands or millions) of grains.

59. Suppose that $a_n = \sin(n)$ for positive integers n and that you place a single grain of sand at each point $x = a_n$ on the x-axis. (a) Describe the likely result after you have placed a few grains. (b) After you have placed a lot (thousands or millions) of grains. (c) Do two grains ever end up on the same point?

60. Suppose that $a_n = \cos(n)$ and $b_n = \sin(n)$ for positive integers n. If you place a single grain of sand at each point $(x, y) = (a_n, b_n)$ on the xy-plane, describe the likely result (a) after you have placed a few grains and (b) after you have placed millions of grains.

9.2 Practice Answers

1. We need to show that, for any positive ϵ, there is a number N so that the distance from $a_n = \dfrac{n+1}{n}$ to $L = 1$, $|a_n - L|$, is less than ϵ whenever n is larger than N. For this particular sequence:

$$|a_n - L| = \left|\frac{n+1}{n} - 1\right| = \left|\frac{n}{n} + \frac{1}{n} - 1\right| = \left|\frac{1}{n}\right|$$

needs to be less than ϵ. To determine how large N needs to be, solve the inequality:

$$\left|\frac{1}{n}\right| < \epsilon \quad \Rightarrow n > \frac{1}{\epsilon}$$

So, for any positive ϵ, we can take $N = \left\lfloor \frac{1}{\epsilon} \right\rfloor + 1$. Then:

We're "rounding down" and then adding 1 to ensure that N is an integer.

$$n > N = \left\lfloor \frac{1}{\epsilon} \right\rfloor + 1 \geq \frac{1}{\epsilon} \quad \Rightarrow \quad \frac{1}{n} < \epsilon \quad \Rightarrow \quad \left|\frac{n+1}{n} - 1\right| < \epsilon$$

2. Replacing n with x and applying L'Hôpital's Rule:

$$\lim_{n\to\infty} \frac{\ln(n)}{n} = \lim_{x\to\infty} \frac{\ln(x)}{x} = \lim_{x\to\infty} \frac{\frac{1}{x}}{1} = 0$$

3. We can show that $a_n = \left\{\sin\left(\frac{\pi n}{2}\right)\right\}$ diverges by finding two subsequences b_n and c_n of a_n that converge to different limits. Let $\{b_n\}$ consist of the terms $\{a_1, a_5, a_9, \ldots, a_{4k-3}, \ldots\}$:

$$\left\{\sin\left(\frac{\pi}{2}\right), \sin\left(\frac{5\pi}{2}\right), \sin\left(\frac{9\pi}{2}\right), \ldots\right\} = \{1, 1, 1, \ldots\}$$

so that $\lim\limits_{n\to\infty} b_n = 1$. Then let $\{c_n\}$ consist of $\{a_2, a_4, a_6, \ldots, a_{2k}, \ldots\}$:

$$\left\{\sin\left(\frac{2\pi}{2}\right), \sin\left(\frac{4\pi}{2}\right), \sin\left(\frac{6\pi}{2}\right), \ldots\right\} = \{0, 0, 0, \ldots\}$$

so that $\lim\limits_{n\to\infty} c_n = 0$. Because the subsequences $\{b_n\}$ and $\{c_n\}$ have different limits, the original sequence $\{a_n\}$ must diverge.

Many other pairs of subsequences will also work in place of the $\{b_n\}$ and $\{c_n\}$ that we used.

4. By showing that:

   - $a_{n+1} \leq a_n$ for all n
   - $a_n > 0$ and that $\dfrac{a_{n+1}}{a_n} \leq 1$ for all n
   - $a_n = f(n)$ for integer values n and some differentiable function f for which $f'(x) \leq 0$ for all $x > 0$

5. Using the second method from Practice 9:

$$\frac{a_{n+1}}{a_n} = \frac{\left(\frac{2}{3}\right)^{n+1}}{\left(\frac{2}{3}\right)^n} = \frac{2}{3} < 1$$

so $\left\{\left(\dfrac{2}{3}\right)^n\right\}$ is monotonically decreasing.

9.3 *Infinite Series*

What does it mean to add together an infinite number of terms of a sequence? We will define that concept carefully in this section. Is the sum of all the terms in an infinite sequence a finite number? Over the next few sections we will examine a variety of techniques for determining whether the sum of an infinite number of terms is a finite number. If we know the sum of an infinite number of terms is finite, can we determine the value of that sum? In certain special situations, this can be very easy, but often it turns out to be very, very difficult.

Bouncing Golf Ball
(This is **not** the ball's path. The actual motion of the ball is straight up and down.)

Example 1. You throw a golf ball 9 feet straight up into the air, and on each bounce it rebounds to two-thirds of its previous height (see margin). Find a sequence whose terms give the distances the ball travels during each successive bounce. Represent the total distance traveled by the ball as a sum.

Solution. The heights of the successive bounces are 9 feet, $\left(\frac{2}{3}\right) \cdot 9$ feet, $\left(\frac{2}{3}\right) \cdot \left(\frac{2}{3}\right) \cdot 9$ feet, $\left(\frac{2}{3}\right)^3 \cdot 9$ feet, and so forth. On each bounce, the ball rises and falls, so the distance traveled is twice the height of that bounce:

$$18 \text{ ft}, \left(\frac{2}{3}\right) \cdot 18 \text{ ft}, \left(\frac{2}{3}\right) \cdot \left(\frac{2}{3}\right) \cdot 18 \text{ ft}, \left(\frac{2}{3}\right)^3 \cdot 18 \text{ ft}, \ldots$$

The total distance traveled is the sum of these bounce-distances:

$$18 + \left(\frac{2}{3}\right) \cdot 18 + \left(\frac{2}{3}\right)^2 \cdot 18 + \left(\frac{2}{3}\right)^3 \cdot 18 + \cdots$$
$$= 18\left[1 + \frac{2}{3} + \left(\frac{2}{3}\right)^2 + \left(\frac{2}{3}\right)^3 + \cdots\right]$$

At the completion of the first bounce, the ball has traveled 18 feet. After the second bounce, it has traveled 30 feet, a total of 38 feet after the third bounce, $\frac{130}{3} \approx 43.67$ feet after the fourth, and so on. With a calculator—and some patience—you can determine that after the 20th bounce the ball has traveled a total of (approximately) 53.996 feet, after the 30th bounce (approximately) 53.99994 feet, and after the 40th bounce (approximately) 53.9999989 feet. ◀

Do these total distances appear to be approaching a limiting value?

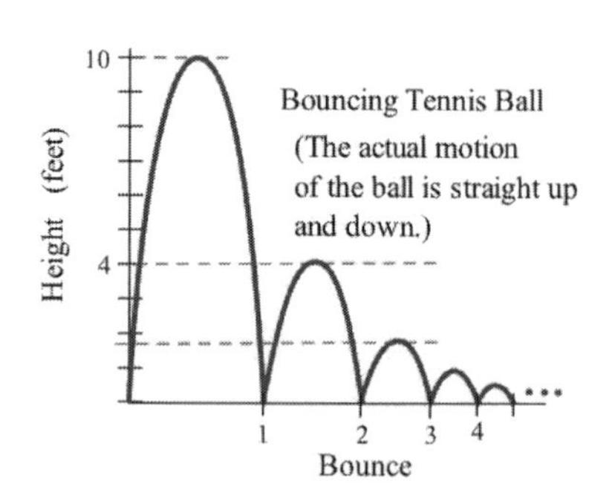

Practice 1. Your friend throws a tennis ball 10 feet straight up, and on each bounce it rebounds to 40% of its previous height (see margin). Represent the total distance traveled by the ball as a sum, and find the total distance traveled by the ball after its third bounce.

Infinite Series

The infinite sums in the preceding Example and Practice problems are called **infinite series**.

Definitions: An **infinite series** is an expression of the form

$$a_1 + a_2 + a_3 + a_4 + \cdots = \sum_{k=1}^{\infty} a_k$$

The numbers $a_1, a_2, a_3, a_4, \ldots$ are called the **terms** of the series.

Although often used interchangeably in everyday language, the words "sequence" and "series" have precise meanings in mathematics: a sequence is a list of numbers and a series is sum of a sequence.

Example 2. Represent the following series using sigma notation.

(a) $1 + \frac{1}{3} + \frac{1}{9} + \frac{1}{27} + \cdots$ (b) $-1 + \frac{1}{2} - \frac{1}{3} + \frac{1}{4} - \frac{1}{5} + \cdots$

(c) $18\left[1 + \frac{2}{3} + \frac{4}{9} + \frac{8}{27} + \frac{16}{81} + \cdots\right]$

(d) $0.777\ldots = \frac{7}{10} + \frac{7}{100} + \cdots$ (e) $0.222\ldots = \frac{2}{10} + \frac{2}{100} + \cdots$

While the plural of "sequence" is "sequences," the plural of "series" is "series," so you might say "the series is convergent" when talking about a single series and "the series are convergent" when discussing more than one series.

Solution. (a) The terms are all of the form $\left(\frac{1}{3}\right)^k$, starting with $k = 0$:

$$1 + \frac{1}{3} + \frac{1}{9} + \frac{1}{27} + \cdots = \sum_{k=0}^{\infty} \left(\frac{1}{3}\right)^k$$

(b) The terms are all of the form $\frac{\pm 1}{k}$ starting with $k = 1$:

$$-1 + \frac{1}{2} - \frac{1}{3} + \frac{1}{4} - \frac{1}{5} + \cdots = \sum_{k=1}^{\infty} \frac{(-1)^k}{k}$$

(c) The terms in the brackets are of the form $\left(\frac{2}{3}\right)^k$, starting with $k = 0$:

$$18\left[1 + \frac{2}{3} + \frac{4}{9} + \frac{8}{27} + \frac{16}{81} + \cdots\right] = 18 \cdot \sum_{k=0}^{\infty} \left(\frac{2}{3}\right)^k$$

(d) The terms are all of the form $\frac{7}{10^k}$, starting with $k = 1$, so:

$$0.777\ldots = \frac{7}{10} + \frac{7}{100} + \frac{7}{1000} + \cdots = \sum_{k=1}^{\infty} \frac{7}{10^k}$$

(e) $0.222\ldots = \sum_{k=1}^{\infty} \frac{2}{10^k}$ ◀

Practice 2. Represent the following series using sigma notation.

(a) $1 + 2 + 3 + 4 + \cdots$ (b) $-1 + 1 - 1 + 1 - \cdots$

(c) $2 + 1 + \frac{1}{2} + \frac{1}{4} + \frac{1}{8} + \cdots$ (d) $\frac{1}{2} + \frac{1}{4} + \frac{1}{6} + \frac{1}{8} + \frac{1}{10} + \cdots$

(e) $0.111\ldots$

Partial Sums

In Example 1, we computed the total distance traveled by the golf ball after successive bounces: the first term in the sequence, then the sum of the first two terms, then the sum of the first three terms, and so forth. We call these numbers **partial sums** of an infinite series.

> **Definition:** The **partial sums** s_n of the infinite series $\sum_{k=1}^{\infty} a_k$ are:
>
> $$\begin{aligned} s_1 &= a_1 \\ s_2 &= a_1 + a_2 \\ s_3 &= a_1 + a_2 + a_3 \\ &\vdots \\ s_n &= \sum_{k=1}^{n} a_k \end{aligned}$$

We can also define the partial sums recursively as $s_n = s_{n-1} + a_n$. The partial sums form a sequence of partial sums $\{s_n\}$.

Example 3. Compute the first four partial sums for the following series.

(a) $1 + \frac{1}{2} + \frac{1}{4} + \frac{1}{8} + \frac{1}{16} + \cdots$ (b) $\sum_{k=1}^{\infty} (-1)^k$ (c) $\sum_{k=1}^{\infty} \frac{1}{k}$

Solution. (a) $s_1 = 1$, $s_2 = 1 + \frac{1}{2} = \frac{3}{2}$, $s_3 = 1 + \frac{1}{2} + \frac{1}{4} = \frac{7}{4}$ and $s_4 = 1 + \frac{1}{2} + \frac{1}{4} + \frac{1}{8} = \frac{15}{8}$. Typically, it is easier to use the recursive formulation of s_n:

$$s_3 = s_2 + a_3 = \frac{3}{2} + \frac{1}{4} = \frac{7}{4} \Rightarrow s_4 = s_3 + a_4 = \frac{7}{4} + \frac{1}{8} = \frac{15}{8} \Rightarrow \cdots$$

(b) Using the recursive formulation: $s_1 = (-1)^1 = -1$, $s_2 = s_1 + a_2 = -1 + (-1)^2 = 0$, $s_3 = s_2 + a_3 = 0 + (-1)^3 = -1$ and $s_4 = 0$.

(c) $s_1 = 1$, $s_2 = \frac{3}{2}$, $s_3 = \frac{11}{6}$ and $s_4 = \frac{25}{12}$ ◀

Practice 3. Compute the first four partial sums for the following series.

(a) $1 - \frac{1}{2} + \frac{1}{4} - \frac{1}{8} + \frac{1}{16} - \cdots$ (b) $\sum_{k=1}^{\infty} \left(\frac{1}{3}\right)^k$ (c) $\sum_{k=1}^{\infty} \frac{(-1)^k}{k}$

If we know the values of the partial sums s_n for an infinite series, we can recover the values of the terms for the series used to build the s_n.

Example 4. If $s_1 = 2.1$, $s_2 = 2.6$, $s_3 = 2.84$ and $s_4 = 2.87$ are the first four partial sums of $\sum_{k=1}^{\infty} a_k$, find the first four terms of $\{a_n\}$.

Solution. $s_1 = a_1$, so $a_1 = 2.1$, while $s_2 = a_1 + a_2$, so $2.6 = 2.1 + a_2 \Rightarrow a_2 = 2.6 - 2.1 = 0.5$. Similarly, $s_3 = a_1 + a_2 + a_3$ so $2.84 = 2.1 + 0.5 + a_3 \Rightarrow a_3 = 2.84 - 2.6 = 0.24$. Finally, $a_4 = 0.03$. ◀

Alternatively, we could use the recursive formula $s_n = s_{n-1} + a_n$ to conclude that $a_n = s_n - s_{n-1}$ and compute:

$$a_2 = s_2 - s_1 = 2.6 - 2.1 = 0.5$$
$$a_3 = s_3 - s_2 = 2.84 - 2.6 = 0.24$$
$$a_4 = s_4 - s_3 = 2.87 - 2.84 = 0.03$$

Practice 4. If $s_1 = 3.2$, $s_2 = 3.6$, $s_3 = 3.5$, $s_4 = 4$, $s_{99} = 7.3$, $s_{100} = 7.6$ and $s_{101} = 7.8$ are partial sums of $\sum_{k=1}^{\infty} a_k$, find a_1, a_2, a_3, a_4 and a_{100}.

Example 5. Graph the first five **terms** of the series $\sum_{k=1}^{\infty} \left(\frac{1}{2}\right)^k$ and then graph the first five **partial sums**.

Solution. The first few terms are $a_1 = \left(\frac{1}{2}\right)^1 = \frac{1}{2}$, $a_2 = \left(\frac{1}{2}\right)^2 = \frac{1}{4}$, $a_3 = \left(\frac{1}{2}\right)^3 = \frac{1}{8}$, $a_4 = \left(\frac{1}{2}\right)^4 = \frac{1}{16}$ and $a_5 = \left(\frac{1}{2}\right)^5 = \frac{1}{32}$, so $s_1 = a_1 = \frac{1}{2}$, $s_2 = a_1 + a_2 = \frac{1}{2} + \frac{1}{4} = \frac{3}{4}$, $s_3 = s_2 + a_3 = \frac{3}{4} + \frac{1}{8} = \frac{7}{8}$, $s_4 = \frac{15}{16}$ and $s_5 = \frac{31}{32}$. See margin for a graph of the terms and partial sums. ◀

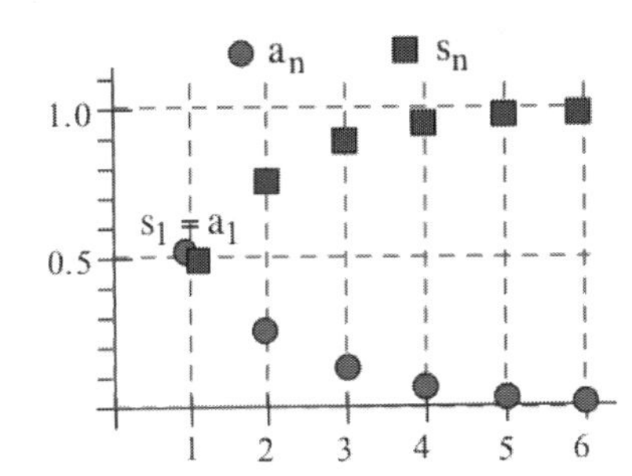

Practice 5. Graph the first five **terms** of the series $\sum_{k=1}^{\infty} \left(-\frac{1}{2}\right)^k$ and then graph the first five **partial sums**.

Convergence of a Series

In Example 1, you may have noticed that the sequence of total distances traveled by the golf ball after each subsequent bounce:

$$18, 30, 38, 43.67, \ldots, 53.996, \ldots, 53.99994, \ldots, 53.9999989, \ldots$$

appeared to approach a limit. If this sequence of partial sums does in fact approach 54 feet (which appears to be true, but remains a fact that we need to prove) then we could write:

$$18 + \left(\frac{2}{3}\right) \cdot 18 + \left(\frac{2}{3}\right)^2 \cdot 18 + \left(\frac{2}{3}\right)^3 \cdot 18 + \cdots = 54$$

and say that the infinite series $\sum_{k=0}^{\infty} 18 \cdot \left(\frac{2}{3}\right)^k$ **converges** to 54.

If the sequence of partial sums for an infinite series—the sequence obtained by adding up more and more of the terms of the series—approaches a finite number, we say the series **converges** to that finite number. If the sequence of partial sums diverges (does not approach a single finite number), we say that the series **diverges**.

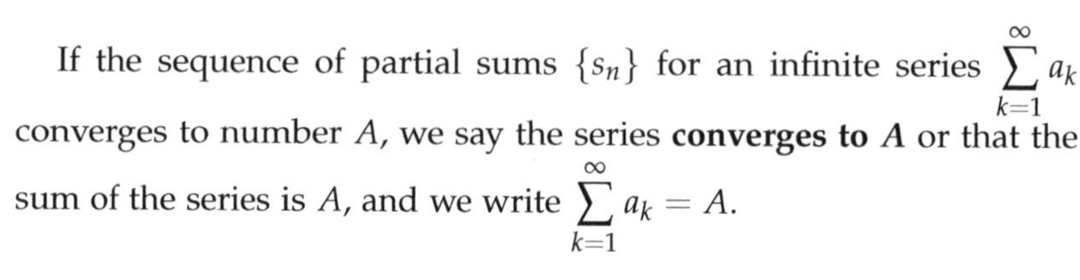

Definition:

Let $\{s_n\}$ be the sequence of partial sums for the series $\sum_{k=1}^{\infty} a_k$.

If the sequence $\{s_n\}$ converges, we say the series $\sum_{k=1}^{\infty} a_k$ **converges**.

If the sequence $\{s_n\}$ diverges, we say the series $\sum_{k=1}^{\infty} a_k$ **diverges**.

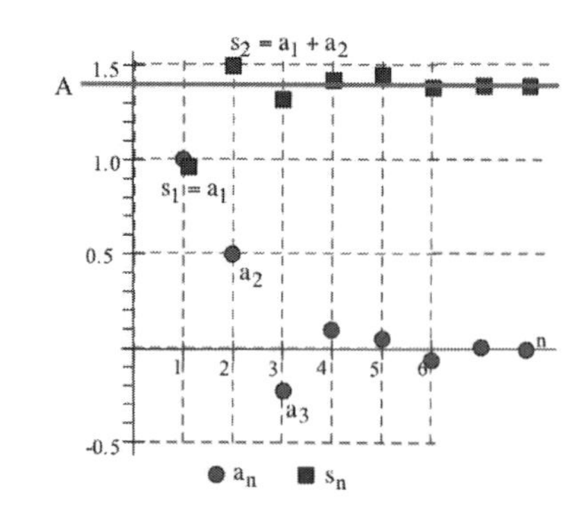

If the sequence of partial sums $\{s_n\}$ for an infinite series $\sum_{k=1}^{\infty} a_k$ converges to number A, we say the series **converges to A** or that the sum of the series is A, and we write $\sum_{k=1}^{\infty} a_k = A$.

Example 6. In Example 3(a), it appears that the partial sums of the infinite series $\sum_{k=0}^{\infty} \left(\frac{1}{2}\right)^k$ are:

$$s_n = \frac{2^{n+1}-1}{2^n} = 2 - \frac{1}{2^n}$$

Use this result to evaluate the limit of the partial sums $\{s_n\}$. Does the series $\sum_{k=0}^{\infty} \left(\frac{1}{2}\right)^k$ converge? If so, to what value?

Solution. $\lim_{n\to\infty} s_n = \lim_{n\to\infty} \left[2 - \frac{1}{2^n}\right] = 2$ so $\sum_{k=0}^{\infty} \left(\frac{1}{2}\right)^k = 2$. ◀

Practice 6. In Practice 5, the partial sums of $\sum_{k=1}^{\infty} \left(-\frac{1}{2}\right)^k$ are:

We will discover how to arrive at this formula for s_n in the next section.

$$s_n = -\frac{1}{3} - \frac{1}{3}\left(-\frac{1}{2}\right)^n$$

Use this result to evaluate the limit of the partial sums $\{s_n\}$. Does the series $\sum_{k=1}^{\infty} \left(-\frac{1}{2}\right)^k$ converge? If so, to what value?

A Test for Divergence

If a series converges, its partial sums s_n approach a finite limit L, so all of the s_n must be close to L for large values of n. This means that s_n and s_{n-1} must be close to each other (why?), so $a_n = s_n - s_{n-1} \longrightarrow 0$.

Theorem:

If the series $\sum_{k=1}^{\infty} a_k$ converges, then $\lim_{k\to\infty} a_k = 0$.

See Problem 47 for a formal proof.

We can **not** use this theorem to conclude that a series converges. If the terms of the series *do* approach 0, then the series may or may not converge—we will need more information to draw a conclusion above the convergence of the series. An alternate form of the above theorem, called the Test for Divergence, can be very useful for quickly concluding that some series diverge.

Test for Divergence:

If the terms of a series do not approach 0 (as $k \to \infty$)
then the series diverges.

In other words, if $\lim_{k\to\infty} a_k \neq 0$ then $\sum_{k=0}^{\infty} a_k$ must diverge.

Example 7. Which of these series must diverge according to the Test for Divergence?

(a) $\sum_{k=1}^{\infty} (-1)^k$ (b) $\sum_{k=1}^{\infty} \left(\frac{3}{4}\right)^k$ (c) $\sum_{k=1}^{\infty} \left(1+\frac{1}{k}\right)^k$ (d) $\sum_{k=1}^{\infty} \frac{1}{k}$

Solution. (a) $a_k = (-1)^k$ oscillates between -1 and $+1$ and does not approach 0, so $\sum_{k=1}^{\infty} (-1)^k$ diverges. (b) $\lim_{k\to\infty} \left(\frac{3}{4}\right)^k = 0$, so $\sum_{k=1}^{\infty} \left(\frac{3}{4}\right)^k$ may or may not converge. (c) $\lim_{k\to\infty} \left(1+\frac{1}{k}\right)^k = e \neq 0$, so $\sum_{k=1}^{\infty} \left(1+\frac{1}{k}\right)^k$ diverges. (d) $\lim_{k\to\infty} \frac{1}{k} = 0$, so $\sum_{k=1}^{\infty} \frac{1}{k}$ may or may not converge. We can be certain the series in (a) and (c) diverge. We don't have enough information to decide about (b) and (d). ◀

In the next section, we will show that the series in (b) converges and that the series in (d) diverges.

Practice 7. Which of these series must diverge according to the Test for Divergence?

(a) $\sum_{k=1}^{\infty} (-0.9)^k$ (b) $\sum_{k=1}^{\infty} (1.1)^k$ (c) $\sum_{k=1}^{\infty} \sin(k \cdot \pi)$ (d) $\sum_{k=1}^{\infty} \frac{1}{\sqrt{k}}$

New Series from Old

If we know whether an infinite series converges or diverges, then we also know whether several related series converge or diverge.

Inserting or deleting an *infinite* number of terms *can* change the convergence or divergence.

- Inserting or deleting a "few" terms (any finite number of terms) does not change the convergence or divergence of a series. The insertions or deletions typically change the sum of the series (the limit of its partial sums), but do not change whether or not it converges.
- Multiplying each term in a series by the same nonzero constant does not change the convergence or divergence of a series. If $c \neq 0$, then $\sum_{k=1}^{\infty} a_k$ converges if and only if $\sum_{k=1}^{\infty} c \cdot a_k$ converges.

Term-by-term multiplication and division of series do not have such nice results.

- Term-by-term addition and subtraction of the terms of two *convergent* series result in a new convergent series.

The proofs of these statements follow directly from the definition of convergence of a series and from results about convergence of sequences (of partial sums). See Problems 44–46.

> **Theorem:**
>
> If $\sum_{k=1}^{\infty} a_k = A$, $\sum_{k=1}^{\infty} b_k = B$ and $c \neq 0$
>
> then $\sum_{k=1}^{\infty} [a_k + b_k] = A + B$, $\sum_{k=1}^{\infty} [a_k - b_k] = A - B$
>
> and $\sum_{k=1}^{\infty} c \cdot a_k = c \cdot A$.

9.3 Problems

In Problems 1–6, rewrite each sum using sigma notation, starting with $k = 1$.

1. $1 + \frac{1}{2} + \frac{1}{3} + \frac{1}{4} + \frac{1}{5} + \cdots$
2. $1 + \frac{1}{4} + \frac{1}{9} + \frac{1}{16} + \frac{1}{25} + \cdots$
3. $\frac{2}{3} + \frac{2}{6} + \frac{2}{9} + \frac{2}{12} + \frac{2}{15} + \cdots$
4. $\sin(1) + \sin(8) + \sin(27) + \sin(64) + \cdots$
5. $-\frac{1}{2} + \frac{1}{4} - \frac{1}{8} + \frac{1}{16} - \cdots$
6. $-\frac{1}{3} + \frac{1}{9} - \frac{1}{27} + \frac{1}{81} - \cdots$

In Problems 7–12, calculate and graph the first four partial sums s_1 through s_4 of the given series.

7. $\sum_{k=1}^{\infty} k^2$
8. $\sum_{k=1}^{\infty} (-1)^k$
9. $\sum_{k=1}^{\infty} \frac{1}{k+2}$
10. $\sum_{k=1}^{\infty} (-\frac{1}{2})^k$
11. $\sum_{k=1}^{\infty} \frac{1}{2^k}$
12. $\sum_{k=2}^{\infty} \left[\frac{1}{k} - \frac{1}{k-1}\right]$

Problems 13–18 give the first five partial sums, s_1 through s_5, of a series $\sum_{k=1}^{\infty} a_k$. Find a_1 through a_4.

13. $s_1 = 3$, $s_2 = 2$, $s_3 = 4$, $s_4 = 5$, $s_5 = 3$
14. $s_1 = 3$, $s_2 = 5$, $s_3 = 4$, $s_4 = 6$, $s_5 = 5$
15. $s_1 = 4$, $s_2 = 4.5$, $s_3 = 4.3$, $s_4 = 4.8$, $s_5 = 5$
16. $s_1 = 4$, $s_2 = 3.7$, $s_3 = 3.9$, $s_4 = 4.1$, $s_5 = 4$
17. $s_1 = 1$, $s_2 = 1.1$, $s_3 = 1.11$, $s_4 = 1.111$, $s_5 = 1.1111$
18. $s_1 = 1$, $s_2 = 0.9$, $s_3 = 0.93$, $s_4 = 0.91$, $s_5 = 0.92$

In Problems 19–28, represent each repeating decimal as a series using sigma notation.

19. 0.888...
20. 0.333...
21. 0.555...
22. 0.111...
23. 0.aaa...
24. 0.232323...
25. 0.171717...
26. 0.838383...
27. 0.070707...
28. 0.ababab...
29. 0.abcabcabc...

30. After you throw a golf ball 20 feet straight up into the air, on each bounce it rebounds to 60% of its previous height. Represent the total distance traveled by the ball as an infinite sum.

31. After your friend throws a "super ball" 15 feet straight up, on each bounce it rebounds to 80% of its previous height. Represent the total distance traveled by the ball as an infinite sum.

32. Each special washing of a pair of coveralls removes 80% of the radioactive particles attached to the coveralls. Represent, as a sequence of numbers, the percent of the original radioactive particles that remain after each washing.

33. Each week, 20% of the argon gas in a container leaks out of the container. Represent, as a sequence of numbers, the percent of the original argon gas that remains in the container at the end of the first, second, third and n-th weeks.

34. Eight researchers depart on an expedition by horseback through desolate wilderness. The researchers and their equipment require 12 horses, and those horses require additional horses to carry food for the horses. Each horse can carry enough food to feed 2 horses for the trip. Represent the number of horses needed to carry food as an infinite sum.

Which of the series in Problems 35–43 definitely diverge by the Test for Divergence? What can you conclude about the other series?

35. $\sum_{k=1}^{\infty} \left(\frac{1}{4}\right)^k$ 36. $\sum_{k=1}^{\infty} \frac{7}{k}$ 37. $\sum_{k=1}^{\infty} \left(\frac{4}{3}\right)^k$

38. $\sum_{k=1}^{\infty} \left(-\frac{7}{4}\right)^2$ 39. $\sum_{k=1}^{\infty} \frac{\sin(k)}{k}$ 40. $\sum_{k=2}^{\infty} \frac{\ln(k)}{k}$

41. $\sum_{k=10}^{\infty} \cos(k)$ 42. $\sum_{k=5}^{\infty} \frac{k^2-20}{k^2+4}$ 43. $\sum_{k=5}^{\infty} \frac{k^2-20}{k^5+4}$

In Problems 44–47, you know that $\sum_{k=1}^{\infty} a_k = A$, $\sum_{k=1}^{\infty} b_k = B$ and $c \neq 0$.

44. Prove that $\sum_{k=1}^{\infty} [a_k + b_k] = A + B$.

45. Prove that $\sum_{k=1}^{\infty} c \cdot a_k = c \cdot A$.

46. Prove that $\sum_{k=1}^{\infty} [a_k - b_k] = A - B$. (Hint: Use the results from the two previous problems, rather than starting over from scratch.)

47. Prove that $\lim_{k \to \infty} a_k = 0$.

9.3 Practice Answers

1. The heights of the bounces are 10, $(0.4)10 = 4$, $(0.4)(0.4)10 = 1.6$, $(0.4)^3 10 = 0.64, \ldots$, so the distances traveled (up and down) by the ball are 20, $(0.4)20 = 8$, $(0.4)(0.4)20 = 3.2$, $(0.4)^3 20 = 1.28, \ldots$. The total distance traveled is:

$$20 + (0.4)20 + (0.4)^2 20 + (0.4)^3 .0 + \cdots$$
$$= 20\left[1 + 0.4 + (0.4)^2 + (0.4)^3 + \cdots\right] = 20 \sum_{k=0}^{\infty} (0.4)^k$$

After three bounces, the ball has traveled $20 + 8 + 3.2 = 31.2$ feet.

2. (a) $\sum_{k=1}^{\infty} k$ (b) $\sum_{k=1}^{\infty} (-1)^k$ (c) $\sum_{k=-1}^{\infty} \left(\frac{1}{2}\right)^k = 2\sum_{k=0}^{\infty} \left(\frac{1}{2}\right)^k$ (d) $\sum_{k=1}^{\infty} \frac{1}{2k}$
(e) We can write $0.111\ldots$ as:

$$0.111\ldots = 0.1 + 0.01 + 0.001 + \cdots = \frac{1}{10} + \frac{1}{100} + \frac{1}{1000} + \cdots$$
$$= \left(\frac{1}{10}\right)^1 + \left(\frac{1}{10}\right)^2 + \left(\frac{1}{10}\right)^3 + \cdots = \sum_{k=1}^{\infty} (\frac{1}{10})^k$$

3. (a) $s_1 = 1$, $s_2 = \frac{1}{2}$, $s_3 = \frac{3}{4}$, $s_4 = \frac{5}{8}$ (b) $s_1 = \frac{1}{3}$, $s_2 = \frac{4}{9}$, $s_3 = \frac{13}{27}$, $s_4 = \frac{40}{81}$
(c) $s_1 = -1$, $s_2 = -1 + \frac{1}{2} = -\frac{1}{2}$, $s_3 = -\frac{1}{2} - \frac{1}{3} = -\frac{5}{6}$, $s_4 = -\frac{7}{12}$

4. $a_1 = s_1 = 3.2$, $a_2 = s_2 - s_1 = 3.6 - 3.2 = 0.4$, $a_3 = s_3 - s_2 = 3.5 - 3.6 = -0.1$, $a_4 = s_4 - s_3 = 4 - 3.5 = 0.5$, $a_{100} = s_{100} - s_{99} = 7.6 - 7.3 = 0.3$

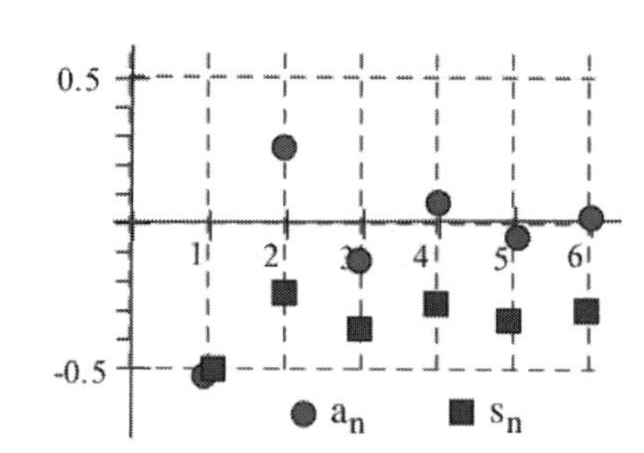

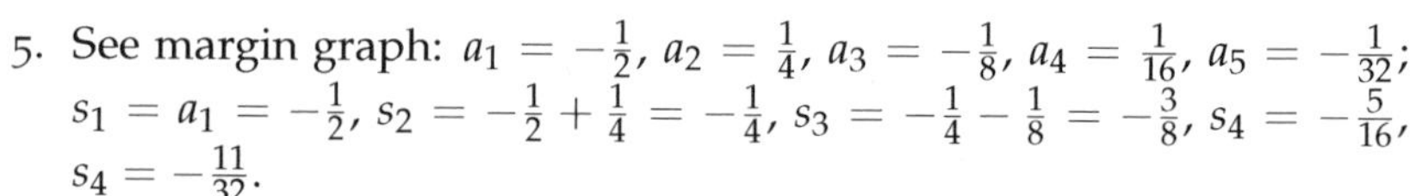

5. See margin graph: $a_1 = -\frac{1}{2}$, $a_2 = \frac{1}{4}$, $a_3 = -\frac{1}{8}$, $a_4 = \frac{1}{16}$, $a_5 = -\frac{1}{32}$; $s_1 = a_1 = -\frac{1}{2}$, $s_2 = -\frac{1}{2} + \frac{1}{4} = -\frac{1}{4}$, $s_3 = -\frac{1}{4} - \frac{1}{8} = -\frac{3}{8}$, $s_4 = -\frac{5}{16}$, $s_4 = -\frac{11}{32}$.

6. Taking the limit of the partial sums of the series:

$$\lim_{n\to\infty} s_n = \lim_{n\to\infty} \left[-\frac{1}{3} - \frac{1}{3}\left(-\frac{1}{2}\right)^n\right] = -\frac{1}{3}$$

so the series $\sum_{k=1}^{\infty} \left(-\frac{1}{2}\right)^k$ converges and its sum is $-\frac{1}{3}$.

7. (a) $\lim_{k\to\infty} (-0.9)^k = 0$, so $\sum_{k=1}^{\infty} (-0.9)^k$ may or may not converge.
(b) $\lim_{k\to\infty} (1.1)^k = \infty \neq 0$ so $\sum_{k=1}^{\infty} (1.1)^k$ must diverge. (c) $\sin(k \cdot \pi) = 0$ for any integer k, so all of the terms of the series are 0; the test for Divergence tells us nothing, but because all of the terms are 0, we know that $\sum_{k=1}^{\infty} \sin(k \cdot \pi) = 0$ (so the series converges). (d) $\lim_{k\to\infty} \frac{1}{\sqrt{k}} = 0$, so $\sum_{k=1}^{\infty} \frac{1}{\sqrt{k}}$ may or may not converge.

In the next section, we will show that the series in (a) converges and the series in (d) diverges.

9.4 Geometric and Harmonic Series

This section investigates three special types of series. Geometric series appear throughout mathematics and arise in a variety of applications. Much of the early work in the 17th century with series focused on geometric series and generalized them. And many of the ideas used later in this chapter (and into the next) originated with geometric series. It is very easy to determine whether a geometric series converges or diverges — and when one does converge, we can easily find its sum.

The harmonic series is important as an example of a *divergent* series whose terms approach 0.

The last part of this section briefly discusses a third type of series, called "telescoping": these are relatively uncommon, but their partial sums exhibit a particularly nice pattern.

Geometric Series

Example 1. Your friend throws a "super ball" 10 feet straight up into the air. On each bounce, it rebounds to 80% of its previous height (see margin) so the sequence of heights the ball attains is 10 feet, 8 feet, 6.4 feet, 5.12 feet, and so on.

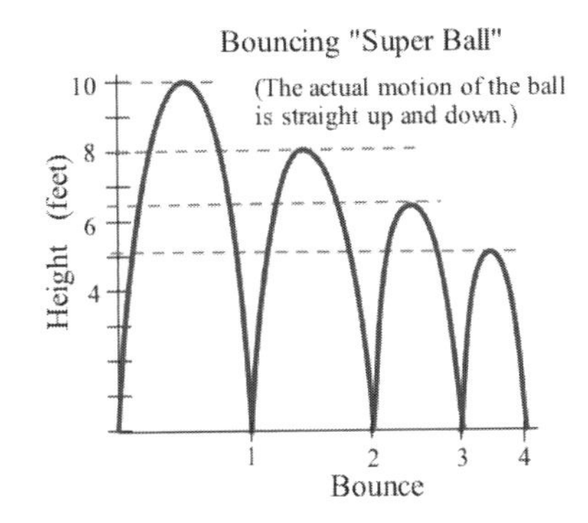

(a) How far does the ball travel (up and down) during its n-th bounce?

(b) Use a sum to represent the total distance traveled by the ball.

Solution. (a) Because the ball travels up and down on each bounce, the distance traveled during each bounce is twice the height the ball attains on that bounce. So the distance the ball travels prior to its first bound is $d_1 = 2\,(10 \text{ feet}) = 20$ feet, the distance it travels between the first and second bounces is $d_2 = (0.8)(20) = 16$ feet, $d_3 = (0.8)(16) = 12.8$ feet and, in general, $d_n = 0.8d_{n-1}$. Looking at these values in another way:

$$d_1 = 20,\ d_2 = 0.8(20),\ d_3 = (0.8)d_2 = (0.8)^2(20),\ d_4 = (0.8)^3(20)$$

and, in general, $d_n = (0.8)^{n-1}(20)$.

(b) In theory, the ball bounces up and down forever, and the total distance traveled by the ball is the sum of the distances traveled during each bounce (an up-and-down flight):

$$\begin{aligned} 20 + (0.8)(20) + (0.8)^2(20) + (0.8)^3(20) + (0.8)^4(20) + \cdots \\ = 20\left[1 + 0.8 + (0.8)^2 + (0.8)^3 + (0.8)^4 + \cdots\right] \\ = 20 \cdot \sum_{k=0}^{\infty} (0.8)^k \end{aligned}$$

(In practice, the ball will eventually top bouncing.) ◀

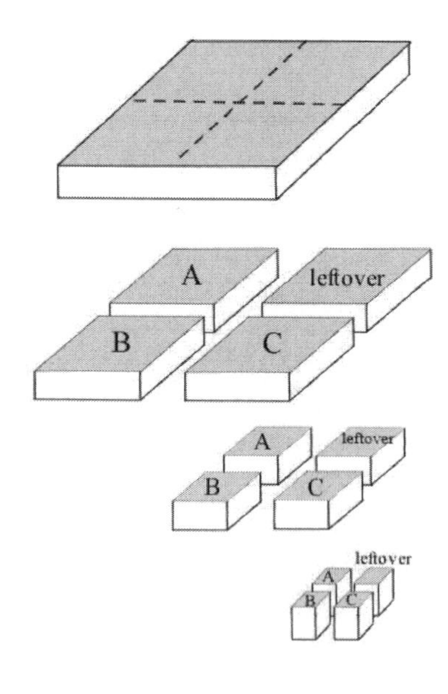

Practice 1. Three calculus students want to share a small square cake equally, but they go about it in a rather strange way. First they cut the cake into four equal square pieces, then each person takes one square, leaving one square (see margin). Then they cut the leftover square piece into four equal square pieces, each person takes one square, leaving one square. And they keep repeating this process.

(a) What fraction of the total cake does each student "eventually" eat?

(b) Represent the amount of cake each person gets as an infinite series.

The infinite series in the previous Example and Practice problems are both **geometric series**, a type of series in which each term is a fixed multiple of the previous term. Geometric series have the form:

$$\sum_{k=0}^{\infty} C \cdot r^k = C + C \cdot r + C \cdot r^2 + C \cdot r^3 + \cdots$$
$$= C\left[1 + r + r^2 + r^3 + \cdots\right] = C \cdot \sum_{k=0}^{\infty} r^k$$

with $C \neq 0$ and $r \neq 0$ representing fixed numbers. Each term in the series is r times the previous term. Geometric series are among the most common series we will encounter, and among the easiest to work with. A simple test determines whether a geometric series converges, and we can even determine the sum of any convergent geometric series.

> **Geometric Series Theorem**
>
> The geometric series $\sum_{k=0}^{\infty} r^k = 1 + r + r^2 + r^3 + \cdots$
>
> - converges to $\dfrac{1}{1-r}$ if $|r| < 1$
> - diverges if $|r| \geq 1$

Proof. If $|r| = 1$, then $\lim_{k\to\infty} \left|r^k\right| = 1$; if $|r| > 1$, then $\lim_{k\to\infty} \left|r^k\right| = \infty$. Either way, $\lim_{k\to\infty} r^k \neq 0$, so $\sum_{k=0}^{\infty} r^k$ diverges by the Test for Divergence.

If $|r| < 1$, then $\lim_{k\to\infty} r^k = 0$, so the Test for Divergence says that $\sum_{k=0}^{\infty} r^k$ may or may not converge.

Examining the partial sums $s_n = 1 + r + r^2 + r^3 + \cdots + r^n$ of the geometric series when $|r| < 1$, a clever insight allows us to find a simple formula for those partial sums:

$$\begin{aligned}
(1-r) \cdot s_n &= (1-r) \cdot \left(1 + r + r^2 + r^3 + \cdots + r^n\right) \\
&= \left(1 + r + r^2 + r^3 + \cdots + r^n\right) - r \cdot \left(1 + r + r^2 + r^3 + \cdots + r^n\right) \\
&= \left(1 + r + r^2 + r^3 + \cdots + r^n\right) - \left(r + r^2 + r^3 + \cdots + r^n + r^{n+1}\right) \\
&= 1 - r^{n+1}
\end{aligned}$$

Because $|r| < 1$, we know that $r \neq 1$ so that $1 - r \neq 0$, meaning we can divide the preceding equality by $1 - r$ to get:

$$s_n = 1 + r + r^2 + r^3 + \cdots + r^n = \frac{1 - r^{n+1}}{1 - r}$$

This formula for the n-th partial sum of a geometric series is sometimes useful; at the moment, we are interested in $\lim\limits_{n\to\infty} s_n$.

Because $|r| < 1$, $\lim\limits_{n\to\infty} r^{n+1} = 0$, so we can conclude that:

$$\sum_{k=0}^{\infty} r^k = \lim_{n\to\infty} \sum_{k=0}^{n} r^k = \lim_{n\to\infty} s_n = \lim_{n\to\infty} \frac{1 - r^{n+1}}{1 - r} = \frac{1}{1 - r}$$

giving a formula for the sum of any (convergent) geometric series. □

More generally, for any $C \neq 0$ and any r with $|r| < 1$ we can write:

$$\sum_{k=0}^{\infty} C \cdot r^k = \frac{C}{1 - r}$$

allowing us to quickly find the sum of any convergent geometric series.

Example 2. How far did the ball in Example 1 travel?

Solution. In Example 1, we expressed the total distance the ball travels as a geometric series, so:

$$20 \cdot \sum_{k=0}^{\infty} (0.8)^k = 20 \cdot \frac{1}{1 - 0.8} = \frac{20}{0.2} = 100$$

so the ball (theoretically) travels a total distance of 100 feet. ◀

Repeating decimal numbers are really geometric series in disguise, so we can now represent their exact values as fractions.

Example 3. Represent the repeating decimals $0.\overline{4}$ and $0.\overline{13}$ as geometric series and find their sums.

Solution. We can rewrite $0.\overline{4} = 0.444\ldots$ as:

$$\begin{aligned} 0.444\ldots &= \frac{4}{10} + \frac{4}{100} + \frac{4}{1000} + \cdots = \frac{4}{10} + \frac{4}{10^2} + \frac{4}{10^3} + \cdots \\ &= \frac{4}{10}\left[1 + \frac{1}{10} + \frac{1}{10^2} + \cdots\right] = \frac{4}{10} \cdot \sum_{k=0}^{\infty} \left(\frac{1}{10}\right)^k \\ &= \frac{4}{10} \cdot \frac{1}{1 - \frac{1}{10}} = \frac{4}{10} \cdot \frac{1}{\frac{9}{10}} = \frac{4}{10} \cdot \frac{10}{9} = \frac{4}{9} \end{aligned}$$

so $0.\overline{4} = \frac{4}{9}$. Proceeding similarly with $0.\overline{13} = 0.131313\ldots$:

$$\begin{aligned} 0.131313\ldots &= \frac{13}{100} + \frac{13}{10000} + \frac{13}{1000000} + \cdots = \frac{13}{100} + \frac{13}{100^2} + \frac{13}{100^3} + \cdots \\ &= \frac{13}{100}\left[1 + \frac{1}{100} + \frac{1}{100^2} + \cdots\right] = \frac{13}{100} \cdot \sum_{k=0}^{\infty} \left(\frac{1}{100}\right)^k \\ &= \frac{13}{100} \cdot \frac{1}{1 - \frac{1}{100}} = \frac{13}{100} \cdot \frac{1}{\frac{99}{100}} = \frac{13}{100} \cdot \frac{100}{99} = \frac{13}{99} \end{aligned}$$

so we can express $0.\overline{13}$ as $\frac{13}{99}$. ◀

Practice 2. Represent the repeating decimals $0.\overline{3}$ and $0.\overline{432}$ as geometric series and find their sums.

Replacing the number r with an expression involving x allows us to create a function defined as an infinite series.

Example 4. Given the functions $f(x) = \sum_{k=0}^{\infty} 3x^k = 3 + 3x + 3x^2 + \cdots$ and $g(x) = \sum_{k=0}^{\infty} (2x-5)^k = 1 + (2x-5) + (2x-5)^2 + \cdots$, find the domains of each function (that is, determine the values of x for which each infinite series converges).

Solution. We know that a geometric series converges if and only if $|r| < 1$, and the series defining $f(x)$ has ratio $r = x$, so it converges if and only if $|x| < 1$. The sum of this first series is:

$$f(x) = \sum_{k=0}^{\infty} 3x^k = 3 \cdot \sum_{k=0}^{\infty} x^k = 3 \cdot \frac{1}{1-x} = \frac{3}{1-x}$$

provided that $|x| < 1$, or, equivalently, that $-1 < x < 1$. Notice that the domain of the function $\frac{3}{1-x}$ consists of all real numbers except $x = 1$, but that $\sum_{k=0}^{\infty} 3x^k$ converges only when $-1 < x < 1$, so $f(x) = \frac{3}{1-x}$ holds only on the interval $(-1, 1)$.

In the series defining $g(x)$, the ratio is $r = 2x - 5$ so the series converges if and only if $|2x - 5| < 1$. Removing the absolute values and solving for x, we get:

$$|2x-5| < 1 \quad \Rightarrow \; -1 < 2x - 5 < 1 \quad \Rightarrow \; 2 < x < 3$$

so this second series converges precisely on the interval $(2, 3)$. The sum of the second series is:

$$g(x) = \sum_{k=0}^{\infty} (2x-5)^k = \frac{1}{1-(2x-5)} = \frac{1}{6-2x}$$

as long as $2 < x < 3$. ◀

Practice 3. Given $F(x) = \sum_{k=0}^{\infty} (2x)^k$ and $G(x) = \sum_{k=0}^{\infty} (3x-4)^k$, determine the values of x for which these series converge.

We call the type of series in the previous Example and Practice problems "**power series**" because they involve powers of the variable x. In the next chapter, we will embark on an extensive investigation of other power series, including many non-geometric series, such as:

$$1 + x + \frac{1}{2}x^2 + \frac{1}{3}x^3 + \cdots$$

For each power series we will attempt to determine the values of x for which the series is guaranteed to converge.

These series are extensions of the MacLaurin polynomials from Section 8.7 to "infinite polynomials" with an unlimited number of terms.

The Harmonic Series: $\sum_{k=1}^{\infty} \frac{1}{k}$

The series $\sum_{k=1}^{\infty} \frac{1}{k}$ ranks among the best-known and most important divergent series. We call it the **harmonic series** because of its ties to music (see the discussion following the Practice Answers for additional background). Because $\lim_{k\to\infty} \frac{1}{k} = 0$, the Test for Divergence tells us nothing about the convergence or divergence of this series. Calculating partial sums of the harmonic series (see margin table) reveals that the partial sums s_n increase very, very slowly. By taking $n \approx 2{,}000{,}000$, we can make $s_n > 15$, but does it ever exceed 16? The answer to that question turns out to be "yes," but in our examination of the divergence of the harmonic series, brain power will prove to be much more effective than lots and lots of computing power.

n	s_n
31	4.0224519544
83	5.00206827268
227	6.00436670835
1,674	8.00048557200
12,367	10.00004300827
1,835,421	15.00000378267

Theorem:

The harmonic series $\sum_{k=1}^{\infty} \frac{1}{k} = 1 + \frac{1}{2} + \frac{1}{3} + \frac{1}{4} + \cdots$ diverges.

Proof. Assume the harmonic series converges, and let S be its sum:

$$S = \sum_{k=1}^{\infty} \frac{1}{k} = 1 + \frac{1}{2} + \frac{1}{3} + \frac{1}{4} + \frac{1}{5} + \frac{1}{6} + \frac{1}{7} + \frac{1}{8} + \cdots$$

This proof is essentially due to Oresme in 1630 (12 years before Newton's birth). In 1821, Cauchy included Oresme's proof in his *Course in Analysis*, after which it became known as "Cauchy's argument."

Next, group the terms of the series as indicated by the brackets:

$$S = 1 + \frac{1}{2} + \left[\frac{1}{3} + \frac{1}{4}\right] + \left[\frac{1}{5} + \frac{1}{6} + \frac{1}{7} + \frac{1}{8}\right] + \left[\frac{1}{9} + \frac{1}{10} + \frac{1}{11} + \frac{1}{12} + \frac{1}{13} + \frac{1}{14} + \frac{1}{15} + \frac{1}{16}\right] + \cdots$$

so each set of brackets includes twice as many terms as the previous set. Notice that:

$$\frac{1}{3} + \frac{1}{4} > \frac{1}{4} + \frac{1}{4} = \frac{2}{4} = \frac{1}{2}$$

Looking at the next set of terms:

$$\frac{1}{5} + \frac{1}{6} + \frac{1}{7} + \frac{1}{8} > \frac{1}{8} + \frac{1}{8} + \frac{1}{8} + \frac{1}{8} = \frac{4}{8} = \frac{1}{2}$$

The sum of the terms in the next set of brackets exceeds $\frac{8}{16} = \frac{1}{2}$, the sum after that exceeds $\frac{16}{32} = \frac{1}{2}$, and so on. We can therefore write:

$$S > 1 + \frac{1}{2} + \frac{1}{2} + \frac{1}{2} + \frac{1}{2} + \cdots$$

By adding enough sets of terms, you can make this lower bound for S arbitrarily large: S cannot be finite, so the harmonic series diverges. □

The Test for Divergence is typically a good first step when investigating the converge or divergence of an infinite series, but merely works as a "screening test" to identify certain series that definitely diverge.

The harmonic series provides an example of a *divergent* series whose terms, $a_k = \frac{1}{k}$, approach 0. Any geometric series with $|r| < 1$ provides an example of a *convergent* series whose terms approach 0. This illustrates why the Test for Divergence says that a series with terms approaching 0 may or may not converge.

Telescoping Series

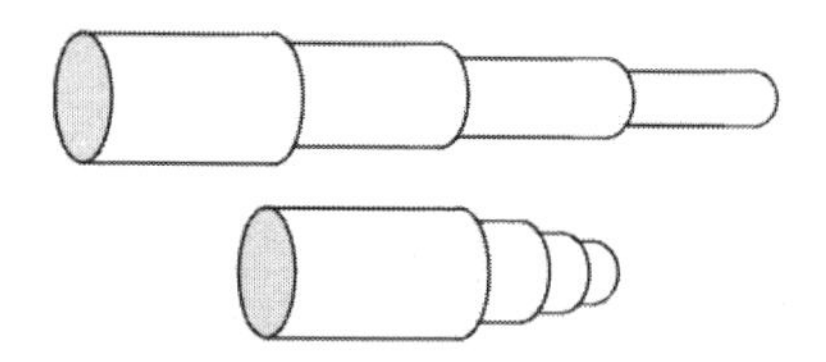

During the 17th and 18th centuries, sailors used telescopes (see margin) that could be extended for viewing and collapsed for storage. Telescoping series get their name because they exhibit a similar "collapsing" property. Telescoping series arise infrequently, but they are easy to analyze and it can be useful to recognize them.

Example 5. Determine a formula for the partial sum s_n of the series $\sum_{k=1}^{\infty}\left[\frac{1}{k} - \frac{1}{k+1}\right]$ and then compute $\lim_{n\to\infty} s_n$.

Solution. It is tempting to rewrite the formula for a_k as a single fraction, but the pattern becomes clearer if you begin writing out all of the terms:

$$s_1 = \left[1 - \frac{1}{2}\right] = 1 - \frac{1}{2}$$

$$s_2 = \left[1 - \frac{1}{2}\right] + \left[\frac{1}{2} - \frac{1}{3}\right] = 1 - \frac{1}{3}$$

$$s_3 = \left[1 - \frac{1}{2}\right] + \left[\frac{1}{2} - \frac{1}{3}\right] + \left[\frac{1}{3} - \frac{1}{4}\right] = 1 - \frac{1}{4}$$

In these partial sums, all terms cancel except the first and last terms, so we can write:

$$s_n = \left[1 - \frac{1}{2}\right] + \left[\frac{1}{2} - \frac{1}{3}\right] + \left[\frac{1}{3} - \frac{1}{4}\right] + \cdots + \left[\frac{1}{n} - \frac{1}{n+1}\right] = 1 - \frac{1}{n+1}$$

It should now be obvious that $\lim_{n\to\infty} s_n = 1$ so that the original series converges and equals 1. ◀

Practice 4. Find the sum of the series $\sum_{k=3}^{\infty}\left[\sin\left(\frac{1}{k}\right) - \sin\left(\frac{1}{k+1}\right)\right]$.

9.4 Problems

In Problems 1–12, calculate the value of the sum or explain why the series diverges.

1. $\sum_{k=0}^{\infty}\left(\frac{2}{7}\right)^k$
2. $\sum_{k=0}^{\infty}\left(\frac{4}{7}\right)^k$
3. $\sum_{k=0}^{\infty}\left(-\frac{4}{7}\right)^k$
4. $\sum_{k=0}^{\infty}\left(-\frac{2}{7}\right)^k$
5. $\sum_{k=1}^{\infty}\left(\frac{2}{7}\right)^k$
6. $\sum_{k=2}^{\infty}\left(\frac{4}{7}\right)^k$
7. $\sum_{k=3}^{\infty}\left(-\frac{7}{4}\right)^k$
8. $\sum_{k=4}^{\infty}\left(-\frac{7}{2}\right)^k$
9. $\sum_{k=5}^{\infty}\left(-\frac{2}{7}\right)^k$

10. $\sum_{k=0}^{\infty}\left(\frac{e}{3}\right)^k$ 11. $\sum_{k=0}^{\infty}\left(\frac{\pi}{3}\right)^k$ 12. $\sum_{k=0}^{\infty}\left(\frac{e}{\pi}\right)^k$

In Problems 13–18, rewrite each geometric series using sigma notation, then compute the sum.

13. $1+\frac{1}{3}+\frac{1}{9}+\frac{1}{27}+\cdots$ 14. $1+\frac{2}{3}+\frac{4}{9}+\frac{8}{27}+\cdots$

15. $\frac{1}{8}+\frac{1}{16}+\frac{1}{32}+\frac{1}{64}+\cdots$

16. $1-\frac{1}{2}+\frac{1}{4}-\frac{1}{8}+\cdots$

17. $-\frac{2}{3}+\frac{4}{9}-\frac{8}{27}+\frac{16}{81}-\cdots$

18. $1+\frac{1}{e}+\frac{1}{e^2}+\frac{1}{e^3}+\cdots$

19. Show that:
   (a) $\frac{1}{2}+\frac{1}{4}+\frac{1}{8}+\cdots=1$
   (b) $\frac{1}{3}+\frac{1}{9}+\frac{1}{27}+\cdots=\frac{1}{2}$
   (c) $\frac{1}{a}+\frac{1}{a^2}+\frac{1}{a^3}+\cdots=\frac{1}{a-1}$ (for $a>1$)

20. You throw a ball 10 feet straight up into the air, and on each bounce it rebounds to 60% of its previous height.
   (a) How far does the ball travel (up and down) during its n-th bounce?
   (b) Use an infinite sum to represent the total distance traveled by the ball.
   (c) Find the total distance traveled by the ball.

21. Your friend throws an old tennis ball 20 feet straight up into the air, and on each bounce it rebounds to 40% of its previous height.
   (a) How far does the ball travel (up and down) during its n-th bounce?
   (b) Use an infinite sum to represent the total distance traveled by the ball.
   (c) Find the total distance traveled by the ball.

22. Eighty people embark on an expedition by horseback through desolate country. The people and their gear require 90 horses, with additional horses needed to carry food for the original 90 horses. Each additional horse can carry enough food to feed three horses for the trip. How many total horses are needed for the trip?

23. Your friend follows a mathematical diet that says he can eat "half of whatever is on the plate," so he bites off half of a cake and puts the other half back on the plate. Then he picks up the remaining half from the plate (it's "on the plate"), bites off half of that and returns the rest to the plate. He continues this silly process of picking up the remaining piece from the plate, biting off half, and returning the rest to the plate.
   (a) Use an infinite sum to represent the total amount of cake he eats.
   (b) How much cake is left after one bite? Two bites? n bites?
   (c) "Eventually," how much does he eat?

24. As indicated in the figure below left, begin with a square with sides of length 1 (so its area is 1). Construct another square inside the first one by connecting the midpoints of the sides of the first square, so the new square has area $\frac{1}{2}$. Continue this process of constructing each new square by connecting the midpoints of the sides of the previous square to get a sequence of squares, each of which has half the area of the previous square. Find the total area of all of the squares.

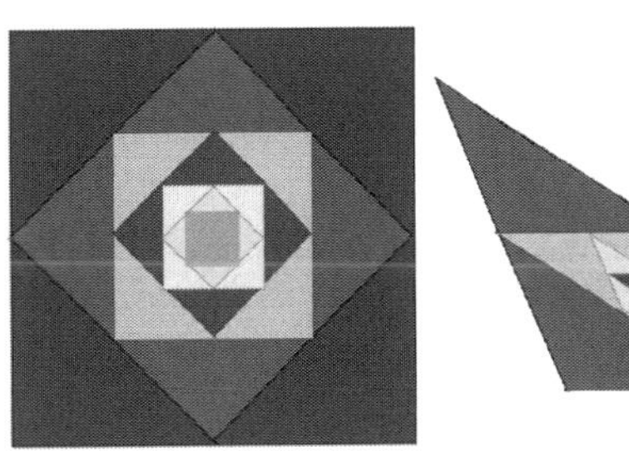

25. As indicated in the figure above right, begin with a triangle with area 1. Construct another triangle inside the first triangle by connecting the midpoints of the sides of the first triangle, so this new triangle will have area $\frac{1}{4}$. Imagine that you continue this construction process "forever" and find the total area of all of the triangles.

26. Begin with a circle of radius 1. Construct two more circles inside the first one, each with radius $\frac{1}{2}$. Continue this process "forever," constructing two new circles inside each previous circle (see

below left). Find the total area of all the circles.

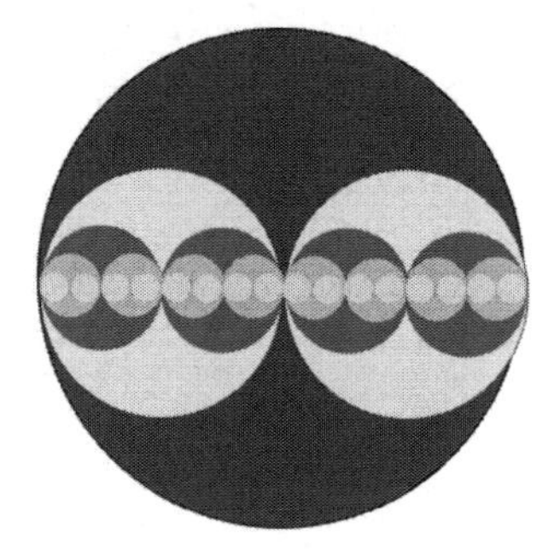

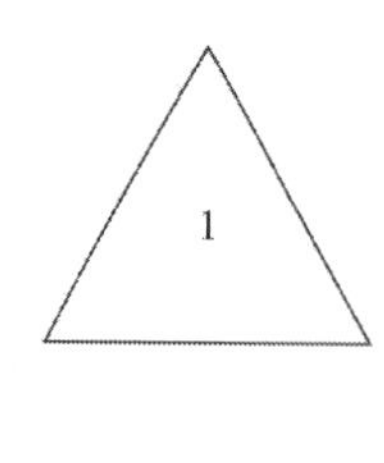

27. Swedish mathematician Helge von Koch (1870-1924) described one of the earliest examples of a fractal, now known as the **Koch snowflake**. Beginning with an equilateral triangle of area 1 (above right), subdivide each edge into three equal lengths, then build three equilateral triangles, each with area $\frac{1}{9}$, on the "middle thirds" of sides of the original triangle, adding a total of $3 \cdot \frac{1}{9} = \frac{1}{3}$ to the original area (below left). Now repeat this process: at the next stage, build $3 \cdot 4$ equilateral triangles, each with area $\frac{1}{81}$ on the new "middle thirds," adding $3 \cdot 4 \cdot \frac{1}{81}$ to the total area.
   (a) Find the total area that results when you repeat this process "forever."
   (b) Express the perimeter of the Koch Snowflake as a geometric series and find its sum.

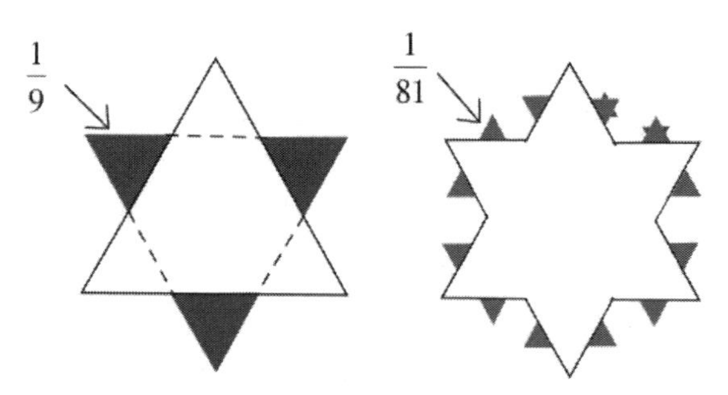

28. The base of a "harmonic tower" is a cube with edges one foot long. Sitting on top of the bases are cubes with edges of length $\frac{1}{2}, \frac{1}{3}, \frac{1}{4}$ and so on.
   (a) Represent the total height of the tower as a series. Is the height finite?
   (b) Represent the total surface area of the cubes as an infinite series.
   (c) Represent the total volume of the cubes as an infinite series.

(See below left for a picture of the harmonic tower. In the next section, we will be able to determine whether its surface area and volume are finite or infinite.)

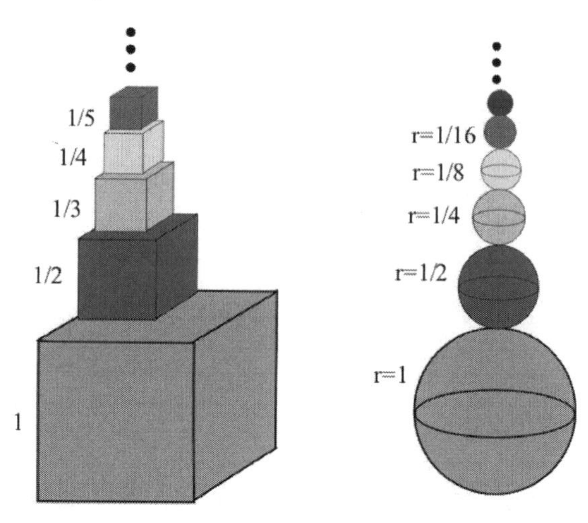

29. The base of a tower is a sphere with radius one foot. On top of each sphere sits another sphere with a radius half the radius of the sphere immediately beneath it (as in the figure above right).
   (a) Represent the total height of the tower as a series and evaluate the sum.
   (b) Represent the total surface area of the spheres as an infinite series and evaluate the sum.
   (c) Represent the total volume of the spheres as an infinite series and evaluate the sum.

30. Represent the repeating decimals $0.\overline{6}$ and $0.\overline{63}$ as geometric series and express the value of each series as a fraction in lowest terms.

31. Represent the repeating decimals $0.\overline{8}$, $0.\overline{9}$ and $0.\overline{285714}$ as geometric series and express the value of each series as a fraction in lowest terms.

32. Represent the repeating decimals $0.\overline{a}$, $0.\overline{ab}$ and $0.\overline{abc}$ as geometric series and express the value of each series as a fraction. What do you think a fractional representation for $0.\overline{abcd}$ would be?

In Problems 33–44, find all values of x for which the geometric series converges.

33. $\sum_{k=1}^{\infty} (2x+1)^k$ 34. $\sum_{k=1}^{\infty} (3-x)^k$ 35. $\sum_{k=1}^{\infty} (1-2x)^k$

36. $\sum_{k=1}^{\infty} 5x^k$ 37. $\sum_{k=1}^{\infty} (7x)^k$ 38. $\sum_{k=1}^{\infty} \left(\frac{x}{3}\right)^k$

39. $1 + \frac{x}{2} + \frac{x^2}{4} + \frac{x^3}{8} + \cdots$

40. $1 + \frac{2}{x} + \frac{4}{x^2} + \frac{8}{x^3} + \cdots$

41. $1 + 2x + 4x^2 + 8x^3 + \cdots$

42. $\sum_{k=1}^{\infty} \left(\frac{2x}{3}\right)^k$ 43. $\sum_{k=1}^{\infty} \sin^k(x)$ 44. $\sum_{k=1}^{\infty} e^{kx}$

45. A student thought she remembered the formula for a geometric series as:

$$1 + x + x^2 + x^3 + \cdots = \frac{1}{1-x}$$

Her friend said, "That can't be right. If we replace x with 2, then the formula says the sum of the positive numbers $1 + 2 + 4 + 8 + \cdots$ is a negative number: $\frac{1}{1-2} = -1$." Who was right? Why?

46. If you have many identical 1-foot-long boards, you can arrange them so that they hang over the edge of a table. One board can extend $\frac{1}{2}$ foot beyond the edge, two boards can extend $\frac{1}{2} + \frac{1}{4}$ feet and, in general, n boards can extend:

$$\frac{1}{2} + \frac{1}{4} + \frac{1}{6} + \cdots + \frac{1}{2n}$$

feet beyond the edge (see below).

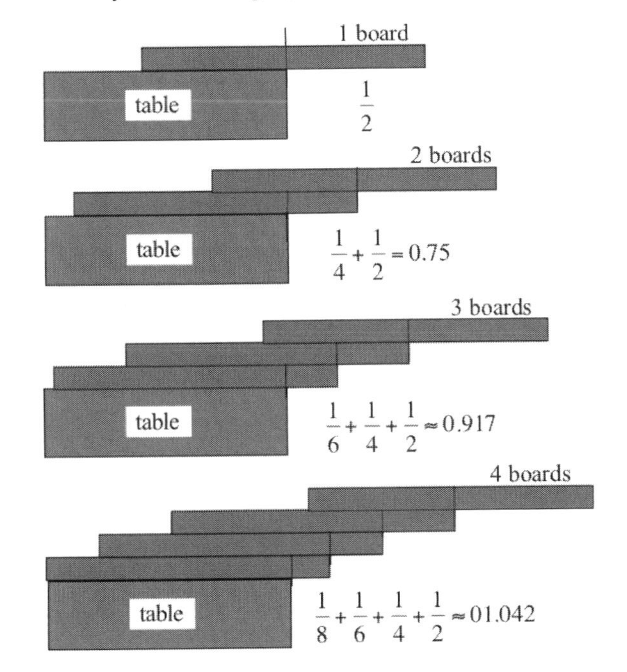

(a) How many boards do you need for an arrangement in which the entire top board sits beyond the edge of the table?

(b) How many boards do you need for an arrangement in which the entire top *two* boards sit beyond the edge of the table?

(c) How far can any such arrangement extend beyond the edge of the table?

In Problems 47–52, compute the value of the partial sums s_4 and s_5 for the given series, then find a formula for s_n. (The patterns may be more obvious if you do not simplify each term.)

47. $\sum_{k=3}^{\infty} \left[\frac{1}{k} - \frac{1}{k+1}\right]$

48. $\sum_{k=1}^{\infty} \left[\cos\left(\frac{1}{k}\right) - \cos\left(\frac{1}{k+2}\right)\right]$

49. $\sum_{k=1}^{\infty} \left[k^3 - (k+1)^3\right]$ 50. $\sum_{k=1}^{\infty} \left[\ln\left(\frac{k}{k+1}\right)\right]$

51. $\sum_{k=1}^{\infty} [f(k) - f(k+1)]$ 52. $\sum_{k=1}^{\infty} [g(k) - g(k+1)]$

In 53–56, compute s_4 and s_5 for each series and then $\lim_{n\to\infty} s_n$. (If the limit is a finite value, it represents the value of the corresponding infinite series.)

53. $\sum_{k=1}^{\infty} \left[\sin\left(\frac{1}{k}\right) - \sin\left(\frac{1}{k+1}\right)\right]$

54. $\sum_{k=2}^{\infty} \left[\cos\left(\frac{1}{k}\right) - \cos\left(\frac{1}{k+1}\right)\right]$

55. $\sum_{k=2}^{\infty} \left[\frac{1}{k^2} - \frac{1}{(k+1)^2}\right]$ 56. $\sum_{k=3}^{\infty} \ln\left(1 - \frac{1}{k^2}\right)$

Problems 57–58 outline two "proofs by contradiction" that the harmonic series diverges. Each proof begins with the assumption that the "sum" of the harmonic series is a finite number and then obtains an obviously false conclusion based on this assumption. Verify that each step follows from the assumption and the previous steps, then explain why the conclusion is false.

57. Assume that $H = 1 + \frac{1}{2} + \frac{1}{3} + \frac{1}{4} + \frac{1}{5} + \cdots$ is a finite number. Let $O = 1 + \frac{1}{3} + \frac{1}{5} + \cdots$ be the sum of the "odd reciprocals" and $E = \frac{1}{2} + \frac{1}{4} + \frac{1}{6} + \cdots$ be the sum of the "even reciprocals." Then:
   - $H = O + E$
   - each O term > the corresponding E term
   - $O > E$
   - $E = \frac{1}{2}\left[1 + \frac{1}{2} + \frac{1}{3} + \frac{1}{4} + \frac{1}{5} + \cdots\right] = \frac{1}{2}H$
   - $H = O + E > 2E > 2 \cdot H > H$

58. Assume that $H = 1 + \frac{1}{2} + \frac{1}{3} + \frac{1}{4} + \frac{1}{5} + \cdots$ is a finite number. Starting with the second term, group the terms into groups of three. Using:

$$\frac{1}{n-1} + \frac{1}{n} + \frac{1}{n+1} > \frac{3}{n}$$

(show this inequality is true) conclude that:

$$\begin{aligned} H &= 1 + \left[\frac{1}{2} + \frac{1}{3} + \frac{1}{4}\right] + \left[\frac{1}{5} + \frac{1}{6} + \frac{1}{7}\right] + \cdots \\ &> 1 + [1] + \left[\frac{1}{2}\right] + \left[\frac{1}{3}\right] + \cdots \\ &= 1 + \left[1 + \frac{1}{2} + \frac{1}{3} + \cdots\right] = 1 + H \end{aligned}$$

so that $H > 1 + H$.

59. Jacob Bernoulli (1654–1705) was a master of understanding and manipulating series by breaking a difficult series into easier pieces. In his 1713 book *Ars Conjectandi*, he used such a technique to find the sum of the non-geometric series:

$$\sum_{k=1}^{\infty} \frac{k}{2^k} = \frac{1}{2} + \frac{2}{4} + \frac{3}{8} + \frac{4}{16} + \frac{5}{32} + \cdots$$

Show that you can write:

$$\begin{aligned} \frac{1}{2} + \frac{1}{4} + \frac{1}{8} + \frac{1}{16} + \frac{1}{32} + \cdots + \frac{1}{2^n} + \cdots &= b_1 \\ \frac{1}{4} + \frac{1}{8} + \frac{1}{16} + \frac{1}{32} + \cdots + \frac{1}{2^n} + \cdots &= b_2 \\ \frac{1}{8} + \frac{1}{16} + \frac{1}{32} + \cdots + \frac{1}{2^n} + \cdots &= b_3 \\ \frac{1}{16} + \frac{1}{32} + \cdots + \frac{1}{2^n} + \cdots &= b_4 \end{aligned}$$

and so forth, so that $\sum_{k=1}^{\infty} \frac{k}{2^k} = \sum_{n=1}^{\infty} b_n$. Find the values of the geometric series b_n, and then find $\sum_{n=1}^{\infty} b_n$ (which will be another geometric series).

60. We can also interpret Bernoulli's approach in the previous problem as a geometric argument for representing the area of an infinitely long region in two different ways.

   (a) Represent the total area in the figure below as a (geometric) sum of areas of side-by-side rectangles, then find the sum of the series.

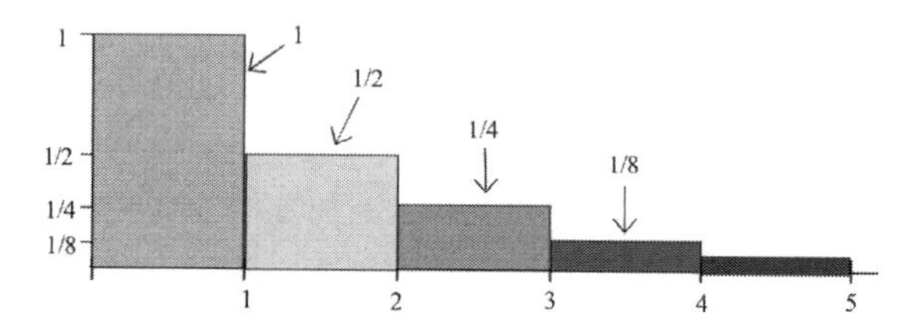

   (b) Represent the total area of the stacked rectangles in the figure below as a sum of the areas of the horizontal slices.

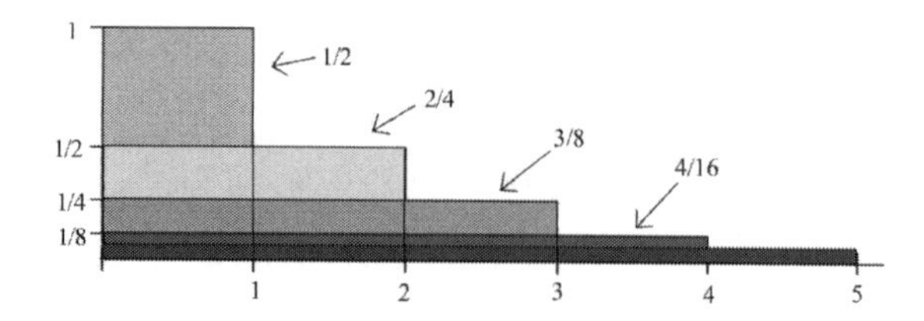

   (c) Explain why the series must be equal.

61. Use the approach of Problem 59 to find:

   (a) the value of the non-geometric series:

$$\sum_{k=1}^{\infty} \frac{k}{3^k} = \frac{1}{3} + \frac{2}{9} + \frac{3}{27} + \frac{4}{81} + \cdots$$

   (b) a formula (when $c > 1$) for the value of:

$$\sum_{k=1}^{\infty} \frac{k}{c^k} = \frac{1}{c} + \frac{2}{c^2} + \frac{3}{c^3} + \frac{4}{c^4} + \cdots$$

9.4 Practice Answers

1. (a) Because each student gets an equal share, and because they eventually eat all of the cake, they each get $\frac{1}{3}$ of the cake. More precisely, after first first step, $\frac{1}{4}$ of the cake remains, with $\frac{3}{4}$ having been eaten by the students. After the second step, $\left(\frac{1}{4}\right)^2$ of the cake remains, with $1-\left(\frac{1}{4}\right)^2$ having been eaten. After the n-th step, $\left(\frac{1}{4}\right)^n$ of the cake remains, with $1-\left(\frac{1}{4}\right)^n$ having been eaten. So after the n-th step, each student has eaten $\frac{1}{3}\left[1-\left(\frac{1}{4}\right)^n\right]$ of the cake. "Eventually," each student gets (almost) $\frac{1}{3}$ of the cake.

   (b) As an infinite series each person gets:

$$\frac{1}{4}+\left(\frac{1}{4}\right)^2+\left(\frac{1}{4}\right)^3+\left(\frac{1}{4}\right)^4+\cdots=\sum_{k=1}^{\infty}\left(\frac{1}{4}\right)^k$$

2. We can rewrite $0.\overline{3}=0.333\ldots$ as:

$$\begin{aligned}\frac{3}{10}+\frac{3}{100}+\frac{3}{1000}&=\frac{3}{10}\left[1+\frac{1}{10}+\frac{1}{100}+\frac{1}{1000}+\cdots\right]\\&=\frac{3}{10}\left[1+\frac{1}{10}+\frac{1}{10^2}+\frac{1}{10^3}+\cdots\right]\end{aligned}$$

   which is a geometric series with $C=\frac{3}{10}$ and $r=\frac{1}{10}$. Because $|r|=\frac{1}{10}<1$, the series converges to:

$$\frac{C}{1-r}=\frac{\frac{3}{10}}{1-\frac{1}{10}}=\frac{\frac{3}{10}}{\frac{9}{10}}=\frac{3}{9}=\frac{1}{3}$$

   Similarly, we can rewrite $0.\overline{432}=0.432432432\ldots$ as:

$$\begin{aligned}\frac{432}{1000}+\frac{432}{1000000}+\frac{432}{1000000000}&=\frac{432}{1000}\left[1+\frac{1}{1000}+\frac{1}{1000000}+\cdots\right]\\&=\frac{432}{1000}\left[1+\frac{1}{1000}+\frac{1}{1000^2}+\cdots\right]\end{aligned}$$

   which is a geometric series with $C=\frac{432}{1000}$ and $r=\frac{1}{1000}$. Becasue $|r|=\frac{1}{1000}<1$, the series converges to:

$$\frac{C}{1-r}=\frac{\frac{432}{1000}}{1-\frac{1}{1000}}=\frac{\frac{432}{1000}}{\frac{999}{1000}}=\frac{432}{999}=\frac{16}{37}$$

3. The ratio for $F(x)$ is $r=2x$, so for the series to converge we need:

$$|2x|<1 \quad\Rightarrow\quad -1<2x<1 \quad\Rightarrow\quad -\frac{1}{2}<x<\frac{1}{2}$$

The series $\sum_{k=0}^{\infty}(2x)^k$ therefore converges to $\dfrac{1}{1-2x}$ when $-\dfrac{1}{2} < x < \dfrac{1}{2}$.
For $G(x)$, the ratio is $r = 3x - 4$, so we need:

$$|3x-4| < 1 \Rightarrow -1 < 3x-4 < 1 \Rightarrow 3 < 3x < 5 \Rightarrow 1 < x < \frac{5}{3}$$

The series $\sum_{k=0}^{\infty}(3x-4)^k$ converges to $\dfrac{1}{1-(3x-4)} = \dfrac{1}{5-3x}$ when $1 < x < \dfrac{5}{3}$.

4. Let $s_n = \sum_{k=3}^{n}\left[\sin\left(\frac{1}{k}\right) - \sin\left(\frac{1}{k+1}\right)\right]$ so that:

$$\begin{aligned} s_n &= \left[\sin\left(\frac{1}{3}\right) - \sin\left(\frac{1}{4}\right)\right] + \left[\sin\left(\frac{1}{4}\right) - \sin\left(\frac{1}{5}\right)\right] \\ &\quad + \left[\sin\left(\frac{1}{5}\right) - \sin\left(\frac{1}{6}\right)\right] + \cdots + \left[\sin\left(\frac{1}{n}\right) - \sin\left(\frac{1}{n+1}\right)\right] \\ &= \left[\sin\left(\frac{1}{3}\right) - \sin\left(\frac{1}{n+1}\right)\right] \end{aligned}$$

allowing us to see that $\lim_{n\to\infty} s_n = \sin\left(\frac{1}{3}\right)$ and hence:

$$\sum_{k=3}^{\infty}\left[\sin\left(\frac{1}{k}\right) - \sin\left(\frac{1}{k+1}\right)\right] = \sin\left(\frac{1}{3}\right) \approx 0.327$$

Background on the Harmonic Series

A taut piece of string, such as a guitar string or piano wire, can only vibrate in such a way that it forms an integer number of waves. The fundamental mode determines the note being played, while the number and intensity of the harmonics (overtones) determine the characteristic quality of the sound. Because of these characteristic qualities, a listener can distinguish between a middle C (264 vibrations per second) played on a piano versus the same note played on a guitar or clarinet.

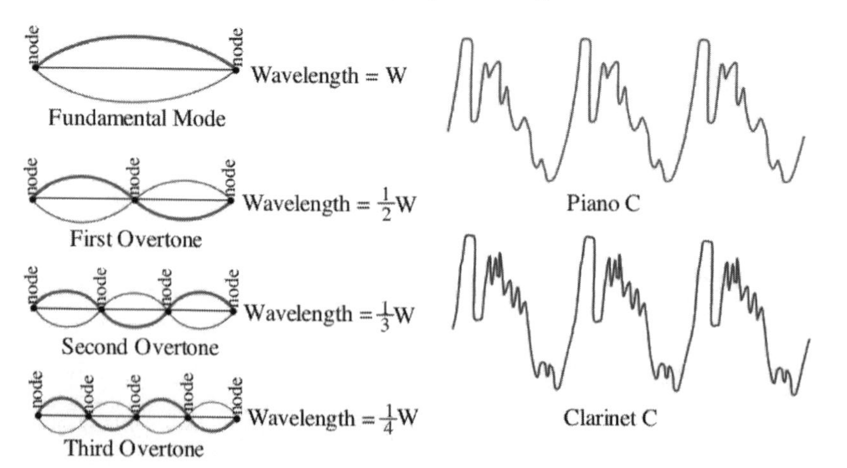

9.5 *An Interlude*

So far in this chapter we have investigated sequences and series, discussed convergence and divergence of infinite series, and examined geometric series in some detail. These concepts, definitions and results will play a fundamental role in understanding the material in the remainder of this chapter—and prove to be important for later work in theoretical and applied mathematics.

The material in the following sections is of a different sort: it is more technical and specialized. In order to work effectively with power series (the subject of Chapter 10) we need to know where (for which values of x) a power series converges. And to determine that convergence we need additional methods. Over the next several sections we will examine several methods (called "convergence tests") for determining whether particular series converge or diverge. For many of the convergent series we will study, we will be unable to determine the exact sum of the series, even if we know that the series converges.

As mentioned in the previous section, geometric series are a very special type of series for which it is easy to find an exact value for the sum of a convergent series.

The Problems in this brief section illustrate some of the reasoning that you will need to use throughout the rest of this chapter (but they do not assume you know any information from those later sections). Practicing this reasoning now will pay dividends as you work through the rest of Chapter 9 and into Chapter 10.

9.5 *Problems*

1. Which shaded region in the figure below has the larger area, the sum or the integral?

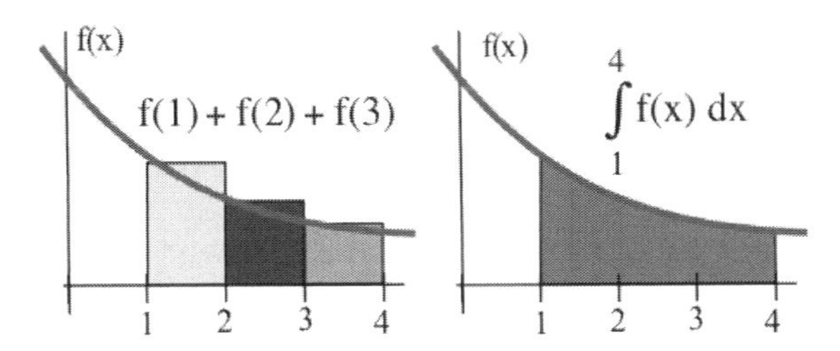

2. Which shaded region in the figure below has the larger area, the sum or the integral?

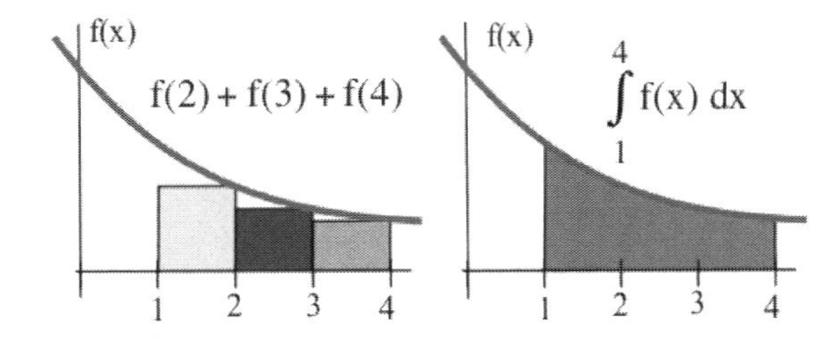

3. Represent the area of the shaded region in the figure below as an infinite series.

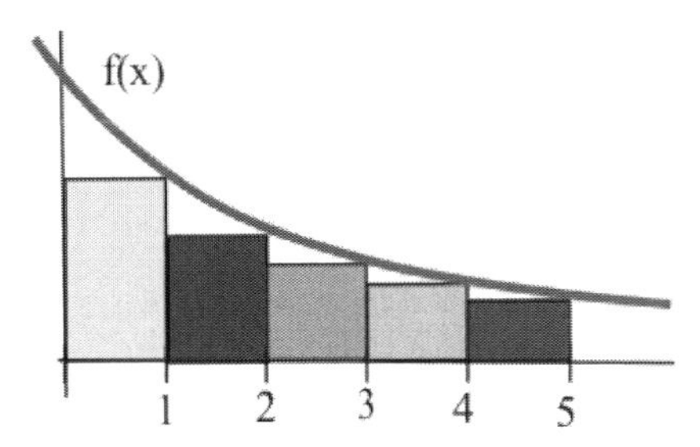

4. Represent the area of the shaded region in the figure below as an infinite series.

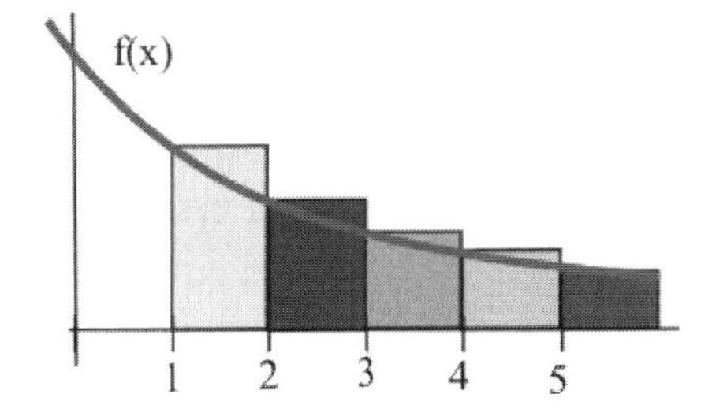

5. Represent the area of the shaded region in the figure below as an infinite series.

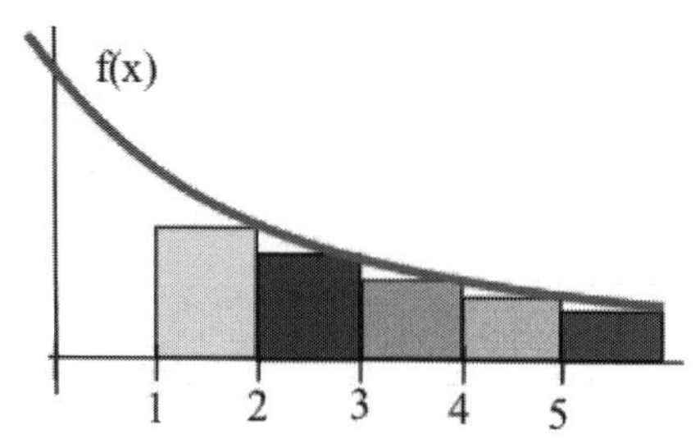

For Problems 6–9, which of the following expressions represents the shaded area in the given figure?

- $f(0)+f(1)$
- $f(1)+f(2)$
- $f(2)+f(3)$
- $f(3)+f(4)$

6. See below left.
7. See below right.

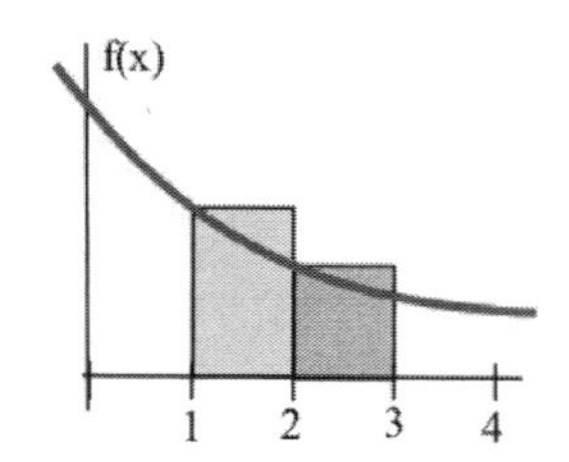

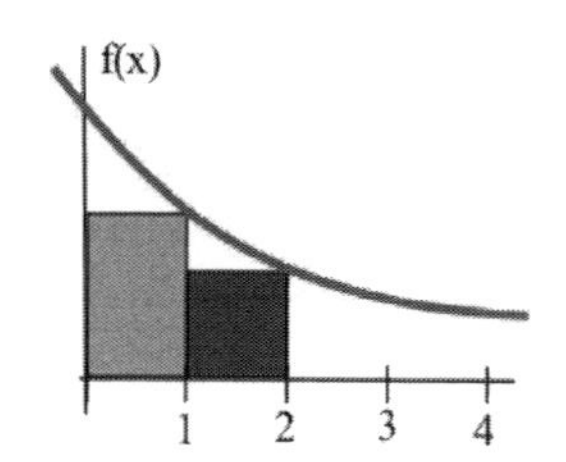

8. See below left.
9. See below right.

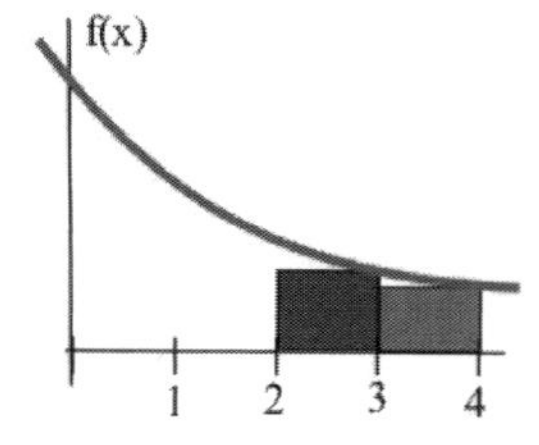

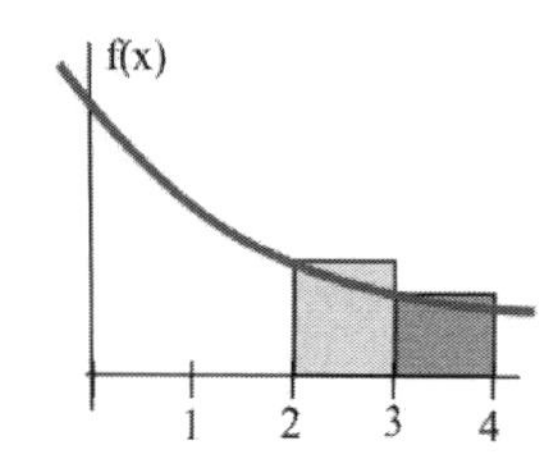

10. Refer to the figure below and arrange the following four values in increasing order:

- $\int_1^3 f(x)\,dx$
- $\int_2^4 f(x)\,dx$
- $f(1)+f(2)$
- $f(2)+f(3)$

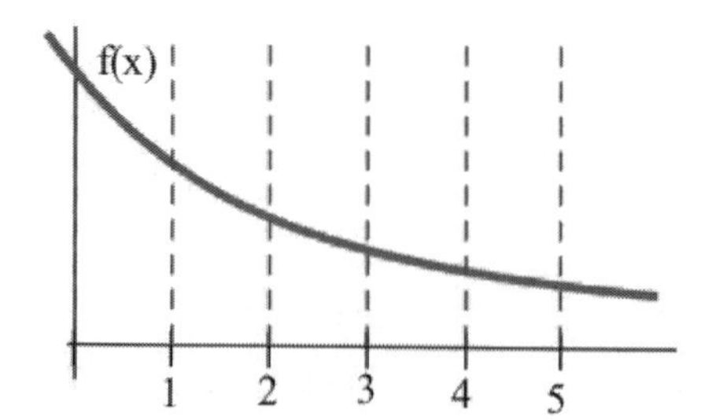

11. Refer to the figure below and arrange the following four values in increasing order:

- $\int_1^4 f(x)\,dx$
- $\int_2^5 f(x)\,dx$
- $f(1)+f(2)+f(3)$
- $f(2)+f(3)+f(4)$

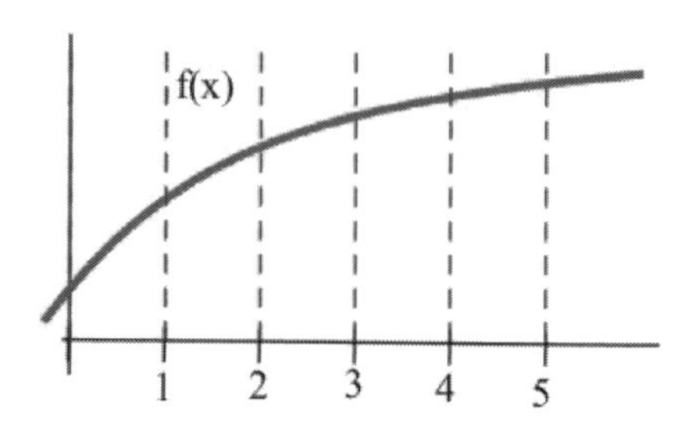

12. You want to get a summer job operating heavy equipment and you know there are certain height requirements in order for the operator to fit safely in the cab of the machine. You don't remember what the requirements are, but three of your friends applied. Tom was rejected as too tall. Sam was rejected as too short. Jenny got a job.

    (a) Should you apply for the job if:

    i. you are taller than Tom? Why?

    ii. you are taller than Sam? Why?

    iii. you are shorter than Sam? Why?

    iv. you are shorter than Jenny? Why?

    (b) List the comparisons that indicate you are the wrong height for the job.

    (c) List the comparisons that do not provide enough information to determine whether or not you are an acceptable height for the job.

13. You know Wendy did well on a recent calculus exam and Paula did poorly, but you haven't received your test back yet. If the instructor violates FERPA regulations and provides you the following information, what can you conclude?

    (a) "You did better than Wendy."

    (b) "You did better than Paula."

    (c) "You did worse than Wendy."

    (d) "You did worse than Paula."

14. Planning to take up mountain climbing, you consider a climb of Mt. Baker. You know that Mt. Index is too easy to be challenging, but that Mt. Liberty Bell is too difficult for you.
    (a) State what you can conclude about your plan to climb Mt. Baker if an experienced climber friend tells you that:
        i. "Baker is easier than Index."
        ii. "Baker is more difficult than Index."
        iii. "Baker is easier than Liberty Bell."
        iv. "Baker is more difficult than Liberty Bell."
    (b) Which comparisons indicate that Baker is appropriate: challenging but not too difficult?
    (c) Which comparisons indicate that Baker is not appropriate?

15. You have previously taken classes taught by Professors Good and Bad, and they each lived up to their names. Now you are considering taking a class taught by Prof. Unknown, with whom you are unfamiliar. State what can you expect about Prof. Unknown's class if:
    (a) a classmate who had Good and Unknown says, "Unknown was better than Good."
    (b) a classmate who had Good and Unknown says, "Good was better than Unknown."
    (c) a classmate who had Bad and Unknown says, "Unknown was better than Bad."
    (d) a classmate who had Bad and Unknown says, "Bad was better than Unknown."

In Problems 16–19, all of the series converge. For each pair, which series has the larger sum? (You shouldn't need to compute any of these sums to determine which is bigger. In fact, it is difficult—if not impossible—to find the exact values of these sums.)

16. $\sum_{k=1}^{\infty} \frac{1}{k^2+1}$ vs. $\sum_{k=1}^{\infty} \frac{1}{k^2}$

17. $\sum_{k=2}^{\infty} \frac{1}{k^3-5}$ vs. $\sum_{k=2}^{\infty} \frac{1}{k^3}$

18. $\sum_{k=1}^{\infty} \frac{1}{k^2+3k-1}$ vs. $\sum_{k=1}^{\infty} \frac{1}{k^2}$

19. $\sum_{k=3}^{\infty} \frac{1}{k^2+5k}$ vs. $\sum_{k=3}^{\infty} \frac{1}{k^3+k-1}$

Problems 20–28 give a formula a_k for the terms of a sequence. For each sequence:

(a) write a formula for a_{k+1}
(b) compute the ratio $\frac{a_{k+1}}{a_k}$
(c) simplify the ratio

20. $a_k = 3k$
21. $a_k = k+3$
22. $a_k = 2k+5$
23. $a_k = \frac{3}{k}$
24. $a_k = k^2$
25. $a_k = 2^k$
26. $a_k = \left(\frac{1}{2}\right)^k$
27. $a_k = x^k$
28. $a_k = (x-1)^k$

In Problems 29–36, state whether the series converges or diverges, then calculate the ratio $\frac{a_{k+1}}{a_k}$.

29. $\sum_{k=1}^{\infty} \left(\frac{1}{2}\right)^k$
30. $\sum_{k=1}^{\infty} \left(\frac{1}{5}\right)^k$
31. $\sum_{k=1}^{\infty} 2^k$
32. $\sum_{k=1}^{\infty} (-3)^k$
33. $\sum_{k=1}^{\infty} 4$
34. $\sum_{k=1}^{\infty} (-1)^k$
35. $\sum_{k=1}^{\infty} \frac{1}{k}$
36. $\sum_{k=1}^{\infty} \frac{7}{k}$

For Problems 37–41, s_n represents the n-th partial sum of the series with terms a_k (for example $s_3 = a_1 + a_2 + a_3$). In each problem, write the appropriate symbol ("$<$" or "$=$" or "$>$") in the box provided to make a true statement.

37. If $a_5 > 0$ then s_4 ☐ s_5
38. If $a_5 = 0$ then s_4 ☐ s_5
39. If $a_5 < 0$ then s_4 ☐ s_5
40. If $a_n > 0$ for all n, then s_n ☐ s_{n+1} for all n.
41. If $a_n < 0$ for all n, then s_n ☐ s_{n+1} for all n.

For Problems 42–45, explain how s_3, s_4, s_5 and s_6 compare, given the information about a_4, a_5 and a_6.

42. $a_4 > 0$, $a_5 > 0$ and $a_6 > 0$
43. $a_4 = 0.2$, $a_5 = -0.1$ and $a_6 = 0.2$
44. $a_4 = -0.3$, $a_5 = 0.2$ and $a_6 = -0.1$
45. $a_4 = -0.3$, $a_5 = -0.2$ and $a_6 = 0.1$

Problems 46–50 list the first eight terms $a_1, a_2, \ldots, a_8$ of a series. Calculate and graph the first eight partial sums $s_1, s_2, \ldots, s_8$ of the series and describe the pattern of the graph of the partial sums.

46. 2, −1, 2, −1, 2, −1, 2, −1
47. 2, −1, 0.9, −0.8, 0.7, −0.6, 0.5, −0.4
48. 2, −1, 1, −1, 1, −1, 1, −1
49. −2, 1.5, −0.8, 0.6, −0.4, 0.2, 2, −0.1
50. 5, 1, −0.6, −0.4, 0.2, 0.1, 0.1, −0.2
51. What condition on the terms a_k guarantees that the graph of the partial sums s_n follows an "up-down-up-down" pattern?
52. What condition on the terms a_k guarantees that the graph of the partial sums s_n follows a "narrowing funnel" pattern (see below)?

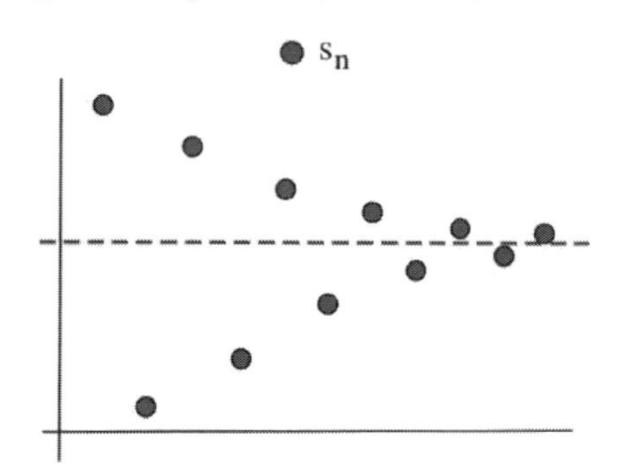

53. The figure below shows the graphs of several partial sums s_n.
    (a) For which do the terms a_k alternate in sign?
    (b) For which do the values of $|a_k|$ decrease?
    (c) For which graphs do the terms a_k alternate in sign and decrease in absolute value?

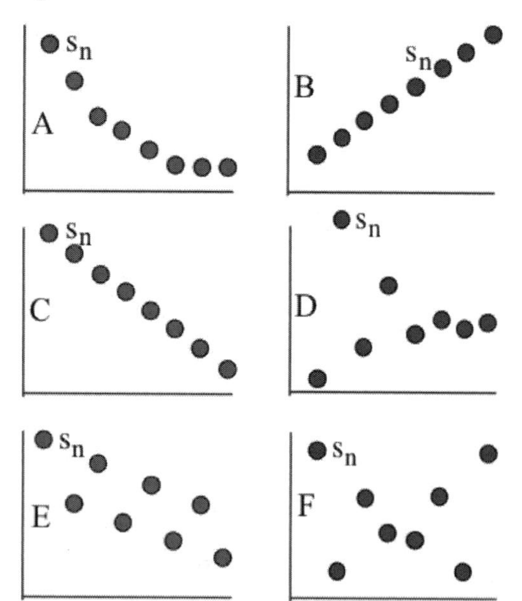

54. Redo Problem 53 for the figure below.

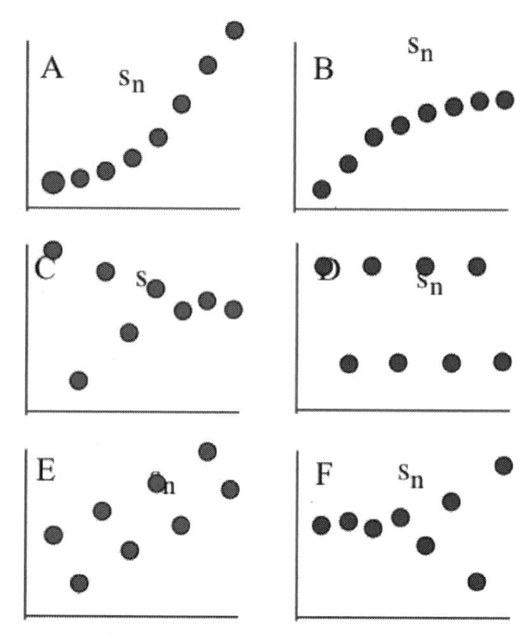

55. The geometric series:

$$\sum_{k=0}^{\infty}\left(-\frac{1}{2}\right)^k = 1 - \frac{1}{2} + \frac{1}{4} - \frac{1}{8} + \cdots$$

converges to:

$$\frac{1}{1-(-\frac{1}{2})} = \frac{1}{\frac{3}{2}} = \frac{2}{3}$$

Graph the horizontal line $y = \frac{2}{3}$ along with the partial sums s_0 through s_8 of the series.

56. The geometric series:

$$\sum_{k=0}^{\infty}(-0.6)^k = 1 - 0.6 + 0.36 - 0.216 + \cdots$$

converges to:

$$\frac{1}{1-(-0.6)} = \frac{1}{1.6} = 0.625$$

Graph the horizontal line $y = 0.625$ along with the partial sums s_0 through s_8 of the series.

57. The geometric series:

$$\sum_{k=0}^{\infty}(-2)^k = 1 - 2 + 4 - 8 + \cdots$$

diverges because $|r| = |-2| = 2 > 1$. Graph the partial sums s_0 through s_8 of the series.

9.6 *Integral Test and P-Test*

This section presents two methods for determining whether certain series converge or diverge. The first of these, the Integral Test, says that a given series converges if and only if a related improper integral converges. This lets us trade a question about the convergence of a series for a question about the convergence of an improper integral.

The second of these convergence tests, the P-Test, says that you can determine the convergence of a series of the form $\sum_{k=1}^{\infty} \frac{1}{k^p}$ by knowing the value of p. Both of these tests apply only to certain classes of series whose terms are positive and (unfortunately) the tests only tell us whether a series converges or diverges — they do *not* tell us the actual sum of the series. The Integral Test is the more fundamental and general of the two, and we will use it to prove the P-Test. The P-Test, however, is easier to apply and is likely to be the test you use more often.

Integral Test

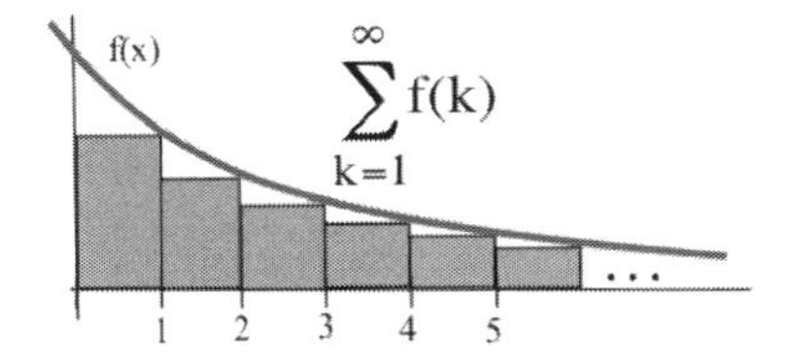

We can represent a series geometrically as a sum of areas of rectangles with a width of one unit and a heights corresponding to the values a_k of the terms of the series (see margin). This area interpretation of series leads to a natural connection between series and integrals — and between the convergence of a series and the convergence of a corresponding improper integral.

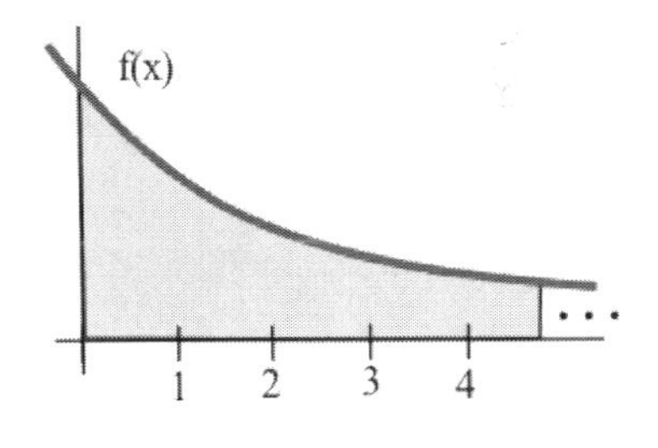

Example 1. If you can paint the shaded region in margin figure using 3 gallons of paint, how much paint do you need to paint all of the shaded rectangles in the lower margin figure?

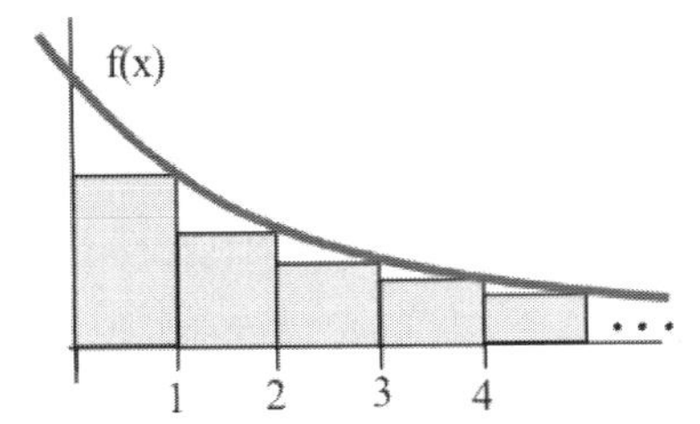

Solution. We don't have enough information to determine the exact amount of paint needed for the rectangles, but the sum of the rectangular areas is smaller than the area in under the graph in the original margin figure, so we *can* say that we need less than 3 gallons of paint are needed for the rectangles. (We just can't say how much less.) ◀

Practice 1. If the area of the shaded region below left is infinite, what can you say about the total area of the rectangular regions below right?

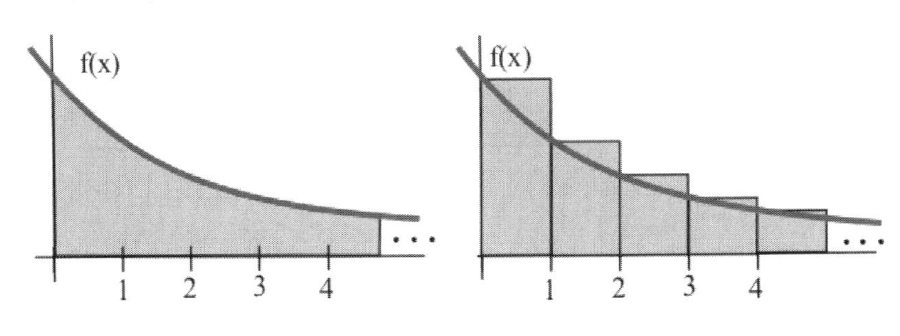

Example 2. Determine which is larger:

$$\sum_{k=2}^{\infty} \frac{1}{k^2} \quad \text{or} \quad \int_2^{\infty} \frac{1}{x^2}\,dx$$

and use the result to show that $\sum_{k=2}^{\infty} \frac{1}{k^2}$ converges.

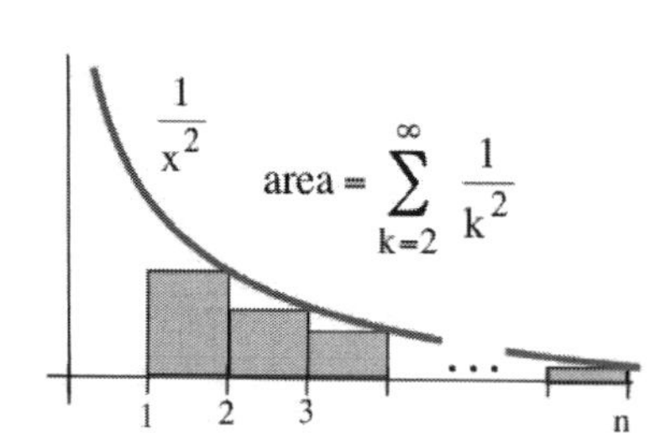

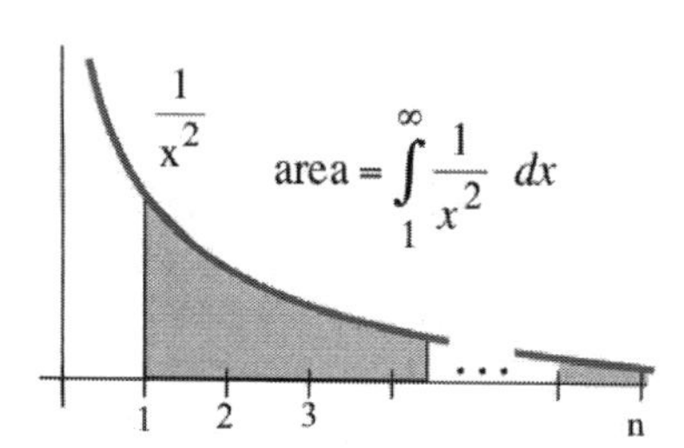

Solution. The margin figure illustrates that the area of the rectangles:

$$\frac{1}{2^2} + \frac{1}{3^2} + \frac{1}{4^2} + \cdots + \frac{1}{n^2} + \cdots$$

is less than the area under the graph of the function $f(x) = \frac{1}{x^2}$ for $1 \le x \le n$ so that:

$$s_n = \sum_{k=2}^{n} \frac{1}{k^2} < \int_1^{n} \frac{1}{x^2}\,dx \le \int_1^{\infty} \frac{1}{x^2}\,dx$$

where the second inequality holds because $f(x) = \frac{1}{x^2} > 0$. Evaluating the improper integral:

$$\int_1^{\infty} \frac{1}{x^2}\,dx = \lim_{M\to\infty} \int_1^{M} x^{-2}\,dx = \lim_{M\to\infty} \left[-\frac{1}{x}\right]_1^M = \lim_{M\to\infty} \left[-\frac{1}{M} + 1\right] = 1$$

tells us that, for any integer n with $n \ge 2$, $s_n = \sum_{k=2}^{n} \frac{1}{k^2} \le 1$, so that the sequence of partial sums $\{s_n\}$ is bounded above. That sequence is also monotonically increasing because:

$$s_{n+1} = s_n + \frac{1}{(n+1)^2} > s_n$$

for any $n \ge 2$. The Monotone Convergence Theorem from Section 9.2 therefore tells us that $\{s_n\}$ converges, so $\sum_{k=2}^{\infty} \frac{1}{k^2}$ converges. ◀

Although we now know that $\sum_{k=2}^{\infty} \frac{1}{k^2}$ converges, the argument in Example 2 tells us nothing about the *sum* of the series (other than it is less than 1). It turns out the actual sum of the series is:

$$\sum_{k=2}^{\infty} \frac{1}{k^2} = \frac{\pi^2}{6} - 1 \approx 0.644934$$

but this is not easy to prove.

The reasoning of Example 2 extends to a more general result comparing the convergence of infinte series to the convergence of certain improper integrals.

Integral Test

If f is a continuous, positive, decreasing function on $[1, \infty)$ and $a_k = f(k)$ for $k \ge 1$, then:

- $\int_1^{\infty} f(x)\,dx$ converges $\Rightarrow \sum_{k=1}^{\infty} a_k$ converges
- $\int_1^{\infty} f(x)\,dx$ diverges $\Rightarrow \sum_{k=1}^{\infty} a_k$ diverges

Proof. Assume that f is a continuous, positive, decreasing function on $[1, \infty)$ and that $a_k = f(k)$ for $k \geq 1$.

To prove the first implication, assume that $\int_1^\infty f(x)\,dx$ converges so that:

The proof of the first statement mirrors the argument used in Example 2.

$$\int_1^\infty f(x)\,dx = \lim_{n \to \infty} \int_1^n f(x)\,dx = L$$

for some finite number L. Because $a_k = f(k)$ for $k \geq 1$ and because $f(x)$ is decreasing, we know that for $1 \leq x \leq 2$:

$$a_2 = f(2) \leq f(x) \;\Rightarrow \int_1^2 a_2\,dx \leq \int_1^2 f(x)\,dx \;\Rightarrow a_2 \leq \int_1^2 f(x)\,dx$$

Using similar reasoning for $k \geq 2$ we can conclude that:

$$a_k \leq \int_{k-1}^k f(x)\,dx \;\Rightarrow \sum_{k=2}^n a_k \leq \sum_{k=2}^n \int_{k-1}^k f(x)\,dx = \int_1^n f(x)\,dx$$

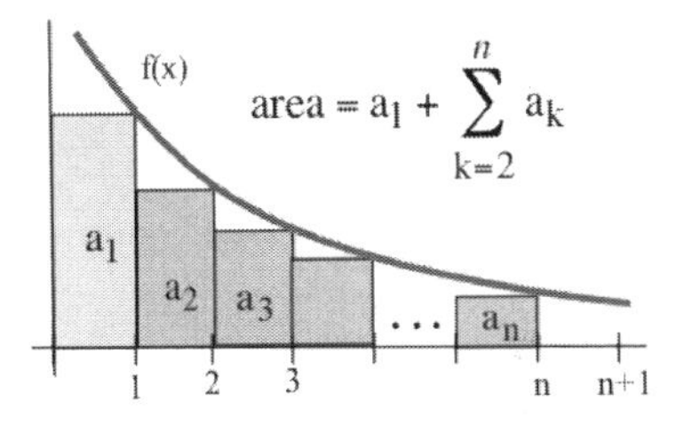

Adding the finite number a_1 to each side of this inequality yields:

$$s_n = \sum_{k=1}^n a_k = a_1 + \sum_{k=2}^n a_k \leq a_1 + \int_1^n f(x)\,dx$$

(see margin). Because $f(x) > 0$, $\int_1^n f(x)\,dx$ increases as n increases, so:

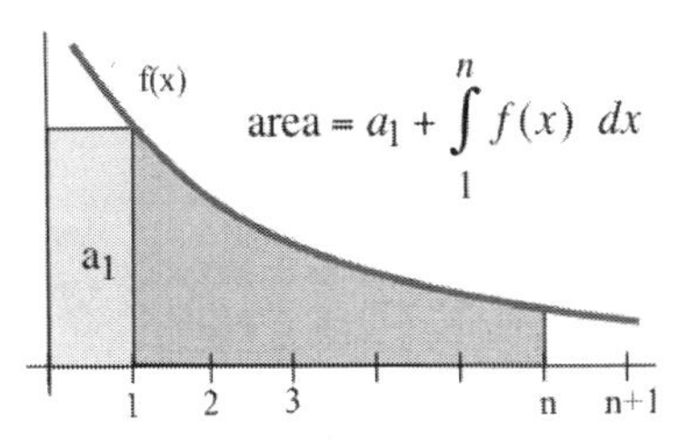

$$s_n \leq a_1 + \int_1^n f(x)\,dx \leq \int_1^\infty f(x)\,dx = L$$

This tells us that the sequence $\{s_n\}$ is bounded above. Because $f(x) > 0$ we know that $a_{n+1} = f(n+1) > 0$ for any $n \geq 1$, so:

$$s_{n+1} = a_{n+1} + s_n > s_n$$

which means that $\{s_n\}$ is an increasing sequence. The Monotone Convergence Theorem then tells us that $\{s_n\}$ is a convergent sequence, so $\sum_{k=1}^{\infty} a_k$ must converge.

Now assume that $\int_1^\infty f(x)\,dx$ diverges. Because $a_k = f(k)$ for $k \geq 1$ and because $f(x)$ is decreasing, we know that for $1 \leq x \leq 2$:

$$a_1 = f(1) \geq f(x) \;\Rightarrow \int_1^2 a_1\,dx \geq \int_1^2 f(x)\,dx \;\Rightarrow a_1 \geq \int_1^2 f(x)\,dx$$

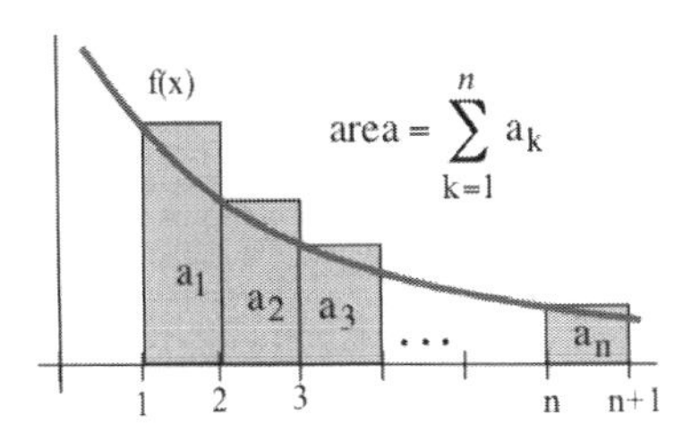

Using similar reasoning for $k \geq 1$ (see margin) we can conclude that:

$$a_k \geq \int_k^{k+1} f(x)\,dx \;\Rightarrow s_n = \sum_{k=1}^n a_k \geq \sum_{k=1}^n \int_k^{k+1} f(x)\,dx = \int_1^{n+1} f(x)\,dx$$

Because $\int_1^\infty f(x)\,dx$ diverges, we know that:

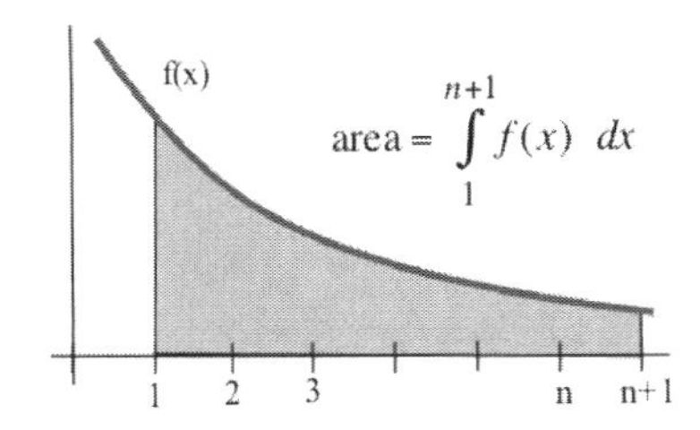

$$\lim_{n \to \infty} s_n \geq \lim_{n \to \infty} \int_1^{n+1} f(x)\,dx = \int_1^\infty f(x)\,dx = \infty$$

so $\{s_n\}$ is a divergent sequence and $\sum_{k=1}^{\infty} a_k$ must diverge. □

Altering the first few terms—or even the first million terms—of an infinite series does not affect whether that series converges or diverges (although changing these terms likely will affect the sum of the series). With this in mind, we can relax the hypotheses of the integral test.

Integral Test Corollary

If f is a continuous, positive, decreasing function on $[N, \infty)$ and $a_k = f(k)$ for $k \geq N$, then

- $\displaystyle\int_N^\infty f(x)\,dx$ converges $\Rightarrow \displaystyle\sum_{k=1}^\infty a_k$ converges
- $\displaystyle\int_N^\infty f(x)\,dx$ diverges $\Rightarrow \displaystyle\sum_{k=1}^\infty a_k$ diverges

Example 3. Use the Integral Test to determine whether the infinite series $\displaystyle\sum_{k=1}^\infty \frac{1}{k^3}$ and $\displaystyle\sum_{k=2}^\infty \frac{1}{k \cdot \ln(k)}$ converge or diverge.

Solution. For the first series, let $f(x) = \dfrac{1}{x^3}$ so that $a_k = \dfrac{1}{k^3} = f(k)$ for $k \geq 1$. When $x \geq 1$, $f(x) = x^{-3} > 0$ and $f'(x) = -3x^{-2} < 0$, so $f(x)$ is continuous, positive and decreasing on $[1, \infty)$, as required by the Integral Test. Furthermore:

$$\int_1^\infty \frac{1}{x^3}\,dx = \lim_{M\to\infty} \int_1^M x^{-3}\,dx = \lim_{M\to\infty} \left[-\frac{1}{2x^2}\right]_1^M = \lim_{M\to\infty} \left[-\frac{1}{2M^2} + \frac{1}{2}\right]$$

This limit is $\dfrac{1}{2}$, so $\displaystyle\int_1^\infty \frac{1}{x^3}\,dx$ converges, hence $\displaystyle\sum_{k=1}^\infty \frac{1}{k^3}$ also converges.

Although the Integral Test tells us that the series $\displaystyle\sum_{k=1}^\infty \frac{1}{k^3}$ converges, it does *not* tell us the sum of that series.

Next, let $g(x) = \dfrac{1}{x \cdot \ln(x)}$ so that $a_k = \dfrac{1}{k \cdot \ln(k)} = g(k)$ for $k \geq 2$. When $x \geq 2$, $g(x) > 0$ and:

$$g'(x) = \frac{[x \cdot \ln(x)] \cdot 0 - 1 \cdot \left[x \cdot \frac{1}{x} + \ln(x) \cdot 1\right]}{[x \cdot \ln(x)]^2} = \frac{-x - \ln(x)}{x^2 \cdot [\ln(x)]^2} < 0$$

so $g(x)$ is continuous, positive and decreasing on $[2, \infty)$, as required. Furthermore, using the substitution $u = \ln(x) \Rightarrow du = \dfrac{1}{x}\,dx$:

We use the Integral Test Corollary because a_1 is undefined for $a_k = \dfrac{1}{k \cdot \ln(k)}$.

$$\int \frac{1}{x \cdot \ln(x)}\,dx = \int \frac{1}{u}\,du = \ln(|u|) + C = \ln(|\ln(x)|) + C$$

so that:

$$\int_2^\infty \frac{1}{x \cdot \ln(x)}\,dx = \lim_{M\to\infty} \Big[\ln(|\ln(x)|)\Big]_2^M = \lim_{M\to\infty} \Big[\ln(\ln(M)) - \ln(\ln(2))\Big]$$

which diverges to ∞, so $\displaystyle\sum_{k=2}^\infty \frac{1}{k \cdot \ln(k)}$ must also diverge. ◀

Practice 2. Determine whether $\sum_{k=4}^{\infty} \frac{1}{\sqrt{k}}$ and $\sum_{k=1}^{\infty} e^{-k}$ converge or diverge.

The P-Test

In the preceding Examples and Practice problems, you have observed that $\sum_{k=2}^{\infty} \frac{1}{k^2}$ and $\sum_{k=1}^{\infty} \frac{1}{k^3}$ both converge, while $\sum_{k=4}^{\infty} \frac{1}{\sqrt{k}} = \sum_{k=4}^{\infty} \frac{1}{k^{\frac{1}{2}}}$ diverges. These series all have the form $\sum_{k=1}^{\infty} \frac{1}{k^p}$ for some power p. Rather than continuing to employ the Integral Test every time we encounter a series of this form, we can instead develop a test that applies to all such series.

> **P-Test**: The "p-series" $\sum_{k=1}^{\infty} \frac{1}{k^p}$:
>
> - converges if $p > 1$
> - diverges if $p \leq 1$

Proof. If $p \leq 0$, $\lim_{k\to\infty} \frac{1}{k^p} \neq 0$, so the Test for Divergence shows $\sum_{k=1}^{\infty} \frac{1}{k^p}$ diverges. If $p = 1$, $\sum_{k=1}^{\infty} \frac{1}{k^p} = \sum_{k=1}^{\infty} \frac{1}{k}$ is the harmonic series, which we already know diverges (from Section 9.4). If $p > 0$, let $f(x) = \frac{1}{x^p} = x^{-p}$ so that $f(k) = \frac{1}{k^p}$ for any integer $k \geq 1$. When $x \geq 1$, $f(x) > 0$ and:

$$f'(x) = -p \cdot x^{-p-1} = \frac{-p}{x^{p+1}} < 0$$

so $f(x)$ is continuous, positive and decreasing, as required by the Integral Test. Furthermore, for $p \neq 1$:

$$\int_1^{\infty} \frac{1}{x^p}\,dx = \lim_{M\to\infty} \int_1^M x^{-p}\,dx = \lim_{M\to\infty} \left[\frac{x^{-p+1}}{-p+1}\right]_1^M$$
$$= \lim_{M\to\infty} \left[\frac{M^{1-p}}{1-p} - \frac{1}{1-p}\right]$$

This limit converges when $p > 1$ and diverges when $0 < p < 1$, so $\sum_{k=1}^{\infty} \frac{1}{k^p}$ also converges when $p > 1$ and diverges when $p \leq 1$. □

Example 4. Do $\sum_{k=1}^{\infty} \frac{1}{k^2}$ and $\sum_{k=4}^{\infty} \frac{1}{\sqrt{k}}$ converge or diverge?

Solution. For $\sum_{k=1}^{\infty} \frac{1}{k^2}$, $p = 2 > 1$ so the series converges; for $\sum_{k=4}^{\infty} \frac{1}{\sqrt{k}}$, $p = \frac{1}{2} \leq 1$, so the series diverges. ◀

Like the Integral Test, the P-Test tells us whether or not a particular series converges, but it does *not* give the value of a convergent infinite series.

Estimating Sums

As mentioned several times so far, neither the Integral Test nor the P-Test can tell us the value of an infinite sum — these tests only tell us whether certain series converge or diverge. But some of the concepts we used to prove the Integral Test *can* help us estimate how close a partial sum comes to approximating the value of an infinite series.

If $f(x) > 0$ is continuous and decreasing on $[1, \infty)$ with $a_k = f(k)$:

$$\int_1^{n+1} f(x)\,dx \le \sum_{k=1}^{n} a_k \le a_1 + \int_1^n f(x)\,dx$$

For $a_k = \frac{1}{k^2}$ and $n = 1000$ we can say that:

Evaluating these improper integrals requires much less arithmetic than computing s_{1000}.

$$\int_1^{1001} \frac{1}{x^2}\,dx \le \sum_{k=1}^{1000} \frac{1}{k^2} \le \frac{1}{1^2} + \int_1^{1000} \frac{1}{x^2}\,dx$$

Evaluating these integrals reveals that:

$$0.999 \le \sum_{k=1}^{1000} \frac{1}{k^2} \le 1.999$$

Unfortunately, this is does not provide a very accurate estimate of s_{1000}. Letting $n \to \infty$ in the first inequality above:

$$\int_1^{\infty} f(x)\,dx \le \sum_{k=1}^{\infty} a_k \le a_1 + \int_1^{\infty} f(x)\,dx$$

For $a_k = \frac{1}{k^2}$, this tells us that

This merely provides us with a "ballpark" estimate for $\sum_{k=1}^{\infty} \frac{1}{k^2}$.

$$1 = \int_1^{\infty} \frac{1}{x^2}\,dx \le \sum_{k=1}^{\infty} \frac{1}{k^2} \le \frac{1}{1^2} + \int_1^{\infty} \frac{1}{x^2}\,dx = 2$$

If we start the approximation at $k = n+1$ instead of $k = 1$:

$$\int_{n+1}^{\infty} f(x)\,dx \le \sum_{k=n+1}^{\infty} a_k \le a_{n+1} + \int_{n+1}^{\infty} f(x)\,dx$$

Adding s_n to each expression in this inequality yields:

$$s_n + \int_{n+1}^{\infty} f(x)\,dx \le \sum_{k=1}^{\infty} a_k \le s_{n+1} + \int_{n+1}^{\infty} f(x)\,dx$$

For $a_k = \frac{1}{k^2}$ and $n = 10$ we can compute:

$$s_{10} = 1 + \frac{1}{2^2} + \frac{1}{3^2} + \cdots \frac{1}{10^2} \approx 1.5497677$$

so that $s_{11} = s_{10} + \frac{1}{11^2} \approx 1.5580322$ and:

$$\int_{11}^{\infty} \frac{1}{x^2}\,dx = \lim_{M \to \infty} \left[-\frac{1}{x} \right]_{11}^{M} = \lim_{M \to \infty} \left[-\frac{1}{M} + \frac{1}{11} \right] = \frac{1}{11}$$

Therefore:

$$1.6407 \approx s_{10} + \int_{11}^{\infty} \frac{1}{x^2}\,dx \le \sum_{k=1}^{\infty} \frac{1}{k^2} \le s_{11} + \int_{11}^{\infty} \frac{1}{x^2}\,dx \approx 1.6489$$

Advanced mathematical techniques can be used to show that:

$$\sum_{k=1}^{\infty} \frac{1}{k^2} = \frac{\pi^2}{6} \approx 1.6449$$

which is consistent with this estimate.

Practice 3. Find an upper and lower bound for $a_k = \frac{1}{k^3}$ using s_{10}.

9.6 Problems

In Problems 1–16, use the Integral Test to determine whether the series converges or diverges. (Be sure to verify the hypotheses of the Integral Test hold.)

1. $\sum_{k=1}^{\infty} \frac{1}{2k+5}$
2. $\sum_{k=1}^{\infty} \frac{1}{(2k+5)^2}$
3. $\sum_{k=1}^{\infty} \frac{1}{(2k+5)^{\frac{3}{2}}}$
4. $\sum_{k=2}^{\infty} \frac{\ln(k)}{k}$
5. $\sum_{k=2}^{\infty} \frac{1}{k \cdot [\ln(k)]^2}$
6. $\sum_{k=1}^{\infty} \frac{1}{k^2} \cdot \sin\left(\frac{1}{k}\right)$
7. $\sum_{k=1}^{\infty} \frac{1}{k^2+1}$
8. $\sum_{k=1}^{\infty} \frac{1}{k^2+100}$
9. $\sum_{k=1}^{\infty} \left[\frac{1}{k} - \frac{1}{k+3}\right]$
10. $\sum_{k=1}^{\infty} \left[\frac{1}{k} - \frac{1}{k+1}\right]$
11. $\sum_{k=1}^{\infty} \frac{1}{k(k+5)}$
12. $\sum_{k=2}^{\infty} \frac{1}{k^2-1}$
13. $\sum_{k=1}^{\infty} k \cdot e^{-k^2}$
14. $\sum_{k=1}^{\infty} k^2 \cdot e^{-k^3}$
15. $\sum_{k=1}^{\infty} \frac{1}{\sqrt{6k+10}}$
16. $\sum_{k=1}^{\infty} \frac{1}{k\sqrt{k}}$

In Problems 17–28, use the P-Test to determine whether the series converges or diverges.

17. $\sum_{k=1}^{\infty} \frac{1}{k^4}$
18. $\sum_{k=1}^{\infty} \frac{1}{\sqrt[4]{k}}$
19. $\sum_{k=1}^{\infty} \frac{1}{\sqrt[5]{k}}$
20. $\sum_{k=1}^{\infty} \frac{1}{k^5}$
21. $\sum_{k=1}^{\infty} \frac{1}{k}$
22. $\sum_{k=1}^{\infty} \frac{1}{k^{\frac{2}{3}}}$
23. $\sum_{k=1}^{\infty} \frac{1}{k^{\frac{3}{2}}}$
24. $\sum_{k=1}^{\infty} \frac{1}{k^e}$
25. $\sum_{k=1}^{\infty} \frac{1}{k\sqrt[3]{k}}$
26. $\sum_{k=1}^{\infty} \frac{1}{\sqrt[3]{k^4}}$
27. $\sum_{k=1}^{\infty} \frac{1}{\sqrt[3]{k^2}}$
28. $\sum_{k=1}^{\infty} \frac{1}{\sqrt[5]{k^4}}$

In Problems 29–34, use the inequality:

$$\int_1^{n+1} f(x)\,dx \le \sum_{k=1}^{n} a_k \le a_1 + \int_1^n f(x)\,dx$$

to estimate s_{10}, s_{100} and $s_{1000000}$ for the series.

29. $\sum_{k=1}^{\infty} \frac{1}{k^3}$
30. $\sum_{k=1}^{\infty} \frac{1}{k^4}$
31. $\sum_{k=1}^{\infty} \frac{1}{k}$
32. $\sum_{k=1}^{\infty} \frac{1}{\sqrt{k}}$
33. $\sum_{k=1}^{\infty} \frac{1}{k^2+1}$
34. $\sum_{k=1}^{\infty} \frac{1}{k^2+100}$

In Problems 35–40, use the inequality:

$$s_n + \int_{n+1}^{\infty} f(x)\,dx \le \sum_{k=1}^{\infty} a_k \le s_{n+1} + \int_{n+1}^{\infty} f(x)\,dx$$

with $n = 10$ and $n = 20$ to estimate the sum.

35. $\sum_{k=1}^{\infty} \frac{1}{k^4}$
36. $\sum_{k=1}^{\infty} \frac{1}{k}$
37. $\sum_{k=1}^{\infty} \frac{1}{k^2+1}$
38. $\sum_{k=1}^{\infty} \frac{1}{k^2+9}$
39. $\sum_{k=1}^{\infty} \frac{1}{k\sqrt{k}}$
40. $\sum_{k=1}^{\infty} \frac{1}{2^k}$

41. Show that $\int_2^{\infty} \frac{1}{x \cdot [\ln(x)]^q}\,dx$ converges for $q > 1$ and diverges for $q \le 1$, then use this result to state a "Q-Test" for the series $\sum_{k=2}^{\infty} \frac{1}{k \cdot [\ln(k)]^q}$.

In Problems 42–45, use the result for Problem 41 to determine whether the series converges or diverges.

42. $\sum_{k=2}^{\infty} \frac{1}{k \cdot \ln(k)}$
43. $\sum_{k=2}^{\infty} \frac{1}{k \cdot [\ln(k)]^3}$
44. $\sum_{k=2}^{\infty} \frac{1}{k\sqrt{\ln(k)}}$
45. $\sum_{k=2}^{\infty} \frac{1}{k \cdot \ln(k^3)}$

46. For $n \geq 1$, define $g_n = \left[\sum_{k=1}^{n} \frac{1}{k}\right] - \ln(n)$ so that:

$$g_1 = 1 - \ln(1) = 1$$
$$g_2 = 1 + \frac{1}{2} - \ln(2) \approx 0.806853$$
$$g_2 = 1 + \frac{1}{2} + \frac{1}{3} - \ln(3) \approx 0.734721$$

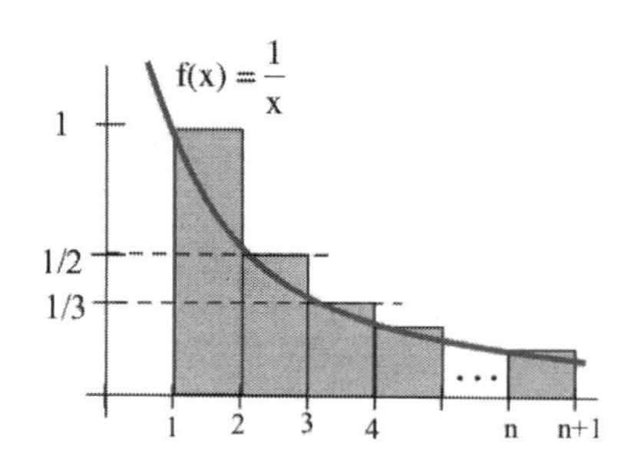

(a) Make several copies of the margin figure and shade the regions represented by g_2, g_3, g_4 and g_n.

(b) Provide a geometric argument that $g_n > 0$ for all $n \geq 1$.

(c) Provide a geometric argument that $\{g_n\}$ is monotonically decreasing; that is: $g_{n+1} < g_n$ for all $n \geq 1$.

(d) Conclude from your geometric results and the Monotone Convergence Theorem that $\{g_n\}$ converges.

The value to which $\{g_n\}$ converges, denoted by γ (the lowercase Greek letter gamma) is called **Euler's constant**. Although $\gamma \approx 0.5772157$ has been approximated to over 100,000 digits, no one (yet) knows whether or not γ is a rational number.

9.6 Practice Answers

1. The area of the rectangles exceeds the area under the curve, which is infinite, so the total area of the rectangles must be infinite.

2. For the first series, let $f(x) = x^{-\frac{1}{2}}$. For $x \geq 1$, $f(x)$ is continuous, $f(x) > 0$ and $f'(x) = -\frac{1}{2}x^{-\frac{3}{2}} < 0$ (so $f(x)$ is decreasing), while $f(k) = \frac{1}{\sqrt{k}}$ for all integers $k \geq 1$. Furthermore:

$$\int_1^\infty x^{-\frac{1}{2}}\,dx = \lim_{M\to\infty} \left[2\sqrt{x}\right]_1^M = \lim_{M\to\infty} \left[2\sqrt{M} - 2\right] = \infty$$

This is a p-series, with $p = \frac{1}{2} < 1$, hence it must diverge.

Because this improper integral diverges, $\sum_{k=1}^{\infty} \frac{1}{\sqrt{k}}$ also diverges.

For the second series, let $g(x) = e^{-x}$. For any x, $g(x)$ is continuous, $g(x) > 0$ and $g'(x) = -e^{-x} < 0$, so $g(x)$ is decreasing, while $g(k) = e^{-k}$ for all integers k. Furthermore:

This is actually a geometric series with ratio $e^{-1} < 1$, hence it converges to:

$$\frac{1}{1-e^{-1}} - 1 = \frac{1}{e-1} \approx 0.582$$

$$\int_1^\infty e^{-x}\,dx = \lim_{M\to\infty} \left[-e^{-x}\right]_1^M = \lim_{M\to\infty} \left[-\frac{1}{e^M} + \frac{1}{e}\right] = \frac{1}{e}$$

Because this improper integral converges, $\sum_{k=1}^{\infty} e^{-k}$ also converges.

3. $1.197532 + \frac{1}{242} \leq \sum_{k=1}^{\infty} \frac{1}{k^3} \leq 1.198283 + \frac{1}{242} \Rightarrow 1.20166 < \sum_{k=1}^{\infty} \frac{1}{k^3} < 1.20242$

9.7 *Comparison Tests*

In the previous section we compared the value of an infinite sum to the value of an improper integral and used the convergence or divergence of the integral to determine whether the series converged or diverged. In this section, we compare an infinite series to another infinite series already known to be convergent or divergent in order to determine the convergence or divergence of the new series. As with the Integral Test, the methods of this section only apply to series with positive terms.

Basic Comparison Test

Consider the series $\sum_{k=1}^{\infty} \frac{1}{k^2}$ and $\sum_{k=1}^{\infty} \frac{1}{k^2+1}$. The first series is a p-series with $p = 2 > 1$, so we know it converges by the P-Test. The second series also converges, but it is not a p-series or a geometric series or a harmonic series, so we need to appeal to the Integral Test. Because:

$$\int_1^{\infty} \frac{1}{x^2+1}\,dx = \lim_{M\to\infty} \int_1^{M} \frac{1}{1+x^2}\,dx = \lim_{M\to\infty} \Big[\arctan(x)\Big]_1^M$$

$$= \lim_{M\to\infty} \left[\arctan(M) - \arctan(1)\right] = \frac{\pi}{2} - \frac{\pi}{4} = \frac{\pi}{4}$$

the series $\sum_{k=1}^{\infty} \frac{1}{k^2+1}$ converges as well. Is there an easier way to see that this second series converges? For large values of k:

$$k^2 + 1 \approx k^2 \;\Rightarrow\; \frac{1}{k^2+1} \approx \frac{1}{k^2}$$

so we might suspect that the convergence of $\sum_{k=1}^{\infty} \frac{1}{k^2}$ and $\sum_{k=1}^{\infty} \frac{1}{k^2+1}$ are related even if these series do not converge to the same sum. Consider a graph of rectangles with areas corresponding to $\sum_{k=1}^{\infty} \frac{1}{k^2}$ (see first margin figure) and another graph with rectangles corresponding to $\sum_{k=1}^{\infty} \frac{1}{k^2+1}$ (second margin figure). Comparing these graphs, it appears that $\sum_{k=1}^{\infty} \frac{1}{k^2+1} \leq \sum_{k=1}^{\infty} \frac{1}{k^2}$. For any $k \geq 1$ and $n \geq 1$:

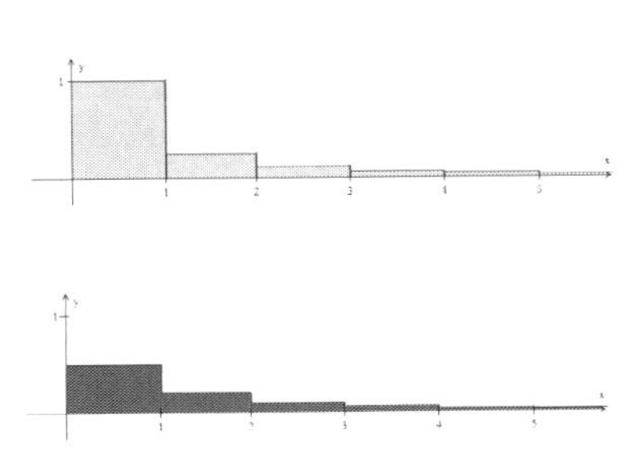

$$0 < \frac{1}{k^2+1} < \frac{1}{k^2} \Rightarrow 0 < \sum_{k=1}^{n} \frac{1}{k^2+1} < \sum_{k=1}^{n} \frac{1}{k^2} \Rightarrow \sum_{k=1}^{\infty} \frac{1}{k^2+1} \leq \sum_{k=1}^{\infty} \frac{1}{k^2}$$

Because we know the series on the right in the last inequality converges (as discussed above), the series on the left, being term-by-term smaller, should converge as well. The following test formalizes and generalizes this result.

To be precise, we know that:

$$\sum_{k=1}^{n} \frac{1}{k^2+1} \leq \sum_{k=1}^{n} \frac{1}{k^2} \leq \sum_{k=1}^{\infty} \frac{1}{k^2}$$

holds for all $n \geq 1$, so the partial sums $\sum_{k=1}^{n} \frac{1}{k^2+1}$ are increasing and bounded above, hence convergent by the Monotone Convergence Theorem.

The Basic Comparison Test (BCT) also works when the hypotheses hold for all $k \geq N$ for some positive integer N.

Basic Comparison Test (BCT)

If $0 < d_k \leq a_k \leq c_k$ for all $k \geq 1$, then:

- $\sum_{k=1}^{\infty} c_k$ converges $\Rightarrow \sum_{k=1}^{\infty} a_k$ converges
- $\sum_{k=1}^{\infty} d_k$ diverges $\Rightarrow \sum_{k=1}^{\infty} a_k$ diverges

Proof. If $0 < a_k \leq c_k$ and $\sum_{k=1}^{\infty} c_k$ converges, then:

$$s_n = \sum_{k=1}^{n} a_k \leq \sum_{k=1}^{n} c_k \leq \sum_{k=1}^{\infty} c_k = C$$

for some finite number C, so the sequence $\{s_n\}$ is bounded above. Because $a_k > 0$, $\{s_n\}$ is monotonically increasing, so by the Monotone Converge Theorem $\{s_n\}$ converges, hence $\sum_{k=1}^{\infty} a_k$ converges.

It's important to note what the BCT does *not* say: if terms of $\sum_{k=1}^{\infty} a_k$ are each smaller than the corresponding terms of a divergent series, or each bigger than the corresponding terms of a convergent series, the BCT does not allow us to conclude anything about the convergence or divergence of $\sum_{k=1}^{\infty} a_k$.

If $0 < d_k \leq a_k$ and $\sum_{k=1}^{\infty} d_k$ diverges, then $s_n = \sum_{k=1}^{n} a_k \geq \sum_{k=1}^{n} d_k = D_n$. Because $d_k > 0$, $\{D_n\}$ is monotonically increasing, and because $\sum_{k=1}^{\infty} d_k$ diverges, the partial sums D_n must not be bounded above, hence the bigger partial sums s_n (which are also increasing) are not bounded above and $\lim_{n\to\infty} s_n = \infty$ so that $\sum_{k=1}^{\infty} a_k$ diverges. □

The Basic Comparison Test requires that we compare a given series to a series whose convergence or divergence we already know. This requires that we have at our disposal a collection of series that converge and some that diverge. Often we select a p-series or a geometric series to compare with the new series, but making this choice quickly requires some experience and practice.

Example 1. Use the Basic Comparison Test to determine the convergence or divergence of $\sum_{k=1}^{\infty} \frac{1}{k^2+3}$ and $\sum_{k=1}^{\infty} \frac{k+1}{k^2}$.

Solution. We can compare each of these series with a p-series for an appropriate value of p. For the first series:

$$k^2 + 3 > k^2 \Rightarrow 0 < \frac{1}{k^2+3} < \frac{1}{k^2}$$

holds for all k. We know that $\sum_{k=1}^{\infty} \frac{1}{k^2}$ converges (it's a p-series with $p = 2 > 1$), so $\sum_{k=1}^{\infty} \frac{1}{k^2+3}$ must also converge.

For the second series, we know that (for all $k \geq 1$):

$$\frac{k+1}{k^2} > \frac{k}{k^2} = \frac{1}{k} > 0$$

The harmonic series $\sum_{k=1}^{\infty} \frac{1}{k}$ diverges, so $\sum_{k=1}^{\infty} \frac{k+1}{k^2}$ must also diverge. ◀

Practice 1. Use the Basic Comparison Test to determine the convergence or divergence of $\sum_{k=3}^{\infty} \frac{1}{\sqrt{k}-2}$ and $\sum_{k=1}^{\infty} \frac{1}{2^k+7}$.

Example 2. Your classmate has shown that $\frac{1}{k^2} < \frac{1}{k^2-1} < \frac{1}{k}$ for all $k \geq 2$. Using this information and the Basic Comparison Test, what can you conclude about the convergence of the series $\sum_{k=2}^{\infty} \frac{1}{k^2-1}$?

Solution. Nothing. The Basic Comparison Test only provides a definitive answer about a series if that series is smaller than a convergent series or larger than a divergent series. In this situation, the series $\sum_{k=2}^{\infty} \frac{1}{k^2-1}$ is larger than a convergent series, $\sum_{k=2}^{\infty} \frac{1}{k^2}$, and smaller than a divergent series, $\sum_{k=2}^{\infty} \frac{1}{k}$, so the Basic Comparison Test does not allow us to conclude anything about the convergence of $\sum_{k=2}^{\infty} \frac{1}{k^2-1}$. ◀

Although the inequalities in the previous Example did not allow us to determine whether $\sum_{k=2}^{\infty} \frac{1}{k^2-1}$ converges or diverges, some clever algebra does allow us to employ the BCT with this series. For $k \geq 2$:

$$k^2 \geq 4 \Rightarrow \frac{1}{4}k^2 \geq 1 \Rightarrow -\frac{1}{4}k^2 \leq -1 \Rightarrow k^2 - \frac{1}{4}k^2 \leq k^2 - 1$$
$$\Rightarrow \frac{3}{4}k^2 \leq k^2 - 1 \Rightarrow \frac{\frac{4}{3}}{k^2} \geq \frac{1}{k^2-1}$$

Because $\sum_{k=2}^{\infty} \frac{1}{k^2}$ converges, the BCT (together with the above inequality) tells us that $\sum_{k=2}^{\infty} \frac{1}{k^2-1}$ must also converge.

We could also have applied the Integral Test, which requires the use of Partial Fraction Decomposition:

$$\int_2^{\infty} \frac{1}{x^2-1}\, dx = \lim_{M\to\infty} \int_2^M \left[\frac{\frac{1}{2}}{x-1} - \frac{\frac{1}{2}}{x+1} \right] dx = \frac{1}{2}\ln(3)$$

We've left out some steps here; you should fill them in.

to show that the series $\sum_{k=2}^{\infty} \frac{1}{k^2-1}$ converges, or we could have used the same partial fraction decomposition to rewrite the series:

$$\sum_{k=2}^{\infty} \frac{1}{k^2-1} = \frac{1}{2}\sum_{k=2}^{\infty}\left[\frac{1}{k-1} - \frac{1}{k+1}\right]$$
$$= \frac{1}{2}\left[\left(1-\frac{1}{3}\right) + \left(\frac{1}{2}-\frac{1}{4}\right) + \left(\frac{1}{3}-\frac{1}{5}\right) + \cdots\right] = \frac{3}{4}$$

as a telescoping series and show that its sum is $\frac{3}{4}$.

Limit Comparison Test

Consider the three infinite series:

$$\sum_{k=2}^{\infty} \frac{1}{k^2}, \quad \sum_{k=2}^{\infty} \frac{1}{k^2+1} \quad \text{and} \quad \sum_{k=2}^{\infty} \frac{1}{k^2-1}$$

All of them converge. Showing that the first series converges is easy: use the P-Test with $p = 2 > 1$. Because $0 < \frac{1}{k^2+1} < \frac{1}{k^2}$, it's relatively easy to show the second series converges (using the BCT and comparing it to the first series). The third series looks quite similar to the first two, but each of the three methods we used in the preceding discussion to show it converges was rather complicated. There must be a better way!

Limit Comparison Test (LCT)

If $a_k > 0$ and $b_k > 0$ for all k and $\lim_{k\to\infty} \frac{a_k}{b_k} = L$ where $0 < L < \infty$, then $\sum_{k=1}^{\infty} a_k$ and $\sum_{k=1}^{\infty} b_k$ both converge or both diverge.

Proof. If the hypotheses hold and $\sum_{k=1}^{\infty} b_k$ converges, then, because $\lim_{k\to\infty} \frac{a_k}{b_k} = L$ and $0 < L < \infty$, there is some integer N so that:

$$k \geq N \Rightarrow \frac{a_k}{b_k} \leq L+1 \Rightarrow a_k \leq (L+1)b_k$$

and the Basic Comparison Test then tells us that $\sum_{k=1}^{\infty} a_k$ converges. If the hypotheses hold and $\sum_{k=1}^{\infty} b_k$ diverges, there is some integer N so that:

$$k \geq N \Rightarrow \frac{a_k}{b_k} \geq \frac{L}{2} > 0 \Rightarrow a_k \geq \frac{L}{2} \cdot b_k$$

The Basic Comparison Test then tells us that $\sum_{k=1}^{\infty} a_k$ diverges. □

Example 3. Use the LCT to show that $\sum_{k=2}^{\infty} \frac{1}{k^2-1}$ converges.

Solution. For $k \geq 2$, let $a_k = \frac{1}{k^2-1}$ and $b_k = \frac{1}{k^2}$ so $a_k > 0$, $b_k > 0$ and:

$$\lim_{k\to\infty} \frac{a_k}{b_k} = \lim_{k\to\infty} \frac{\frac{1}{k^2-1}}{\frac{1}{k^2}} = \lim_{k\to\infty} \frac{k^2}{k^2-1} = 1$$

Because $0 < 1 < \infty$ and $\sum_{k=2}^{\infty} \frac{1}{k^2}$ (a p-series, with $p = 2$) converges, the LCT tells us that $\sum_{k=2}^{\infty} \frac{1}{k^2-1}$ converges. ◀

Practice 2. Use the Limit Comparison Test to show whether the series $\sum_{k=1}^{\infty} \frac{k^2+5k}{k^3+k^2+7}$ and $\sum_{k=3}^{\infty} \frac{5}{\sqrt{k^4-11}}$ converge or diverge.

The Limit Comparison Test allows us to "ignore" some parts of the terms of a series that cause algebraic difficulties when using the BCT, but which have no effect on the convergence of the series.

Using "Dominant Terms"

To use the Limit Comparison Test with a new series, we need to find another, simpler series that we already know converges or diverges to compare with our new series. When the new series involves a rational expression, one effective method to find an appropriate simpler series is ignore everything but the largest power of the variable (the dominant term) in the numerator and denominator of the new series. The Limit Comparison Test will then allow us to conclude that the new series converges if and only if the "dominant term" series converges.

Example 4. For each of series below, form a new series using only the dominant terms from the numerator and the denominator. Does the "dominant term" series converge?

(a) $\sum_{k=3}^{\infty} \frac{5k^2-3k+2}{17+2k^4}$ (b) $\sum_{k=1}^{\infty} \frac{1+4k}{\sqrt{k^3+5k}}$ (c) $\sum_{k=1}^{\infty} \frac{k^{23}+1}{5k^{10}+k^{26}+3}$

Solution. (a) The dominant terms of the numerator and denominator are $5k^2$ and $2k^4$, respectively, so the "dominant term" series is $\sum_{k=3}^{\infty} \frac{5k^2}{2k^4} = \frac{5}{2} \cdot \sum_{k=1}^{\infty} \frac{1}{k^2}$ (a p-series with $p = 2$), which converges.

(b) The dominant terms are $4k$ and $\sqrt{k^3} = k^{\frac{3}{2}}$, so the "dominant term" series is $\sum_{k=1}^{\infty} \frac{4k}{k^{\frac{3}{2}}} = 4 \cdot \sum_{k=1}^{\infty} \frac{1}{k^{\frac{1}{2}}}$ (a p-series with $p = \frac{1}{2}$), which diverges.

(c) The dominant terms are k^{23} and k^{26}, so the "dominant term" series is $\sum_{k=1}^{\infty} \frac{k^{23}}{k^{26}} = \sum_{k=1}^{\infty} \frac{1}{k^3}$ (a p-series with $p = 3$), which converges.

Using the Limit Comparison Test to compare each of the given series with their corresponding "dominant term" series, we can conclude that the first and third converge and that the second diverges. ◀

Practice 3. For each of series below, form a new series using only the dominant terms from the numerator and the denominator. Does the "dominant term" series converge? Does the given series converge?

(a) $\sum_{k=1}^{\infty} \frac{3k^4 - 5k + 2}{1 + 17k^2 + 9k^5}$ (b) $\sum_{k=1}^{\infty} \frac{\sqrt{1+9k}}{k^2 + 5k - 2}$ (c) $\sum_{k=1}^{\infty} \frac{k^{25}+1}{5k^{10} + k^{26} + 3}$

Experienced calculus students commonly use "dominant terms" to make quick and accurate judgments about the convergence or divergence of a series. With practice, so can you.

9.7 Problems

In Problems 1–12, use the Basic Comparison Test to determine whether the series converges or diverges.

1. $\sum_{k=1}^{\infty} \frac{\cos^2(k)}{k^2}$
2. $\sum_{k=1}^{\infty} \frac{3}{k^3+7}$
3. $\sum_{n=3}^{\infty} \frac{5}{n-1}$
4. $\sum_{k=1}^{\infty} \frac{2+\sin(k)}{k^3}$
5. $\sum_{m=1}^{\infty} \frac{3+\cos(m)}{m}$
6. $\sum_{k=1}^{\infty} \frac{\arctan(k)}{k^{\frac{3}{2}}}$
7. $\sum_{k=2}^{\infty} \frac{\ln(k)}{k}$
8. $\sum_{k=2}^{\infty} \frac{k-1}{k \cdot 1.5^k}$
9. $\sum_{k=1}^{\infty} \frac{k+9}{k \cdot 2^k}$
10. $\sum_{n=2}^{\infty} \frac{n^3+7}{n^4-1}$
11. $\sum_{k=1}^{\infty} \frac{1}{k!}$
12. $\sum_{n=1}^{\infty} \frac{1}{1+2+3+\cdots+n}$

In Problems 13–22 use the Limit Comparison Test (or the Test for Divergence) to determine whether the given series converges or diverges.

13. $\sum_{k=3}^{\infty} \frac{k+1}{k^2+4}$
14. $\sum_{k=1}^{\infty} \frac{7}{\sqrt{k^3+3}}$
15. $\sum_{w=1}^{\infty} \frac{5}{w+1}$
16. $\sum_{n=1}^{\infty} \frac{7n^3-4n+3}{3n^4+7n^3+9}$
17. $\sum_{k=1}^{\infty} \frac{k^3}{(1+k^2)^3}$
18. $\sum_{k=1}^{\infty} \left(\frac{\arctan(k)}{k}\right)^2$
19. $\sum_{n=1}^{\infty} \frac{5-\frac{1}{n}}{n}$
20. $\sum_{w=1}^{\infty} \left(1+\frac{1}{w}\right)^w$
21. $\sum_{k=2}^{\infty} \left(\frac{1-\frac{1}{k}}{k}\right)^3$
22. $\sum_{k=2}^{\infty} \sqrt{\frac{k^3-4}{k^5+1}}$

In 23–32, use a "dominant term" series to determine whether the given series converges or diverges.

23. $\sum_{n=3}^{\infty} \frac{n+100}{n^3+4}$
24. $\sum_{n=3}^{\infty} \frac{n+100}{n^2-4}$
25. $\sum_{k=1}^{\infty} \frac{7k}{\sqrt{k^3+5}}$
26. $\sum_{k=1}^{\infty} \frac{5}{k+1}$
27. $\sum_{k=2}^{\infty} \frac{k^3-4k+3}{2k^4+7k^6+9}$
28. $\sum_{n=1}^{\infty} \frac{5n^3+7n^2+9}{(1+n^3)^2}$
29. $\sum_{k=1}^{\infty} \left(\frac{\arctan(3k)}{2k}\right)^2$
30. $\sum_{n=1}^{\infty} \left(\frac{3-\frac{1}{n}}{n}\right)^2$
31. $\sum_{k=1}^{\infty} \frac{\sqrt{k^3+4k^2}}{k^2+3k-2}$
32. $\sum_{k=2}^{\infty} \frac{\arcsin\left(1-\frac{1}{k^2}\right)}{k}$

In 33–78, use any method from this or previous sections to determine whether the series converges or diverges. Include reasoning for your answers.

33. $\displaystyle\sum_{n=2}^{\infty} \frac{n^2+10}{n^3-3}$

34. $\displaystyle\sum_{k=1}^{\infty} \frac{3k}{\sqrt{k^5+7}}$

35. $\displaystyle\sum_{k=1}^{\infty} \frac{3}{2k+1}$

36. $\displaystyle\sum_{n=2}^{\infty} \frac{n^2-n+1}{3n^4+2n^2+1}$

37. $\displaystyle\sum_{n=1}^{\infty} \frac{2n^3+n^2+6}{(3+n^2)^2}$

38. $\displaystyle\sum_{k=1}^{\infty} \left(\frac{\arctan(k)}{3}\right)^k$

39. $\displaystyle\sum_{k=3}^{\infty} \sqrt{\frac{1-\frac{2}{k}}{k}}$

40. $\displaystyle\sum_{n=3}^{\infty} \frac{\sqrt{n^2+4n}}{(n-2)^3}$

41. $\displaystyle\sum_{k=1}^{\infty} \frac{(k+5)^2}{k^2\cdot 3^k}$

42. $\displaystyle\sum_{n=1}^{\infty} \frac{1+\sin(n)}{n^2+4}$

43. $\displaystyle\sum_{k=1}^{\infty} \frac{k+2}{\sqrt{k^2+1}}$

44. $\displaystyle\sum_{k=1}^{\infty} \frac{\sin(k\pi)}{k+1}$

45. $\displaystyle\sum_{k=0}^{\infty} \frac{3}{e^k+k}$

46. $\displaystyle\sum_{n=0}^{\infty} \frac{(2+3n)^2+9}{(1+n^3)^2}$

47. $\displaystyle\sum_{n=1}^{\infty} \left(\frac{\tan(3)}{2+n}\right)^2$

48. $\displaystyle\sum_{n=1}^{\infty} n\cdot\sin\left(\frac{1}{n}\right)$

49. $\displaystyle\sum_{k=1}^{\infty} \sin\left(\frac{1}{k}\right)$

50. $\displaystyle\sum_{n=1}^{\infty} \sin^2\left(\frac{1}{n}\right)$

51. $\displaystyle\sum_{n=1}^{\infty} \sin^3\left(\frac{1}{n}\right)$

52. $\displaystyle\sum_{n=1}^{\infty} \cos^2\left(\frac{1}{n}\right)$

53. $\displaystyle\sum_{n=1}^{\infty} \cos^3\left(\frac{1}{n}\right)$

54. $\displaystyle\sum_{n=1}^{\infty} \tan^2\left(\frac{1}{n}\right)$

55. $\displaystyle\sum_{k=1}^{\infty} \left(1-\frac{2}{k}\right)^k$

56. $\displaystyle\sum_{k=1}^{\infty} \left(1+\frac{2}{k}\right)^k$

57. $\displaystyle\sum_{k=1}^{\infty} \frac{5}{3^k}$

58. $\displaystyle\sum_{n=1}^{\infty} \frac{5+\cos(n^3)}{n^2}$

59. $\displaystyle\sum_{n=1}^{\infty} \frac{2}{3+\sin(n^3)}$

60. $\displaystyle\sum_{k=1}^{\infty} \frac{5}{\left(\frac{1}{3}\right)^k}$

61. $\displaystyle\sum_{k=0}^{\infty} e^{-k}$

62. $\displaystyle\sum_{k=0}^{\infty} \left(\frac{\pi}{e}\right)^k$

63. $\displaystyle\sum_{k=0}^{\infty} \left(\frac{\pi^2}{e^3}\right)^k$

64. $\displaystyle\sum_{k=1}^{\infty} \cos\left(\frac{1}{k^3}\right)$

65. $\displaystyle\sum_{n=1}^{\infty} \frac{5+\cos(n^2)}{n^3}$

66. $\displaystyle\sum_{k=1}^{\infty} \frac{1}{k\cdot[3+\ln(k)]}$

67. $\displaystyle\sum_{m=1}^{\infty} \frac{1}{m\cdot[3+\ln(m)]^2}$

68. $\displaystyle\sum_{n=1}^{\infty} \frac{4}{n\cdot\arctan(n)}$

69. $\displaystyle\sum_{n=1}^{\infty} \frac{4\arctan(n)}{n}$

70. $\displaystyle\sum_{k=2}^{\infty} \frac{\ln(k)}{k^3}$

71. $\displaystyle\sum_{k=2}^{\infty} \frac{\ln(k)}{k^2}$

72. $\displaystyle\sum_{n=1}^{\infty} \left(\frac{n}{2n+3}\right)^n$

73. $\displaystyle\sum_{n=1}^{\infty} \frac{1+n}{1+n^2}$

74. $\displaystyle\sum_{k=2}^{\infty} \left[\sin(k)-\sin(k+1)\right]$

75. $\displaystyle\sum_{k=1}^{\infty} \sqrt{\frac{k^3+5}{k^5+3}}$

76. $\displaystyle\sum_{k=1}^{\infty} \frac{1}{k^2}$

77. $\displaystyle\sum_{n=1}^{\infty} n^{\frac{1}{n}}$

78. $\displaystyle\sum_{k=1}^{\infty} \sqrt[3]{\frac{k^3+7}{k^8+5}}$

9.7 Practice Answers

1. Considering the first series, for any integer $k \geq 3$:

$$k-2<k \quad\Rightarrow\quad \sqrt{k-2}<\sqrt{k} \quad\Rightarrow\quad \frac{1}{\sqrt{k}}<\frac{1}{\sqrt{k-2}}$$

Because $\displaystyle\sum_{k=3}^{\infty} \frac{1}{\sqrt{k}}$ (a p-series with $p=\frac{1}{2}\leq 1$) diverges and each term of $\displaystyle\sum_{k=3}^{\infty} \frac{1}{\sqrt{k-2}}$ is bigger, the BCT says $\displaystyle\sum_{k=3}^{\infty} \frac{1}{\sqrt{k}}$ also diverges.

Considering the second series, for any k:

$$0 < 7 \quad \Rightarrow \quad 2^k < 2^k + 7 \quad \Rightarrow \quad \frac{1}{2^k} > \frac{1}{2^k + 7}$$

Because $\sum_{k=3}^{\infty} \left(\frac{1}{2}\right)^k$ (a geometric series with ratio $|r| = \frac{1}{2} < 1$) converges and each term of $\sum_{k=3}^{\infty} \frac{1}{2^k + 7}$ is even smaller, the BCT tells us that $\sum_{k=3}^{\infty} \frac{1}{2^k + 7}$ also converges.

2. For large values of k, the terms of the series $\sum_{k=1}^{\infty} \frac{k^2 + 5}{k^3 + k^2 + 7}$ behave like $\frac{k^2}{k^3} = \frac{1}{k}$ so we will compare the given series to $\sum_{k=1}^{\infty} \frac{1}{k}$, which we know diverges (it's the harmonic series). Computing the limit of the ratio of the corresponding terms of these series:

$$\lim_{k \to \infty} \frac{\frac{k^2+5}{k^3+k^2+7}}{\frac{1}{k}} = \lim_{k \to \infty} \frac{k^3 + 5k}{k^3 + k^2 + 7} = \lim_{k \to \infty} \frac{1 + \frac{5}{k^2}}{1 + \frac{1}{k} + \frac{7}{k^3}} = 1$$

Because $0 < 1 < \infty$, the LCT says $\sum_{k=1}^{\infty} \frac{k^2 + 5}{k^3 + k^2 + 7}$ also diverges.

For large values of k, the terms of the series $\sum_{k=3}^{\infty} \frac{5}{\sqrt{k^4 - 11}}$ behave like $\frac{5}{\sqrt{k^4}} = \frac{5}{k^2}$ so we will compare the given series to $\sum_{k=3}^{\infty} \frac{5}{k^2}$, which we know converges (it's a constant multiple of a p-series with $p = 2 > 1$). The limit of the ratio of the corresponding terms of these series is:

$$\lim_{k \to \infty} \frac{\frac{5}{\sqrt{k^4 - 11}}}{\frac{5}{k^2}} = \lim_{k \to \infty} \frac{k^2}{\sqrt{k^4 - 11}} = \lim_{k \to \infty} \frac{1}{\sqrt{1 - \frac{11}{k^4}}} = 1$$

Because $0 < 1 < \infty$, the LCT says $\sum_{k=3}^{\infty} \frac{5}{\sqrt{k^4 - 11}}$ also converges.

3. (a) $\sum_{k=1}^{\infty} \frac{3k^4}{9k^5} = \frac{1}{3} \sum_{k=1}^{\infty} \frac{1}{k}$ diverges (it's a multiple of the harmonic series), so $\sum_{k=1}^{\infty} \frac{3k^4 - 5k + 2}{1 + 17k^2 + 9k^5}$ also diverges (by the LCT).

(b) $\sum_{k=1}^{\infty} \frac{\sqrt{9k}}{k^2} = 3 \sum_{k=1}^{\infty} \frac{1}{k^{\frac{3}{2}}}$ converges (it's a multiple of a p-series with $p = \frac{3}{2} > 1$), so $\sum_{k=1}^{\infty} \frac{\sqrt{1 + 9k}}{k^2 + 5k - 2}$ also converges (by the LCT).

(c) $\sum_{k=1}^{\infty} \frac{k^{25}}{k^{26}} = \sum_{k=1}^{\infty} \frac{1}{k}$ diverges, so $\sum_{k=1}^{\infty} \frac{k^{25} + 1}{5k^{10} + k^{26} + 3}$ diverges (LCT).

Why does $\sum_{k=1}^{\infty} \frac{1}{k}$ diverge?

9.8 *Alternating Series*

The Integral Test, P-Test and comparison tests each apply only to series with positive terms. We now examine some series with both positive and negative terms, focusing on series with terms that alternate between positive and negative values.

Examples of Alternating Series

An **alternating series** is a series with terms that alternate between positive and negative. Each of the following are alternating series:

$$\sum_{k=1}^{\infty} (-1)^{k+1}\frac{1}{k} = 1 - \frac{1}{2} + \frac{1}{3} - \frac{1}{4} + \frac{1}{5} - \frac{1}{6} + \cdots$$

$$\sum_{k=1}^{\infty} (-1)^{k}\frac{k}{k+2} = -\frac{1}{3} + \frac{2}{4} - \frac{3}{5} + \frac{4}{6} - \frac{5}{7} + \cdots$$

$$\sum_{k=1}^{\infty} (-1)^{k}\frac{1}{\sqrt{2k+1}} = -\frac{1}{\sqrt{3}} + \frac{1}{\sqrt{5}} - \frac{1}{\sqrt{7}} + \frac{1}{\sqrt{9}} - \frac{1}{\sqrt{11}} + \cdots$$

We call this first example the **alternating harmonic series**.

The margin graphs show values of several partial sums s_n for each of these series. As n increases, the partial sums s_n alternately get larger and smaller, a typical pattern for the partial sums of alternating series. This same pattern appears in tables of the partial sums:

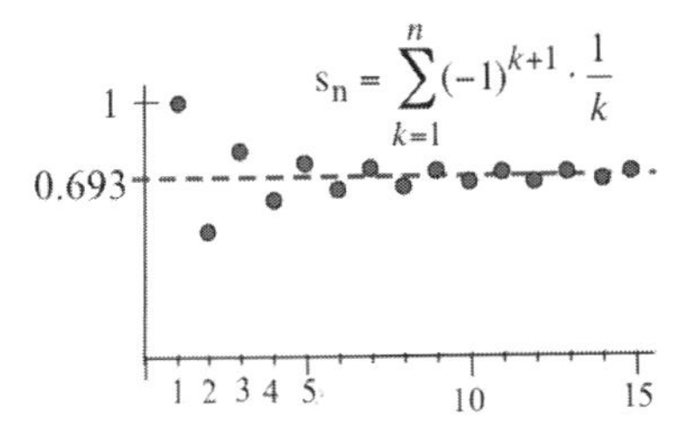

$\sum_{k=1}^{\infty} \frac{(-1)^{k+1}}{k}$

n	s_n
1	1.0
2	0.5
3	0.8333
4	0.5833
5	0.7833
6	0.6166
7	0.7595
8	0.6345
9	0.7456
10	0.6456

$\sum_{k=1}^{\infty} \frac{(-1)^{k}\cdot k}{k+2}$

n	s_n
1	-0.3333
2	0.1667
3	-0.4333
4	0.2333
5	-0.4810
6	0.2690
7	-0.5087
8	0.2913
9	-0.5269
10	0.3064

$\sum_{k=1}^{\infty} \frac{(-1)^{k}}{\sqrt{2k+1}}$

n	s_n
1	-0.5774
2	-0.1301
⋮	⋮
50	-0.2828
51	-0.3813
⋮	⋮
1000	-0.3211
1001	-0.3435

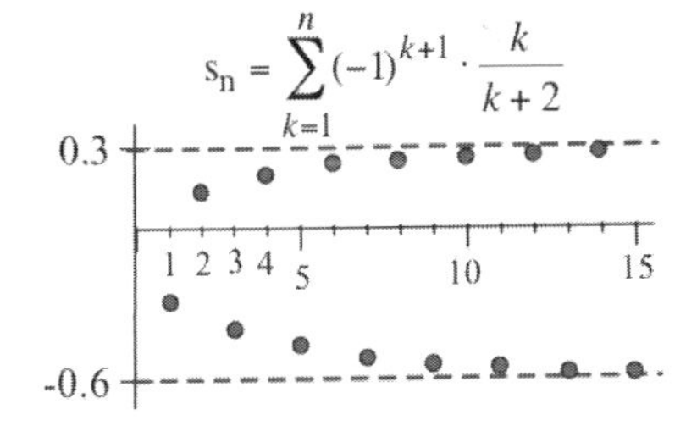

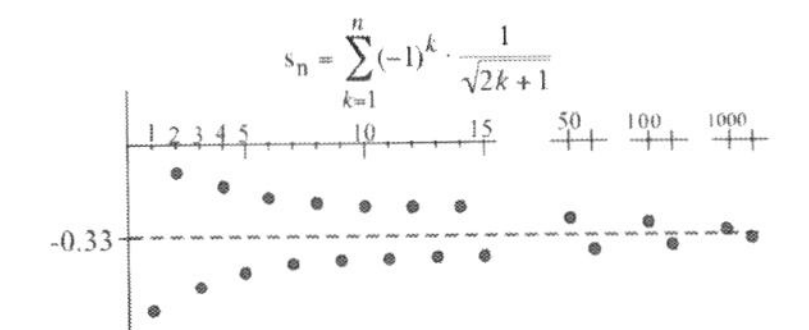

For the first and third series, the partial sums appear to be converging to a limit. For the second series, the partial sums appear to jump back and forth between values near 0.3 and values near -0.6: because $\lim_{k\to\infty} \frac{(-1)^k \cdot k}{k+2}$ does not exist (and, in particular, does not equal 0), the Test for Divergence tells us that this series diverges. The next result will help us determine that the other two series converge.

Alternating Series Test

The next result provides an easy way to determine that *some* alternating series converge: if the absolute values of the terms of an alternating series decrease monotonically to 0, then the series converges.

Alternating Series Test (AST)

If the numbers a_k satisfy the three conditions:

- $a_k > 0$ for all k (each a_k is positive)
- $a_k > a_{k+1}$ ($\{a_k\}$ is monotonically decreasing)
- $\lim_{k\to\infty} a_k = 0$

then the alternating series $\sum_{k=0}^{\infty} (-1)^k a_k$ converges.

Proof. To show that the alternating series converges, we need to show that the sequence of partial sums approaches a finite limit. We do so by showing that the subsequences of even partial sums $\{s_2, s_4, s_6, \ldots\}$ and odd partial sums $\{s_1, s_3, s_5, \ldots\}$ each approach the same value.

Because $\{a_k\}$ is a decreasing sequence, we know that $a_1 > a_2$, $a_3 > a_4$ and so forth, so that for the even partial sums:

$$\begin{aligned} s_2 &= a_1 - a_2 > 0 \\ s_4 &= s_2 + (a_3 - a_4) > s_2 \\ s_6 &= s_4 + (a_5 - a_6) > s_4 \end{aligned}$$

In general, the sequence of even partial sums is positive and increasing:

$$s_{2n+2} = s_{2n} + (a_{2n+1} - a_{2n+2}) > s_{2n} > 0$$

because $\{a_k\}$ is decreasing, so that $a_{2n+1} > a_{2n+2}$ for all n. Furthermore:

$$\begin{aligned} s_{2n} &= a_1 - a_2 + a_3 - a_4 + a_5 - \cdots - a_{2n-2} + a_{2n-1} - a_{2n} \\ &= a_1 - (a_2 - a_3) - (a_4 - a_5) - \cdots - (a_{2n-2} - a_{2n-1}) - a_{2n} < a_1 \end{aligned}$$

so the sequence of even partial sums is bounded above by a_1. Because the sequence $\{s_2, s_4, s_6, \ldots\}$ of even partial sums is increasing and bounded above, the Monotone Convergence Theorem tells us the sequence of even partial sums converges to some finite limit: $\lim_{n\to\infty} s_{2n} = L$

We can write any odd partial sum as $s_{2n+1} = s_{2n} + a_{2n+1}$ so that:

$$\lim_{n\to\infty} s_{2n+1} = \lim_{n\to\infty} s_{2n} + \lim_{n\to\infty} a_{2n+1} = L + 0 = L$$

Because the sequence of even partial sums and the sequence of odd partial sums both approach the same limit, L, we can conclude that the limit of the sequence consisting of all partial sums is L and that the alternating series $\sum_{n=0}^{\infty} (-1)^n a_n$ converges (to L). □

Example 1. Show that each alternating series below satisfies the three conditions in the hypotheses of the Alternating Series Test, allowing you to conclude that each of them converges.

(a) $\sum_{k=1}^{\infty} (-1)^{k+1}\frac{1}{k} = 1 - \frac{1}{2} + \frac{1}{3} - \frac{1}{4} + \frac{1}{5} - \frac{1}{6} + \cdots$

(b) $\sum_{k=1}^{\infty} (-1)^{k+1}\frac{3}{\sqrt{k}} = \frac{3}{1} - \frac{3}{\sqrt{2}} + \frac{3}{\sqrt{3}} - \frac{3}{\sqrt{4}} + \cdots$

(c) $\sum_{k=2}^{\infty} (-1)^{k}\frac{7}{k\cdot\ln(k)} = \frac{7}{2\ln(2)} - \frac{7}{3\ln(3)} + \frac{7}{4\ln(4)} - \frac{7}{5\ln(5)} + \cdots$

Solution. (a) Here $a_k = \frac{1}{k}$, so $a_k > 0$ for all $k \geq 1$. The function $f(x) = x^{-1}$ satisfies $f(k) = a_k$ for $k \geq 1$, and $f'(x) = -x^{-2} < 0$, which tells us that $f(x)$ is decreasing, hence a_k is a decreasing sequence. Finally, $\lim_{k\to\infty}\frac{1}{k} = 0$, so all three conditions are satisfied.

In each part of this Example we use the third technique from Practice 4 in Section 9.2 to show that a sequence is monotonically decreasing. This is not the only possible method, but turns out to be convenient for the sequences under consideration here.

(b) Here $a_k = \frac{3}{\sqrt{k}}$, so $a_k > 0$ for all $k \geq 1$. The function $f(x) = 3x^{-\frac{1}{2}}$ satisfies $f(k) = a_k$ for $k \geq 1$, and $f'(x) = -\frac{3}{2}x^{-\frac{3}{2}} < 0$, which tells us that $f(x)$ is decreasing, hence a_k is a decreasing sequence. Finally, $\lim_{k\to\infty}\frac{3}{\sqrt{k}} = 0$, so all three conditions are satisfied.

(c) Here $a_k = \frac{1}{k\cdot\ln(k)}$, so $a_k > 0$ for all $k \geq 2$. If $f(x) = x\cdot\ln(x)$, $a_k = \frac{1}{f(k)}$ for $k \geq 2$, and $f'(x) = 1 + \ln(x) > 0$ for $x \geq 2$, which tells us that $f(x)$ is increasing, hence a_k is a decreasing sequence. Finally, $\lim_{k\to\infty}\frac{1}{k\cdot\ln(k)} = 0$, so all three conditions are satisfied. The Alternating Series Test therefore tells us that all three series converge. ◄

Practice 1. Show that each alternating series below satisfies the three conditions in the hypotheses of the Alternating Series Test, allowing you to conclude that each of them converges.

(a) $\sum_{k=1}^{\infty} (-1)^{k+1}\frac{1}{k^2} = 1 - \frac{1}{4} + \frac{1}{9} - \frac{1}{16} + \frac{1}{25} - \cdots$

(b) $\sum_{k=2}^{\infty} (-1)^{k}\frac{3}{\ln(k)} = \frac{3}{\ln(2)} - \frac{3}{\ln(3)} + \frac{3}{4\ln(4)} - \frac{3}{\ln(5)} + \cdots$

Examples of Divergent Alternating Series

If $a_k > 0$ fails to hold for a series written in the form $\sum_{k=0}^{\infty} (-1)^k a_k$, then the series does not alternate and the AST fails to apply. Such a series may converge or diverge.

For examples, think of any convergent or divergent series will all positive terms.

If $\lim_{k\to\infty} a_k = 0$ fails to hold for *any* series (not just alternating series), then the Test for Divergence tells us that the series diverges.

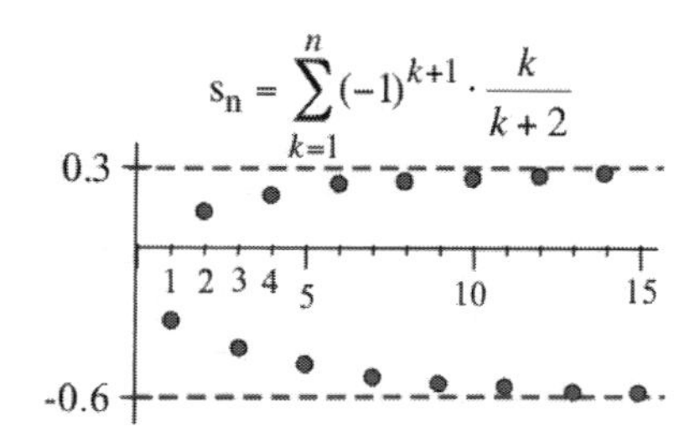

Example 2. Does $\sum_{k=1}^{\infty} (-1)^{k+1} \frac{k}{k+2}$ converge?

Solution. Here $a_k = \frac{k}{k+2} > 0$ but $\lim_{k\to\infty} \frac{k}{k+2} = 1 \neq 0$, so the Test For Divergence tell us the series diverges. (The margin figure shows some of the partial sums for this series. You should notice that the even and the odd partial sums are approaching two different values.) ◀

Practice 2. Does $\sum_{k=1}^{\infty} (-1)^{k+1} k$ converge? Does $\sum_{k=1}^{\infty} \frac{(-1)^{k+1}}{\sqrt{2k+1}}$ converge?

If $a_k > 0$ and $\lim_{k\to\infty} a_k = 0$ for an alternating series $\sum_{k=0}^{\infty} (-1)^k a_k$ but the terms $\{a_k\}$ are not monotonically decreasing, the AST does not apply and the series may converge or may diverge. For example:

$$\frac{3}{2} - \frac{1}{2} + \frac{3}{4} - \frac{1}{4} + \frac{3}{6} - \frac{1}{6} + \frac{3}{8} - \frac{1}{8} + \cdots$$

is an alternating series whose terms approach 0, but the even partial sums of this series are:

$$s_2 = \frac{3}{2} - \frac{1}{2} = 1$$
$$s_4 = \left(\frac{3}{2} - \frac{1}{2}\right) + \left(\frac{3}{4} - \frac{1}{4}\right) = 1 + \frac{1}{2}$$
$$s_6 = \left(\frac{3}{2} - \frac{1}{2}\right) + \left(\frac{3}{4} - \frac{1}{4}\right) + \left(\frac{3}{6} - \frac{1}{6}\right) = 1 + \frac{1}{2} + \frac{1}{3}$$

and, in general:

$$s_{2n} = 1 + \frac{1}{2} + \frac{1}{3} + \frac{1}{4} + \cdots + \frac{1}{n} = \sum_{k=1}^{n} \frac{1}{k}$$

Can you think of an alternating series with terms that approach 0 non-monotonically that converges?

You should recognize these even partial sums as partial sums of the harmonic series, which diverges, so the (even) partial sums of our given series diverge, hence the sequence of all partial sums of that series diverges, so the given series diverges.

Approximating the Sum of an Alternating Series

In Section 9.6, we used a consequence of the Integral Test to help find upper and lower bounds for $\sum_{k=1}^{\infty} a_k$, but this only applied to series with positive terms for which we could find a continuous, positive, decreasing function $f(x)$ with $f(k) = a_k$.

If you know that a series converges and if you add up the first "many" terms, then you should expect that the resulting partial sum is reasonably "close" to the value S obtained by adding all the terms together. Generally, however, we do not know how close the partial sum is to S (because we don't know the value of S). The situation with many alternating series is much nicer.

Alternating Series Estimation Bound

If $S = \sum_{k=0}^{\infty} (-1)^k a_k$ and the numbers a_k satisfy:

- $a_k > 0$ for all k (each a_k is positive)
- $a_k > a_{k+1}$ ($\{a_k\}$ is monotonically decreasing)
- $\lim_{k\to\infty} a_k = 0$

then $|S - s_n| < a_{n+1}$ for any partial sum $s_n = \sum_{k=0}^{n} (-1)^k a_k$.

This estimation bound only applies to *alternating* series. It is often tempting—but wrong—to use it with other series.

The geometric idea behind this estimation bound exploits the fact that when an alternating series satisfies the hypotheses of the AST, then the graph of the sequence $\{s_n\}$ of partial sums is "trumpet-shaped" or "funnel-shaped" (see margin). The partial sums alternately fall above and below the value S and "squeeze" in on the value S. Because the distance from s_n to S is less than the distance between the successive terms s_n and s_{n+1} (see second margin figure) and $|s_{n+1} - s_n| = a_{n+1}$.

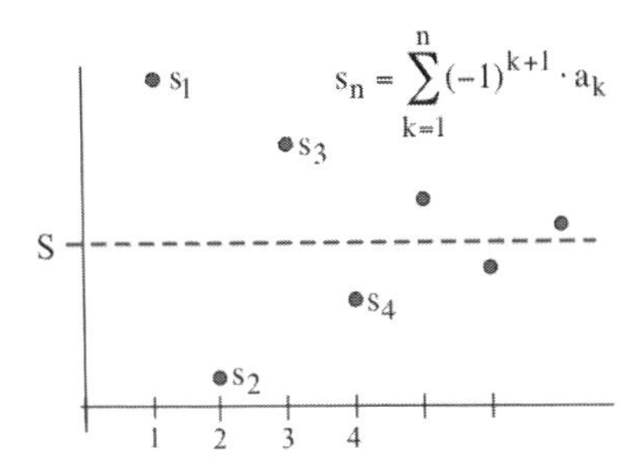

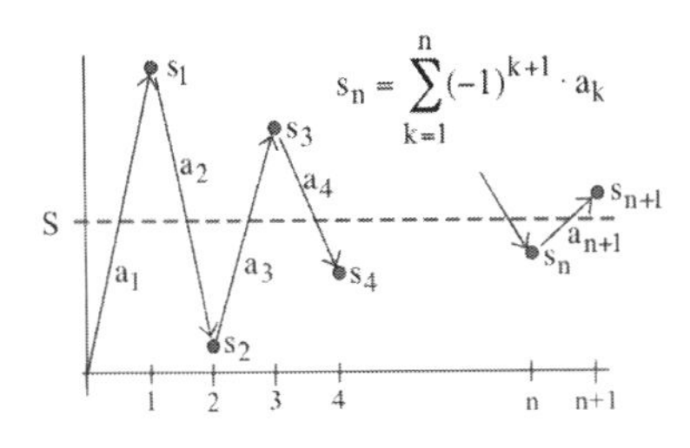

Proof. The distance between the sum S and the n-th partial sum s_n is:

$$
\begin{aligned}
|S - s_n| &= \left| \left[\sum_{k=0}^{\infty} (-1)^k a_k \right] - \left[\sum_{k=0}^{n} (-1)^k a_k \right] \right| = \left| \sum_{k=n+1}^{\infty} (-1)^k a_k \right| \\
&= \left| (-1)^{n+1} a_{n+1} + (-1)^{n+2} a_{n+2} + (-1)^{n+3} a_{n+3} + \cdots \right| \\
&= \left| (-1)^{n+1} \left[a_{n+1} - a_{n+2} + a_{n+3} - a_{n+4} + \cdots \right] \right| \\
&= \left| a_{n+1} - a_{n+2} + a_{n+3} - a_{n+4} + \cdots \right| \\
&= \left| (a_{n+1} - a_{n+2}) + (a_{n+3} - a_{n+4}) + (a_{n+5} - a_{n+6}) + \cdots \right| \\
&= (a_{n+1} - a_{n+2}) + (a_{n+3} - a_{n+4}) + (a_{n+5} - a_{n+6}) + \cdots \\
&= a_{n+1} - (a_{n+2} - a_{n+3}) - (a_{n+4} - a_{n+5}) - \cdots < a_{n+1}
\end{aligned}
$$

The last two equalities hold because $\{a_k\}$ is monotonically decreasing, so that $a_m > a_{m+1} \Rightarrow a_m - a_{m+1} > 0$ for any $m \geq 1$. □

We typically use this estimation bound in two different ways. Sometimes you know the value of n and you want to know how close s_n is to S. Other times you know how close you need s_n to be to S and want to find a value of n to ensure that level of closeness. The next two Examples illustrate these two different uses of the Alternating Series Estimation Bound.

Example 3. How close is $s_4 = \sum_{k=1}^{4} \frac{(-1)^{k+1}}{k^2}$ to $S = \sum_{k=1}^{\infty} \frac{(-1)^{k+1}}{k^2}$?

Solution. The series $\sum_{k=1}^{\infty} \frac{(-1)^{k+1}}{k^2}$ satisfies the hypotheses of the AST with $a_k = \frac{1}{k^2}$: for $k \geq 1$, $a_k > 0$ and $\{a_k\}$ is a decreasing sequence with limit 0. Computing the partial sum:

$$s_4 = \sum_{k=1}^{4} \frac{(-1)^{k+1}}{k^2} = 1 - \frac{1}{4} + \frac{1}{9} - \frac{1}{16} \approx 0.79861$$

so that:

$$|S - s_4| < a_5 = \frac{1}{25} = 0.04 \Rightarrow |S - 0.79861| < 0.04$$

hence $-0.04 < S - 0.79861 < 0.04 \Rightarrow 0.75861 < S < 0.83861$. ◄

Practice 3. Evaluate s_4 and s_9 for $S = \sum_{k=1}^{\infty} \frac{(-1)^{k+1}}{k^3}$ and determine bounds for $|S - s_4|$ and $|S - s_9|$.

Example 4. Find a value N so that s_N will be within 0.001 of the exact value of $S = \sum_{k=1}^{\infty} \frac{(-1)^{k+1}}{k!}$ and evaluate s_N.

Solution. The given series satisfies the hypotheses of the AST with $a_k = \frac{1}{k!}$ because $\frac{1}{k!}$ decreases monotonically to 0, so we know that $|S - s_N| < a_{N+1} = \frac{1}{(N+1)!}$. We need to find N so that $\frac{1}{(N+1)!} \leq 0.001 = \frac{1}{1000}$. With a little experimentation using a calculator, you can see that $6! = 720$, which doesn't quite work, but that $7! = 5040 > 1000$ so that $\frac{1}{7!} = \frac{1}{5040} \approx 0.000198 < 0.001$. With $N + 1 = 7 \Rightarrow N = 6$, $s_6 \approx 0.631944$ is the first partial sum *guaranteed* to be within 0.001 of S. In fact, s_6 is guaranteed to be within 0.000198 of S, so $0.631746 < S < 0.632142$. ◄

Practice 4. Find a value N so that s_N will be within 0.001 of the exact value of $S = \sum_{k=1}^{\infty} \frac{(-1)^{k+1}}{k^3 + 5}$ and evaluate s_N.

Wrap-Up

The Alternating Series Estimation Bound guarantees that s_n will be within a_{n+1} of S. In fact, s_n is often much closer to S.

Because the first finite number of terms do not affect the convergence or divergence of a series (they *do* affect its sum, S) you can use the Alternating Series Test and the Alternating Series Estimation Bound as long as the terms of a series "eventually" satisfy the required hypotheses. More precisely, "eventually" means that there is a value N so that for $n \geq N$ the series is an alternating series and the absolute value of the terms of that series decrease monotonically to 0.

If a series has some positive terms and some negative terms but those terms do **not** "eventually" alternate in sign, then you can **not** use the Alternating Series Test: it simply does not apply to such series. A result from the next section will allow us to show that *some* series of this type converge, but in general you will need more advanced methods to investigate the convergence and divergence of series with both positive and negative terms that do not (eventually) alternate.

9.8 Problems

Problems 1–6 give the values of the first four terms of a series. For each series, (a) calculate and graph the first four partial sums and (b) Determine whether or not it is an alternating series.

1. $1, -0.8, 0.6, -0.4$
2. $-1, 1.5, -0.7, 1$
3. $-1, 2, -3, 4$
4. $2, -1, -0.5, 0.3$
5. $-1, -0.6, 0.4, 0.2$
6. $2, -1, 0.5, -0.3$

Problems 7–12 give the values of the first five partial sums of a series. Which of the series are not alternating series. Why?

7. $2, 1, 3, 2, 4$
8. $2, 1, 1.8, 1.4, 1.6$
9. $2, 3, 2.1, 2.9, 2.8$
10. $-3, -1, -2.5, -1.5, -2$
11. $-1, 1, -0.8, -0.6, -0.4$
12. $-2.3, -1.6, -1.4, -1.8, -1.7$

Problems 13–16 shows the graphs of the partial sums of three series. Which is/are not the partial sums of alternating series? Why?

13.

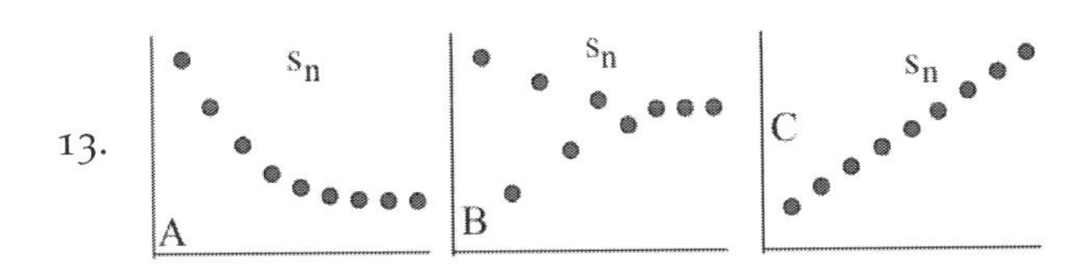

14.

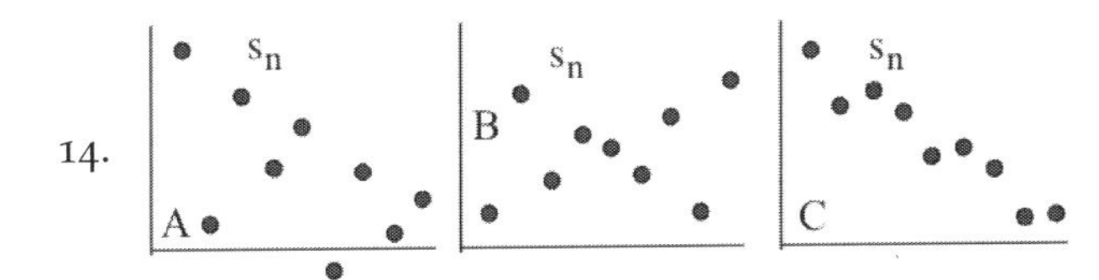

15. A B C s_n

16.

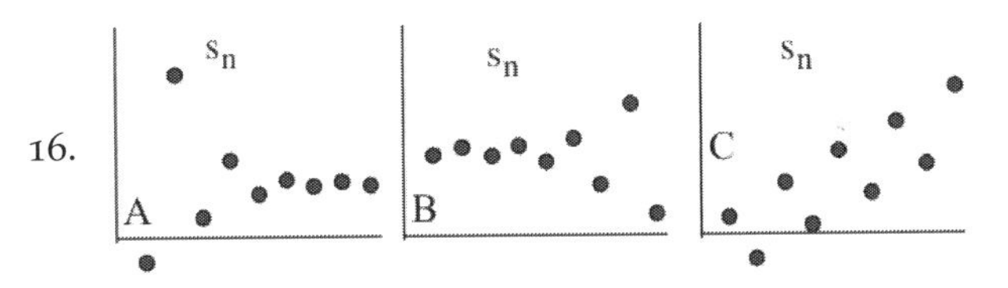

In Problems 17–31, determine whether the given series converges or diverges.

17. $\sum_{k=0}^{\infty} (-1)^k \cdot \frac{1}{k+5}$
18. $\sum_{k=3}^{\infty} (-1)^k \cdot \frac{1}{\ln(k)}$
19. $\sum_{k=1}^{\infty} (-1)^{k+1} \cdot \frac{k}{k^2+3}$
20. $\sum_{k=5}^{\infty} (-0.99)^{k+1}$
21. $\sum_{k=1}^{\infty} (-1)^k \cdot \frac{k+3}{k+7}$
22. $\sum_{k=1}^{\infty} (-1)^{k+1} \cdot \sin\left(\frac{1}{k}\right)$
23. $\sum_{k=4}^{\infty} \frac{\cos(k\pi)}{k}$
24. $\sum_{k=3}^{\infty} \frac{\sin(k\pi)}{k}$
25. $\sum_{k=2}^{\infty} (-1)^k \cdot \frac{\ln(k)}{k}$
26. $\sum_{k=3}^{\infty} (-1)^{k+1} \cdot \frac{\ln(k)}{\ln(k^3)}$
27. $\sum_{m=2}^{\infty} (-1)^m \cdot \frac{\ln(m^3)}{\ln(m^{10})}$
28. $\sum_{k=1}^{\infty} (-1)^{k+1} \cdot \frac{5}{\sqrt{k+7}}$
29. $\sum_{k=1}^{\infty} \frac{(-2)^{k+1}}{1+(-3)^k}$
30. $\sum_{m=1}^{\infty} \frac{(-2)^{m+1}}{1+3^m}$
31. $\sum_{k=1}^{\infty} \cos(k\pi)\sin(k\pi)$
32. $\sum_{k=1}^{\infty} \frac{(-1)^{k+1}}{k^3}$

In Problems 33–42, (a) calculate s_4 for each series, (b) determine an upper bound for the distance between s_4 and the exact value S of the infinite series, then (c) use s_4 to find lower and upper bounds for the value of S.

33. $\sum_{k=1}^{\infty} (-1)^{k+1} \cdot \frac{1}{k^2}$

34. $\sum_{k=1}^{\infty} (-1)^{k+1} \cdot \frac{1}{k+6}$

35. $\sum_{k=1}^{\infty} \frac{(-1)^{k+1}}{\ln(k+1)}$

36. $\sum_{k=1}^{\infty} (-1)^{k+1} \cdot \frac{2}{\sqrt{k+21}}$

37. $\sum_{k=1}^{\infty} (-0.8)^{k+1}$

38. $\sum_{n=1}^{\infty} \left(-\frac{1}{3}\right)^n$

39. $\sum_{k=1}^{\infty} (-1)^{k+1} \cdot \sin\left(\frac{1}{k}\right)$

40. $\sum_{k=1}^{\infty} (-1)^{k+1} \cdot \frac{1}{k^4}$

41. $\sum_{k=1}^{\infty} (-1)^{k} \cdot \frac{1}{k^3}$

42. $\sum_{k=1}^{\infty} \frac{\cos(k\pi)}{k+\ln(k)}$

In Problems 43–52, find the number of terms N needed to guarantee that s_N is within the specified distance D of the exact value S of the sum of the series from the specified Problem.

43. Problem 34, $D = 0.01$

44. 35, $D = 0.01$

45. 36, $D = 0.01$

46. 37, $D = 0.003$

47. 38, $D = 0.002$

48. 39, $D = 0.06$

49. 40, $D = 0.001$

50. 41, $D = 0.0001$

51. 42, $D = 0.04$

52. 33, $D = 0.00005$

Alternating Power Series

Problems 53–63 ask you to use the series $S(x)$, $C(x)$ and $E(x)$ given below as infinite series depending on a continuous variable x:

$$S(x) = x - \frac{1}{3 \cdot 2}x^3 + \frac{1}{5 \cdot 4 \cdot 3 \cdot 2}x^5 + \cdots + \frac{(-1)^k}{(2k+1)!}x^{2k+1} + \cdots$$

$$C(x) = 1 - \frac{1}{2}x^2 + \frac{1}{4 \cdot 3 \cdot 2}x^4 - \frac{1}{6 \cdot 5 \cdot 4 \cdot 3 \cdot 2}x^6 + \cdots + \frac{(-1)^k}{(2k)!}x^{2k} + \cdots$$

$$E(x) = 1 + x - \frac{1}{2}x^2 + \frac{1}{3 \cdot 2}x^3 + \frac{1}{4 \cdot 3 \cdot 2}x^4 + \cdots + \frac{1}{k!}x^k + \cdots$$

You may recognize these series as infinite extensions of MacLaurin polynomials for $\sin(x)$, $\cos(x)$ and e^x from Section 8.7 (which we will study further in Chapter 10). In Problems 53–63, (a) substitute the given value for x in the series, (b) evaluate s_3, the sum of the first three terms of the series, and (c) determine an upper bound on the distance between s_3 and the sum of the entire infinite series.

53. $x = 0.3$ in $S(x)$

54. $x = 0.5$ in $S(x)$

55. $x = 0.1$ in $S(x)$

56. $x = 1$ in $S(x)$

57. $x = 1$ in $C(x)$

58. $x = -0.3$ in $S(x)$

59. $x = -0.2$ in $C(x)$

60. $x = 0.5$ in $C(x)$

61. $x = -1$ in $E(x)$

62. $x = -0.3$ in $C(x)$

63. $x = -0.2$ in $E(x)$

64. $x = -0.5$ in $E(x)$

9.8 Practice Answers

1. (a) For $k \geq 1$, $a_k = \frac{1}{k^2} > 0$. Taking $f(x) = \frac{1}{x^2}$, $f(k) = a_k$ for $k \geq 1$ and $f'(x) = -\frac{2}{x^3} < 0$ for $x \geq 1$, so $f(x)$ is decreasing when $x \geq 1$, telling us that $\{a_k\}$ is a decreasing sequence for $k \geq 1$. Finally, $\lim_{k\to\infty} \frac{1}{k^2} = 0$.

   (b) For $k \geq 2$, $a_k = \frac{3}{\ln(k)} > 0$. Because $\ln(x)$ is an increasing function for all x, $\{a_k\}$ is a decreasing sequence for $k \geq 2$. Finally, $\lim_{k\to\infty} \frac{3}{\ln(k)} = 0$.

2. Because $\lim_{k\to\infty} (-1)^{k+1}k \neq 0$, $\sum_{k=1}^{\infty} (-1)^{k+1}k$ diverges. For the second series, take $a_k = \frac{1}{\sqrt{2k+1}}$ so that $a_k > 0$ for $k \geq 1$. Furthermore, $\sqrt{2x+1}$ is an increasing function, so $\left\{\sqrt{2k+1}\right\}$ is an increasing sequence, hence $\{a_k\}$ is a decreasing sequence for $k \geq 1$. Finally, $\lim_{k\to\infty} \frac{1}{\sqrt{2k+1}} = 0$, so the series satisfies all three conditions of the Alternating Series Test, hence the series converges.

3. For $S = \sum_{k=1}^{\infty} \frac{(-1)^{k+1}}{k^3}$, $a_k = \frac{1}{k^3}$ and the partial sums are:

$$s_4 = 1 - \frac{1}{8} + \frac{1}{27} - \frac{1}{64} = \frac{1549}{1728} \approx 0.896412$$

$$s_9 = s_4 + \frac{1}{125} - \frac{1}{216} + \frac{1}{343} - \frac{1}{512} + \frac{1}{729} \approx 0.9021165$$

   A bound for $|S - s_4|$ is a_5 and a bound for $|S - s_9|$ is a_{10} so:

$$|S - s_4| < a_5 = \frac{1}{5^3} = \frac{1}{125} = 0.008 \Rightarrow s_4 - 0.008 < S < s_4 + 0.008$$
$$\Rightarrow 0.888412 < S < 0.904412$$

$$|S - s_9| < a_{10} = \frac{1}{10^3} = \frac{1}{1000} = 0.001 \Rightarrow s_9 - 0.001 < S < s_9 + 0.001$$
$$\Rightarrow 0.9020165 < S < 0.9022165$$

4. We need to find an integer N so that $|a_{N+1}| < 0.001$:

$$\frac{1}{(N+1)^3+5} < \frac{1}{1000} \Rightarrow (N+1)^3 + 5 > 1000 \Rightarrow (N+1)^3 > 995$$
$$\Rightarrow N+1 > \sqrt[3]{995} \approx 9.98 \Rightarrow N > 8.98$$

   so $N = 9$ should work:

$$s_9 = \frac{1}{6} - \frac{1}{13} + \frac{1}{32} - \frac{1}{69} + \frac{1}{130} - \frac{1}{221} + \frac{1}{348} - \frac{1}{517} - \frac{1}{734} \approx 0.111970$$

9.9 *Absolute Convergence and the Ratio Test*

The series we examined so far have generally behaved very regularly with regard to the signs of the terms: the signs of the terms were typically either all positive or they alternated between $+$ and $-$. Yet the signs of terms in a series need not behave in such regular ways, and in this section we examine techniques for determining whether some of those series converge or diverge.

Two Examples

Consider a series with terms of magnitudes $\frac{1}{k}$ but unknown signs:

$$\sum_{k=1}^{\infty} \frac{\star}{k} = \frac{\star}{1} + \frac{\star}{2} + \frac{\star}{3} + \frac{\star}{4} + \cdots$$

where $\star$ represents $+1$ or -1 in each term. Does this series converge? Replacing $\star$ with 1 in *all* terms results in:

$$\sum_{k=1}^{\infty} \frac{1}{k} = 1 + \frac{1}{2} + \frac{1}{3} + \frac{1}{4} + \cdots$$

which diverges. Replacing $\star$ with -1 in *all* terms results in:

$$\sum_{k=1}^{\infty} \frac{-1}{k} = -1 - \frac{1}{2} - \frac{1}{3} - \frac{1}{4} - \cdots$$

or -1 times the harmonic series, which also diverges. But replacing $\star$ alternately with $+1$ and -1 results in:

$$\sum_{k=1}^{\infty} \frac{(-1)^{k+1}}{k} = 1 - \frac{1}{2} + \frac{1}{3} - \frac{1}{4} + \cdots$$

or the alternating harmonic series, which converges (by the Alternating Series Test). The answer to the question above ("Does this series converge?") is: "It depends on the signs of the terms."

Now consider a series with terms of magnitudes $\frac{1}{k^2}$:

$$\sum_{k=1}^{\infty} \frac{\star}{k^2} = \frac{\star}{1} + \frac{\star}{4} + \frac{\star}{9} + \frac{\star}{16} + \cdots$$

where $\star$ represents $+1$ or -1 in each term. Does this series converge? Replacing $\star$ with 1 in *all* terms results in:

$$\sum_{k=1}^{\infty} \frac{1}{k^2} = 1 + \frac{1}{4} + \frac{1}{9} + \frac{1}{16} + \cdots$$

or a p-series with $p = 2 > 1$, which converges (by the P-Test) to a value S. Replacing $\star$ with -1 in *all* terms results in:

$$\sum_{k=1}^{\infty} \frac{-1}{k^2} = -1 - \frac{1}{4} - \frac{1}{9} - \frac{1}{16} - \cdots = -S$$

so this series also converges. Replacing $\star$ alternately with $+1$ and -1 results in:

$$\sum_{k=1}^{\infty} \frac{(-1)^{k+1}}{k^2} = 1 - \frac{1}{4} + \frac{1}{9} - \frac{1}{16} + \cdots$$

which convereges (by the Alternating Series Test) to some value between $-S$ and S. Replacing $\star$ with $+1$ and -1 in any other way results in a series with partial sums all between $-S$ and S: although we know the partial sums are bounded, we don't (yet) know that they converge (they could bounce back and forth between distinct finite values).

The answer to the question above ("Does this series converge?") appears to be: "Yes, regardless of the signs of the terms." This is in fact true, but will require a careful proof.

Absolute Convergence and Conditional Convergence

If a series $\sum_{k=1}^{\infty} a_k$ converges no matter how the signs of each term are chosen, then $\sum_{k=1}^{\infty} |a_k|$ must converge (just choose all signs to be $+$).

> **Definition**:
> A series $\sum_{k=1}^{\infty} a_k$ is **absolutely convergent** if $\sum_{k=1}^{\infty} |a_k|$ converges.

Any series of the form $\sum_{k=1}^{\infty} \frac{\star}{k^2}$ (where $\star$ represents $+1$ or -1) is absolutely convergent because $\sum_{k=1}^{\infty} \left|\frac{\star}{k^2}\right| = \sum_{k=1}^{\infty} \frac{1}{k^2}$, which is convergent by the P-Test (with $p = 2 > 1$).

The alternating harmonic series $\sum_{k=1}^{\infty} \frac{(-1)^{k+1}}{k}$ converges but is **not** absolutely convergent because $\sum_{k=1}^{\infty} \left|\frac{(-1)^{k+1}}{k}\right| = \sum_{k=1}^{\infty} \frac{1}{k}$ diverges. We call such a series **conditionally convergent**.

> **Definition**:
> A series $\sum_{k=1}^{\infty} a_k$ is **conditionally convergent** if $\sum_{k=1}^{\infty} a_k$ converges but $\sum_{k=1}^{\infty} |a_k|$ diverges.

Example 1. Determine whether these series are absolutely convergent, conditionally convergent or divergent.

(a) $\sum_{k=1}^{\infty} \frac{(-1)^{k+1}}{\sqrt{k}}$ (b) $\sum_{k=1}^{\infty} \frac{\sin(k)}{k^2}$ (c) $\sum_{m=1}^{\infty} (-1)^{k+1} \frac{m^2}{m+1}$

Solution. (a) $\sum_{k=1}^{\infty} \left| \frac{(-1)^{k+1}}{\sqrt{k}} \right| = \sum_{k=1}^{\infty} \frac{1}{\sqrt{k}}$, which diverges (by the P-Test with $p = \frac{1}{2} < 1$) so $\sum_{k=1}^{\infty} \frac{(-1)^{k+1}}{\sqrt{k}}$ does not converge absolutely, but this series does converge (by the Alternating Series Test), so it is conditionally convergent.

See Example 1(b) in Section 9.8.

(b) For all k, $|\sin(k)| < 1$, so $\left| \frac{\sin(k)}{k^2} \right| < \frac{1}{k^2}$. Because $\sum_{k=1}^{\infty} \frac{1}{k^2}$ converges (by the P-Test with $p = 2 > 1$), $\sum_{k=1}^{\infty} \left| \frac{\sin(k)}{k^2} \right|$ converges by the Basic Comparison Test, hence $\sum_{k=1}^{\infty} \frac{\sin(k)}{k^2}$ is absolutely convergent.

(c) $\lim_{m \to \infty} (-1)^{k+1} \frac{m^2}{m+1} \neq 0$, so $\sum_{m=1}^{\infty} (-1)^{k+1} \frac{m^2}{m+1}$ diverges (by the Test for Divergence). ◀

Practice 1. Determine whether each series is absolutely convergent, conditionally convergent or divergent.

(a) $\sum_{k=2}^{\infty} (-1)^k \frac{5}{\ln(k)}$ (b) $\sum_{n=1}^{\infty} \frac{\cos(n\pi)}{n^2}$

We now know that $\sum_{k=1}^{\infty} \frac{\star}{k^2}$ (where $\star$ represents $+1$ or -1) is absolutely convergent no matter how you choose the values of $\star$, and we know that $\sum_{k=1}^{\infty} \frac{\star}{k^2}$ converges for certain choices of $\star$, but does this series converge for *all* possible choices of $\star$? The following important result provides the answer to this question.

This theorem tells us that any absolutely convergent series is convergent.

Absolute Convergence Theorem

If $\sum_{k=1}^{\infty} |a_k|$ converges then $\sum_{k=1}^{\infty} a_k$ converges.

Proof. If $a_k \geq 0$ then $a_k = |a_k|$ and if $a_k < 0$ then $a_k = -|a_k|$ so for all k:

$$-|a_k| \leq a_k \leq |a_k| \quad \Rightarrow \quad 0 \leq |a_k| + a_k \leq 2|a_k|$$

Because $\sum_{k=1}^{\infty} 2|a_k|$ converges, the BCT says $\sum_{k=1}^{\infty} \left[|a_k| + a_k \right]$ converges. Hence the difference between this series and $\sum_{k=1}^{\infty} |a_k|$ converges and:

$$\sum_{k=1}^{\infty} \left[|a_k| + a_k \right] - \sum_{k=1}^{\infty} |a_k| = \sum_{k=1}^{\infty} \left[|a_k| + a_k - |a_k| \right] = \sum_{k=1}^{\infty} a_k$$

must converge. □

The contrapositive form of the Absolute Convergence Theorem can be useful when showing that a series is *not* absolutely convergent.

Corollary

If $\sum_{k=1}^{\infty} a_k$ diverges then $\sum_{k=1}^{\infty} |a_k|$ diverges.

This result tells us that no divergent series can be absolutely convergent.

The Ratio Test

If $|r| \geq 1$, then $\sum_{k=0}^{\infty} r^k$ diverges; if $|r| < 1$, $\sum_{k=0}^{\infty} |r|^k$ converges, so $\sum_{k=0}^{\infty} r^k$ converges absolutely. The following test extends these results about geometric series to a more general class of infinite series that behave like geometric series in a certain way: when the absolute value of the ratio of successive terms (eventually) exceeds 1, the series diverges; when this ratio is (eventually) less than 1, the series converges absolutely.

Ratio Test

If $L = \lim_{k\to\infty} \left| \frac{a_{k+1}}{a_k} \right|$ then:

- $L < 1 \Rightarrow \sum_{k=1}^{\infty} a_k$ converges absolutely
- $L > 1 \Rightarrow \sum_{k=1}^{\infty} a_k$ diverges
- $L = 1 \Rightarrow \sum_{k=1}^{\infty} a_k$ may converge or may diverge

This is the last major convergence test we'll study. You'll use it often in Chapter 10 when you want to determine the interval on which a power series converges.

The Ratio Test can also be inconclusive if the limit in the hypothesis does not exist.

Proof. If $L < 1$, let r be any number so that $L < r < 1$. Because $L = \lim_{k\to\infty} \left| \frac{a_{k+1}}{a_k} \right|$, the ratio $\left| \frac{a_{k+1}}{a_k} \right|$ must get close to L (and be less than r). More precisely, there is an N so that $k \geq N$ guarantees:

$$\left| \frac{a_{k+1}}{a_k} \right| < r \quad \Rightarrow \quad |a_{k+1}| < r \cdot |a_k|$$

Applying this result repeatedly, we know that:

$$|a_{N+1}| < r\,|a_N| \Rightarrow |a_{N+2}| < r\,|a_{N+1}| < r^2\,|a_N| \Rightarrow |a_{N+3}| < r^3\,|a_N|$$

and so forth, so that $|a_{N+j}| < r^j\,|a_N|$ for all $j \geq 0$. Hence:

$$\begin{aligned}
\sum_{k=1}^{\infty} |a_k| &= \sum_{k=1}^{N-1} |a_k| + \sum_{k=N}^{\infty} |a_k| \leq \sum_{k=1}^{N-1} |a_k| + \sum_{j=0}^{\infty} r^j\,|a_N| \\
&= \sum_{k=1}^{N-1} |a_k| + |a_N| \sum_{j=0}^{\infty} r^j = \sum_{k=1}^{N-1} |a_k| + \frac{|a_N|}{1-r}
\end{aligned}$$

We have shown that $\sum_{k=1}^{\infty} |a_k|$ is less than a finite sum plus the sum of a geometric series with ratio r where $|r| < 1$. The Basic Comparison Test tells us that $\sum_{k=1}^{\infty} |a_k|$ converges, hence $\sum_{k=1}^{\infty} a_k$ converges absolutely.

If $L > 1$, the ratio $\left|\frac{a_{k+1}}{a_k}\right|$ exceeds 1 for all $k \geq N$ for some integer N. Therefore $|a_{k+1}| > |a_k|$ when $k \geq N$ so that:

$$|a_N| < |a_{N+1}| < |a_{N+2}| < \cdots < |a_k|$$

for any $k > N$. This means that $\lim_{k\to\infty} |a_k| \geq |a_N| > 0$ so $\lim_{k\to\infty} a_k \neq 0$ and the Test for Divergence tells us that $\sum_{k=1}^{\infty} a_k$ diverges.

The following Practice problem provides an example of a convergent series with $L = 1$ and a divergent series with $L = 1$. □

Practice 2. Show that $\lim_{k\to\infty} \left|\frac{a_{k+1}}{a_k}\right| = 1$ for both $\sum_{k=1}^{\infty} \frac{1}{k}$ (the harmonic series, which diverges) and $\sum_{k=1}^{\infty} \frac{1}{k^2}$ (a convergent p-series with $p = 2$).

A powerful aspect of the Ratio Test is that it is very "mechanical": you simply calculate a particular limit and the value of this limit (often) tells you whether the series converges or diverges.

If the terms of a series involve factorials or k-th powers, the Ratio Test will often be the best test to use, so it will typically be the first test you try.

Example 2. Use the Ratio Test to determine whether each series converges absolutely:

$$\text{(a)} \sum_{k=1}^{\infty} \frac{2^k \cdot k}{5^k} \qquad \text{(b)} \sum_{n=1}^{\infty} \frac{n^2}{n!}$$

Solution. (a) If $a_k = \frac{2^k \cdot k}{5^k}$ then $a_{k+1} = \frac{2^{k+1}\cdot(k+1)}{5^{k+1}}$ so that:

$$\left|\frac{a_{k+1}}{a_k}\right| = \frac{\frac{2^{k+1}\cdot(k+1)}{5^{k+1}}}{\frac{2^k \cdot k}{5^k}} = \frac{2^{k+1}}{2^k} \cdot \frac{5^k}{5^{k+1}} \cdot \frac{k+1}{k} = \frac{2}{5} \cdot \frac{k+1}{k} \to \frac{2}{5} = L$$

Because $L < 1$, the Ratio Test says $\sum_{k=1}^{\infty} \frac{2^k \cdot k}{5^k}$ converges absolutely.

(b) If $a_n = \frac{n^2}{n!}$ then $a_{n+1} = \frac{(n+1)^2}{(n+1)!}$ so that:

$$\left|\frac{a_{n+1}}{a_n}\right| = \frac{\frac{(n+1)^2}{(n+1)!}}{\frac{n^2}{n!}} = \frac{(n+1)^2}{n^2} \cdot \frac{n!}{(n+1)!}$$

$$= \left(\frac{n+1}{n}\right)^2 \cdot \frac{n!}{(n+1)\cdot n!} = \left(1 + \frac{1}{n}\right)^2 \cdot \frac{1}{n+1} \to 1 \cdot 0 = L$$

Because $L < 1$, the Ratio Test says $\sum_{n=1}^{\infty} \frac{n^2}{n!}$ converges absolutely. ◀

Practice 3. Use the Ratio Test to determine whether each series converges absolutely:

$$\text{(a)} \sum_{k=1}^{\infty} (-1)^{k+1} \frac{e^k}{k!} \qquad \text{(b)} \sum_{n=1}^{\infty} \frac{n^5}{3^n}$$

The Ratio Test is often very useful for determining values of a variable that guarantee the absolute convergence (hence convergence) of a power series. You will use this method often in Chapter 10.

Example 3. On what interval does $\sum_{k=1}^{\infty} \frac{(x-3)^k}{k}$ converge absolutely?

Solution. If $a_k = \frac{(x-3)^k}{k}$ then $a_{k+1} = \frac{(x-3)^{k+1}}{k+1}$ so that:

$$\left|\frac{a_{k+1}}{a_k}\right| = \left|\frac{\frac{(x-3)^{k+1}}{k+1}}{\frac{(x-3)^k}{k}}\right| = \left|\frac{(x-3)^{k+1}}{(x-3)^k} \cdot \frac{k}{k+1}\right| = |x-3| \cdot \frac{k}{k+1} \to |x-3|$$

We need $L = |x-3| < 1$, which means that:

$$|x-3| < 1 \quad \Rightarrow \quad -1 < x-3 < 1 \quad \Rightarrow \quad 2 < x < 4$$

If $x < 2$ or $x > 4$, $L = |x-3| > 1$ so the series diverges. If $x = 4$, $x - 3 = 1$ and the series becomes the harmonic series, which diverges; if $x = 2$, $x - 3 = -1$ and the series becomes the alternating harmonic series, which converges conditionally. So the series converges when $2 \le x < 4$ but converges absolutely only when $2 < x < 4$. ◀

Practice 4. On what interval does $\sum_{k=1}^{\infty} \frac{(x-5)^k}{k^2}$ converge absolutely?

9.9 Problems

In Problems 1–30, determine whether the given series converges absolutely, converges conditionally or diverges, giving reasons for your conclusion.

1. $\sum_{k=1}^{\infty} \frac{(-1)^{k+1}}{k+2}$

2. $\sum_{k=1}^{\infty} \frac{(-1)^{k+1}}{\sqrt{k}}$

3. $\sum_{n=1}^{\infty} (-1)^{n+1} \cdot \frac{5}{n^3}$

4. $\sum_{n=1}^{\infty} (-1)^n \cdot \frac{1}{1+\ln(n)}$

5. $\sum_{k=0}^{\infty} (-0.5)^k$

6. $\sum_{k=0}^{\infty} (-0.5)^{-k}$

7. $\sum_{k=1}^{\infty} \frac{(-1)^{k+1}}{k^2}$

8. $\sum_{k=1}^{\infty} \frac{(-1)^{k+1}}{3+k^2}$

9. $\sum_{n=2}^{\infty} (-1)^n \cdot \frac{\ln(n)}{n}$

10. $\sum_{n=2}^{\infty} (-1)^n \cdot \frac{\ln(n)}{n^2}$

11. $\sum_{k=1}^{\infty} \frac{(-1)^k}{k+\ln(k)}$

12. $\sum_{n=0}^{\infty} (-1)^n \cdot \frac{5}{\sqrt{n+7}}$

13. $\sum_{k=1}^{\infty} (-1)^k \cdot \sin\left(\frac{1}{k}\right)$

14. $\sum_{k=1}^{\infty} (-1)^k \cdot \sin\left(\frac{1}{k^2}\right)$

15. $\sum_{k=1}^{\infty} (-1)^k \sqrt{k} \sin\left(\frac{1}{k^2}\right)$

16. $\sum_{k=1}^{\infty} \frac{\cos(k\pi)}{k}$

17. $\sum_{m=2}^{\infty} (-1)^m \cdot \frac{\ln(m)}{\ln(m^3)}$

18. $\sum_{n=0}^{\infty} (-1)^n \cdot \frac{n^2+7}{n^2+10}$

19. $\sum_{n=0}^{\infty} (-1)^n \cdot \frac{n^2+7}{n^3+10}$

20. $\sum_{n=0}^{\infty} (-1)^n \cdot \frac{n^2+7}{n^4+10}$

21. $\sum_{n=0}^{\infty} (-1)^n \cdot \frac{(n^2+7)^2}{n^2+10}$

22. $\sum_{k=1}^{\infty} \frac{\sin(k)}{k^2}$

23. $\sum_{k=1}^{\infty} \frac{\sin(k\pi)}{k}$

24. $\sum_{k=1}^{\infty} (-k)^{-k}$

25. $\sum_{n=0}^{\infty} (-1)^n \cdot \frac{1+\sqrt{3n}}{n+2}$

26. $\sum_{k=1}^{\infty} \frac{(-2)^k}{k^2}$

27. $\sum_{k=1}^{\infty} \frac{(-3)^k}{k^3}$

28. $\sum_{n=2}^{\infty} (-1)^n \left(\frac{\ln(n)}{\ln(n^5)}\right)^2$

29. $\sum_{k=1}^{\infty} \frac{(-2)^k}{k \cdot 3^k}$

30. $\sum_{k=1}^{\infty} \frac{3^k}{k^3}$

The Ratio Test often arises with series that involve factorials. To help you prepare for these situations, Problems 31–39 ask you to simplify the factorial expression on the left side of the equality to arrive at the expression on the right side. Recall that: $0! = 1$, $1! = 1$, $2! = 2 \cdot 1 = 2$, $3! = 3 \cdot 2 \cdot 1 = 6$, $4! = 24$ and, in general, $n! = n \cdot (n-1) \cdots 3 \cdot 2 \cdot 1$.

31. $\frac{n!}{(n+1)!} = \frac{1}{n+1}$

32. $\frac{n!}{(n+3)!} = \frac{1}{(n+1)(n+2)(n+3)}$

33. $\frac{(n-1)!}{(n+1)!} = \frac{1}{n(n+1)}$

34. $\frac{(2n)!}{(2n+1)!} = \frac{1}{2n+1}$

35. $\frac{n!}{(n+2)!} = \frac{1}{(n+1)(n+2)}$

36. $\frac{(n+1)!}{(n+2)!} = \frac{1}{n+2}$

37. $\frac{2n!}{(2n)!} = \frac{2}{(n+1)(n+2)\cdots(2n)}$

38. $\frac{(2n)!}{(2(n+1)!} = \frac{1}{(2n+1)(2n+2)}$

39. $\frac{n^n}{n!} = \frac{n}{1} \cdot \frac{n}{2} \cdot \frac{n}{3} \cdots \frac{n}{n-1} \cdot \frac{n}{n}$

40. For $n > 0$, which is larger, $7!$ or $\frac{(n+7)!}{n!}$?

In Problems 41–58, apply the Ratio Test to the series. If the Ratio Test is inconclusive, use some other method to determine whether the series converges absolutely, converges conditionally or diverges.

41. $\sum_{k=1}^{\infty} \frac{1}{k}$

42. $\sum_{k=1}^{\infty} \frac{1}{k^2}$

43. $\sum_{k=1}^{\infty} \frac{1}{k^3}$

44. $\sum_{k=1}^{\infty} \frac{1}{\sqrt{k}}$

45. $\sum_{k=0}^{\infty} \left(\frac{1}{2}\right)^k$

46. $\sum_{k=0}^{\infty} \left(\frac{1}{3}\right)^k$

47. $\sum_{n=0}^{\infty} 1^n$

48. $\sum_{n=0}^{\infty} (-2)^n$

49. $\sum_{k=0}^{\infty} \frac{1}{k!}$

50. $\sum_{k=0}^{\infty} \frac{5}{k!}$

51. $\sum_{k=0}^{\infty} \frac{2^k}{k!}$

52. $\sum_{k=0}^{\infty} \frac{5^k}{k!}$

53. $\sum_{k=0}^{\infty} \left(\frac{1}{2}\right)^{3k}$

54. $\sum_{k=0}^{\infty} \left(\frac{1}{3}\right)^{2k}$

55. $\sum_{k=5}^{\infty} (0.9)^{2k+1}$

56. $\sum_{k=100}^{\infty} (1.1)^{\frac{k}{2}}$

57. $\sum_{k=100}^{\infty} (-1.1)^k$

58. $\sum_{k=5}^{\infty} (-0.8)^{2k+1}$

In 59–76, apply the Ratio Test to determine the values of x for which the series converges absolutely.

59. $\sum_{k=0}^{\infty} (x-5)^k$

60. $\sum_{k=1}^{\infty} \frac{(x-5)^k}{k}$

61. $\sum_{k=1}^{\infty} \frac{(x-5)^k}{k^2}$

62. $\sum_{k=1}^{\infty} \frac{(x-2)^k}{k^2}$

63. $\sum_{k=0}^{\infty} \frac{(x-2)^k}{k!}$

64. $\sum_{n=0}^{\infty} \frac{(x-10)^n}{n!}$

65. $\sum_{k=1}^{\infty} \frac{(2x-12)^k}{k^2}$

66. $\sum_{k=1}^{\infty} \frac{(4x-12)^k}{k^2}$

67. $\sum_{k=1}^{\infty} \frac{(6x-12)^k}{k!}$

68. $\sum_{k=0}^{\infty} (x-3)^{2k}$

69. $\sum_{k=1}^{\infty} \frac{(x+1)^{2k}}{k}$

70. $\sum_{k=1}^{\infty} \frac{(x+1)^{2k+1}}{k^2}$

71. $\sum_{n=1}^{\infty} \frac{(x-5)^{3n+1}}{n^2}$

72. $\sum_{n=0}^{\infty} \frac{(x+4)^{2n+1}}{n!}$

73. $\sum_{n=0}^{\infty} \frac{(x+3)^{2n+1}}{(n+1)!}$

74. $\sum_{k=0}^{\infty} (-1)^k \frac{x^{2k+1}}{(2k+1)!}$

75. $\sum_{k=0}^{\infty} (-1)^k \frac{x^{2k}}{(2k)!}$

76. $\sum_{k=0}^{\infty} \frac{x^k}{k!}$

77. $\sum_{k=0}^{\infty} \frac{x^{2k}}{(2k)!}$

78. $\sum_{k=0}^{\infty} \frac{x^{2k+1}}{(2k+1)!}$

The Root Test

A geometric series $\sum_{k=0}^{\infty} r^k$ diverges if $|r| \geq 1$ and converges (absolutely) if $|r| < 1$. In a geometric series:

$$a_k = r^k \quad \Rightarrow \quad |a_k| = \left|r^k\right| \quad \Rightarrow \quad \sqrt[k]{|a_k|} = |r|$$

The following test extends this result about a geometric series to a more general class of infinite series that behave like geometric series in a certain way: when the k-th root of the absolute value of the k-th term (eventually) exceeds 1, the series diverges; when this k-th root is (eventually) less than 1, the series converges absolutely.

Root Test

If $L = \lim_{k\to\infty} \sqrt[k]{|a_k|}$ then:

- $L < 1 \Rightarrow \sum_{k=1}^{\infty} a_k$ converges absolutely
- $L > 1 \Rightarrow \sum_{k=1}^{\infty} a_k$ diverges
- $L = 1 \Rightarrow \sum_{k=1}^{\infty} a_k$ may converge or may diverge

The Root Test comes in handy less often than the Ratio Test, but can be useful on occasion. If the Ratio Test is conclusive, the Root Test will be also (but may be more difficult to apply); there are some series, however, for which the Ratio Test is inconclusive but the Root Test is not.

The Root Test can also be inconclusive if the limit in the hypothesis does not exist.

Proof. If $L < 1$, choose any number r with $L < r < 1$. Then there is an integer N so that for $k \geq N$:

$$\sqrt[k]{|a_k|} < r \quad \Rightarrow \quad |a_k| < r^k$$

Therefore:

$$\sum_{k=1}^{\infty} |a_k| = \sum_{k=1}^{N-1} |a_k| + \sum_{k=N}^{\infty} |a_k| \leq \sum_{k=1}^{N-1} |a_k| + \sum_{k=N}^{\infty} r^k$$

This last quantity is the sum of a finite sum and the tail end of a convergent geometric series (because $0 \leq r < 1$). The Basic Comparison Test tells us that $\sum_{k=1}^{\infty} |a_k|$ converges, hence $\sum_{k=1}^{\infty} a_k$ converges absolutely.

If $L > 1$, then there is an integer N so that for $k \geq N$:

$$\sqrt[k]{|a_k|} > 1 \quad \Rightarrow \quad |a_k| > 1 \quad \Rightarrow \quad \lim_{k\to\infty} |a_k| \neq 0 \quad \Rightarrow \quad \lim_{k\to\infty} a_k \neq 0$$

The Test for Divergence then tell us that $\sum_{k=1}^{\infty} a_k$ diverges.

The series $\sum_{k=1}^{\infty} 1$ diverges while $\sum_{k=1}^{\infty} \frac{1}{k^2}$ converges; applying the Root Test to each results in $L = 1$. □

Example 6 in Section 3.7 shows:

$$\lim_{k\to\infty} \left(\frac{1}{k}\right)^{\frac{1}{k}} = \lim_{x\to 0^+} x^x = 1$$

In Problems 79–94, apply the Root Test to the series. If the Root Test is inconclusive, use some other method to determine whether the series converges absolutely, converges conditionally or diverges.

79. $\sum_{k=0}^{\infty} \left(\frac{2}{7}\right)^k$ 80. $\sum_{k=0}^{\infty} \left(\frac{8}{7}\right)^k$ 81. $\sum_{k=1}^{\infty} \frac{1}{k^3}$

82. $\sum_{k=1}^{\infty} \frac{1}{k}$ 83. $\sum_{k=1}^{\infty} \frac{1}{k^k}$ 84. $\sum_{k=1}^{\infty} \frac{3^k}{k^k}$

85. $\sum_{k=5}^{\infty} \left(\frac{1}{2} - \frac{2}{k}\right)^k$ 86. $\sum_{k=1}^{\infty} \left(\frac{1}{2} + \frac{2}{k}\right)^k$

87. $\sum_{k=1}^{\infty} \left(\frac{2+k}{k}\right)^k$ 88. $\sum_{k=1}^{\infty} \left(\frac{k}{2+k}\right)^k$

89. $\sum_{k=1}^{\infty} \cos^k(k\pi)$ 90. $\sum_{k=1}^{\infty} \sin^k\left(\frac{(6k+1)\pi}{6}\right)$

91. $\sum_{k=0}^{\infty} \left(\frac{2k^3+1}{3k^3+2}\right)^k$ 92. $\sum_{k=0}^{\infty} \left(\frac{3k^2+1}{2k^2+3}\right)^k$

93. $\sum_{k=1}^{\infty} \frac{(2k)^k}{k^{2k}}$ 94. $\sum_{k=1}^{\infty} \left(1 - \frac{1}{k}\right)^{k^2}$

Rearrangements

See Problems 103–104 for a proof of this statement.

Absolutely convergent series share an important property with finite sums: no matter what order you add the numbers, the sum will always be the same.

Conditionally convergent series do not possess this property: the order in which you add the terms *does* matter. If you reorder the terms of a conditionally convergent series, the sum after the rearrangement may be different than the sum before the rearrangement. A rather amazing fact is that you can rearrange the terms of a conditionally convergent series to obtain *any* sum you want!

We illustrate this strange result by showing that you can rearrange the alternating harmonic series:

$$\sum_{k=1}^{\infty} \frac{(-1)^k}{k} = 1 - \frac{1}{2} + \frac{1}{3} - \frac{1}{4} + \frac{1}{5} - \frac{1}{6} + \frac{1}{7} - \frac{1}{8} + \cdots$$

which converges conditionally to a value $S \approx 0.69$, so that the sum of the rearranged series is 2.

First, note that the sum of the positive terms of the alternating harmonic series:

$$1 + \frac{1}{3} + \frac{1}{5} + \frac{1}{7} + \cdots = \sum_{n=1}^{\infty} \frac{1}{2n+1}$$

diverges (by the Limit Comparison Test with the harmonic series), so the partial sums of this series of all-positive terms must exceed any positive number you pick (if n becomes large enough).

Similarly, the sum of the negative terms of the harmonic series:

$$-\frac{1}{2} - \frac{1}{4} - \frac{1}{6} - \frac{1}{8} - \cdots = -\frac{1}{2} \cdot \sum_{m=1}^{\infty} \frac{1}{m}$$

diverges (because it's a nonzero multiple of the harmonic series), so the partial sums of this series of all-negative terms must eventually be lower than any negative number you pick (if m becomes large enough).

Begin the rearrangement process by adding up just enough positive terms so that the sum of those terms exceeds 2:

$$s_7 = 1 + \frac{1}{3} + \frac{1}{5} + \frac{1}{7} + \frac{1}{9} + \frac{1}{11} + \frac{1}{13} \approx 1.955133755$$

$$s_8 = 1 + \frac{1}{3} + \frac{1}{5} + \frac{1}{7} + \frac{1}{9} + \frac{1}{11} + \frac{1}{13} + \frac{1}{15} \approx 2.021800422$$

This procedure will still work if you "overshoot" the target by adding on more positive terms than necessary, but not doing so makes the process more efficient.

so the first eight positive terms will do the trick. Now add on enough negative terms to bring the total below 2:

$$s_9 = 1 + \frac{1}{3} + \cdots + \frac{1}{15} - \frac{1}{2} \approx 1.521800422$$

(a single negative term does the job). Then add on more positive terms until the total once again exceeds 2:

$$s_{21} \approx 1.97967321 \Rightarrow s_{22} = s_{21} + \frac{1}{41} \approx 2.004063454$$

Now add enough negative terms to bring the total back down below 2:

$$s_{23} = s_{22} - \frac{1}{4} \approx 1.754063454$$

We will always have enough terms at our disposal to exceed the target number: if not, the partial sums of the positive terms would be bounded above and the sum of the positive terms would be finite (and likewise for the negative terms).

As you continue to repeat this process, you will "eventually" use all of the terms of the original conditionally convergent series, while the partial sums of the new "rearranged" series become—and remain—arbitrarily close to the target number, 2.

This method can be used to rearrange the terms of the alternating harmonic series—or any conditionally convergent series—to get a sum of 0.3, 3, 30 or any positive target number you want.

How do you think the strategy needs to be modified to rearrange a conditionally convergent series to add up to a negative target number?

In Problems 95–100, use the strategy outlined above to find the first 15 terms of a rearrangement of the given conditionally convergent series so that the rearranged series converges to the given target T.

95. $\sum_{k=1}^{\infty} \frac{(-1)^k}{k}$, $T = 0.3$

96. $\sum_{k=1}^{\infty} \frac{(-1)^k}{k}$, $T = 1$

97. $\sum_{k=1}^{\infty} \frac{(-1)^k}{\sqrt{k}}$, $T = 1$

98. $\sum_{k=1}^{\infty} \frac{(-1)^k}{k}$, $T = 0.7$

99. $\sum_{k=1}^{\infty} \frac{(-1)^k}{\sqrt{k}}$, $T = 0.4$

100. $\sum_{k=1}^{\infty} \frac{(-1)^k}{k}$, $T = -1$

101. Show that the Ratio Test applied to the series:

$$\frac{1}{2} + \frac{1}{3} + \frac{1}{2^2} + \frac{1}{3^2} + \frac{1}{2^3} + \frac{1}{3^3} + \cdots$$

is inconclusive. Then apply the Root Test.

102. Show that the Ratio Test applied to the series:

$$\frac{1}{2} + \frac{1}{3^2} + \frac{1}{2^3} + \frac{1}{3^4} + \frac{1}{2^5} + \cdots$$

is inconclusive. Then apply the Root Test.

103. Given an absolutely convergent series $\sum_{k=1}^{\infty} a_k$, let $\{b_n\}$ be any rearrangement of $\{a_k\}$ (so that these sequences both have the same terms, just listed in a different order). Show that the partial sums of $\sum_{k=1}^{\infty} a_k$ are bounded by some number $C > 0$. Then show that the partial sums of $\sum_{n=1}^{\infty} b_n$ are also increasing and bounded by the same number C. Conclude that $\sum_{n=1}^{\infty} b_n$ is absolutely convergent.

104. Refer to Problem 103. Let $a = \sum_{k=1}^{\infty} a_k$, $A = \sum_{k=1}^{\infty} |a_k|$ $b = \sum_{n=1}^{\infty} b_n$, $s_K = \sum_{k=1}^{K} a_k$ and $S_N = \sum_{n=1}^{N} a_n$.

(a) Given $\epsilon > 0$, show that there is some integer K so that $|a - s_K| < \frac{\epsilon}{2}$ and that this inequality holds for any larger value of K.

(b) Show that by choosing a (possibly larger) value of K:

$$\left| A - \sum_{k=1}^{K} |a_k| \right| < \frac{\epsilon}{2}$$

holds for all larger values of K.

(c) Choose N to be large enough so that all of the terms $a_1, a_2, \ldots, a_K$ appear among the terms $b_1, b_2, \ldots, b_N$ and show this holds true for even larger values of N.

(d) Show that $|S_N - s_K| \le \left| A - \sum_{k=1}^{K} |a_k| \right|$.

(e) Use the triangle inequality to conclude that $|a - S_N| < \epsilon$, showing that the partial sums of $\sum_{n=1}^{\infty} b_n$ have limit $\sum_{k=1}^{\infty} a_k$.

9.9 *Practice Answers*

1. (a) $\sum_{k=1}^{\infty} \left| (-1)^k \frac{5}{\ln(k)} \right| = \sum_{k=1}^{\infty} \frac{5}{\ln(k)}$ diverges (use the BCT with $\sum_{k=1}^{\infty} \frac{5}{k}$ to show this) so $\sum_{k=1}^{\infty} (-1)^k \frac{5}{\ln(k)}$ does not converge absolutely, but it does converge (by the AST—you should check this), so it converges conditionally.

Note that: $\cos(k\pi) = \pm 1$

(b) $\sum_{k=1}^{\infty} \left| \frac{\cos(k\pi)}{k^2} \right| = \sum_{k=1}^{\infty} \frac{1}{k^2}$ and this series converges (by the P-Test, with $p = 2 > 1$), so $\sum_{k=1}^{\infty} \frac{\cos(k\pi)}{k^2}$ converges absolutely.

2. For the harmonic series, $a_k = \frac{1}{k} \Rightarrow a_{k+1} = \frac{1}{k+1}$ so:

$$\lim_{k\to\infty} \left| \frac{a_{k+1}}{a_k} \right| = \lim_{k\to\infty} \frac{\frac{1}{k+1}}{\frac{1}{k}} = \lim_{k\to\infty} \frac{k}{k+1} = 1$$

For the other series, $a_k = \frac{1}{k^2} \Rightarrow a_{k+1} = \frac{1}{(k+1)^2}$ so:

$$\lim_{k\to\infty} \left| \frac{a_{k+1}}{a_k} \right| = \lim_{k\to\infty} \frac{\frac{1}{(k+1)^2}}{\frac{1}{k^2}} = \lim_{k\to\infty} \frac{k^2}{(k+1)^2} = \lim_{k\to\infty} \left[\frac{k}{k+1} \right]^2 = 1$$

3. (a) Applying the Ratio Test:

$$\lim_{k\to\infty} \left| \frac{a_{k+1}}{a_k} \right| = \lim_{k\to\infty} \left| \frac{(-1)^{k+2} \frac{e^{k+1}}{(k+1)!}}{(-1)^{k+1} \frac{e^k}{k!}} \right| = \lim_{k\to\infty} e \cdot \frac{k!}{(k+1)!}$$

$$= \lim_{k\to\infty} \frac{e}{k+1} = 0$$

Because $0 < 1$, the series converges absolutely.

(b) Applying the Ratio Test:

$$\lim_{n\to\infty}\left|\frac{a_{n+1}}{a_n}\right| = \lim_{n\to\infty}\left|\frac{\frac{(n+1)^5}{3^{n+1}}}{\frac{n^5}{3^n}}\right| = \lim_{n\to\infty}\frac{1}{3}\cdot\frac{(n+1)^5}{n^5}$$

$$= \lim_{n\to\infty}\frac{1}{3}\cdot\left[\frac{n+1}{n}\right]^5 = \frac{1}{3}\cdot 1 = \frac{1}{3}$$

Because $\frac{1}{3} < 1$, the series converges absolutely.

4. If $a_k = \dfrac{(x-5)^k}{k^2}$ then $a_{k+1} = \dfrac{(x-5)^{k+1}}{(k+1)^2}$ so that:

$$\left|\frac{a_{k+1}}{a_k}\right| = \left|\frac{\frac{(x-5)^{k+1}}{(k+1)^2}}{\frac{(x-5)^k}{k^2}}\right| = \left|\frac{(x-5)^{k+1}}{(x-5)^k}\cdot\frac{k^2}{(k+1)^2}\right| = |x-5|\cdot\left[\frac{k}{k+1}\right]^2$$

which has limit $L = |x-5|$. We need $L < 1$, which means that:

$$|x-5| < 1 \quad\Rightarrow\quad -1 < x-5 < 1 \quad\Rightarrow\quad 4 < x < 6$$

If $x < 4$ or $x > 6$, $L = |x-5| > 1$ so the series diverges. If $x = 4$, $x - 5 = -1$ and the series becomes $\sum_{k=1}^{\infty}\frac{(-1)^2}{k^2}$, which converges absolutely by the P-Test; if $x = 6$, $x - 5 = 1$ and the series becomes $\sum_{k=1}^{\infty}\frac{1}{k^2}$, which also converges absolutely by the P-Test. The series converges absolutely on $[4,6]$ and diverges outside of this interval.

10
Power Series

The preceding chapter focused on the convergence (and divergence) of infinite series, such as the geometric series. If $|r| < 1$, we know that:

$$\sum_{k=0}^{\infty} r^k = 1 + r + r^2 + r^3 + \cdots = \frac{1}{1-r}$$

Replacing the constant r with a variable x yields:

$$G(x) = \sum_{k=0}^{\infty} x^k = 1 + x + x^2 + x^3 + \cdots = \frac{1}{1-x}$$

which defines a function of x with domain $|x| < 1$.

Or, equivalently, $-1 < x < 1$.

We can define other functions of the form $s(x) = \sum_{k=0}^{\infty} a_k x^k$ and ask:

In the geometric series example above, $a_k = 1$ for all k.

- What is the domain of $s(x)$? (Where does the series converge?)
- On that domain, does the series converge to a known function $f(x)$?
- On what interval does $s(x) = f(x)$?

In Section 8.7, we learned how to obtain a MacLaurin polynomial that approximated e^x near $x = 0$:

$$e^x \approx 1 + x + \frac{1}{2!}x^2 + \frac{1}{3!}x^3 + \cdots + \frac{1}{n!}x^n = \sum_{k=0}^{n} \frac{1}{k!}x^k$$

and observed that this approximation appeared to become better as n increases. We now have the tools to investigate what happens when we let $n \to \infty$ to obtain the series $E(x) = \sum_{k=0}^{\infty} \frac{1}{k!}x^k$. To do so, we will ask:

- What is the domain of $E(x)$? (Where does the series converge?)
- On that domain, does the series always converge to e^x?
- Given another known function $f(x)$, can we find a series $s(x)$ that converges to $f(x)$ on some interval?

To answer all of these questions, we will rely on techniques developed in the preceding chapter, especially the Ratio Test.

Finally, we will ask how many terms of the series $s(b)$ we will need to add together in order to approximate $f(b)$ within a specified tolerance.

10.1 *Power Series:* $\sum_{k=0}^{\infty} a_k x^k$

The following functions are examples of **power series**:

$$G(x) = 1 + x + x^2 + x^3 + x^4 + \cdots = \sum_{k=0}^{\infty} x^k$$

$$E(x) = 1 + x + \frac{1}{2!}x^2 + \frac{1}{3!}x^3 + \cdots + = \sum_{k=0}^{\infty} \frac{1}{k!}x^k$$

$$S(x) = x - \frac{1}{3!}x^3 + \frac{1}{5!}x^5 - \cdots + = \sum_{k=0}^{\infty} \frac{(-1)^k}{(2k+1)!}x^{2k+1}$$

We can also approximate these functions using partial sums of the power series, which turn out to be the MacLaurin polynomials discussed in Section 8.7.

A power series looks like an "infinite polynomial." Power series play particularly important roles in mathematics and applications because we can represent many important functions, such as $\sin(x)$, $\cos(x)$, e^x and $\ln(1+x)$, using power series.

Definition:
A **power series** is an expression of the form:

$$\sum_{k=0}^{\infty} a_k x^k = a_0 + a_1 x + a_2 x^2 + a_3 x^3 + \cdots$$

where $a_0, a_1, a_2, a_3, \ldots$ are constants, called the **coefficients** of the series, and x is a variable.

For $k = 0$ we use the convention for power series that $x^0 = 1$ even when $x = 0$. This convention—which ignores the fact that 0^0 is an indeterminate form (as discussed in Section 3.7)—simply makes it easier for us to represent the series using summation notation.

Replacing x with any number, the power series simply becomes a numerical series that may converge or diverge. If the power series does converge, the value of the function is the sum of the series. The domain of the function is the set of x-values for which the series converges.

Finding Where a Power Series Converges

Any power series $f(x) = \sum_{k=0}^{\infty} a_k x^k$ must converge at $x = 0$:

$$f(0) = \sum_{k=0}^{\infty} a_k \cdot 0^k = a_0 + a_1 \cdot 0 + a_2 \cdot 0^2 + a_3 \cdot 0^3 + \cdots = a_0$$

To find which other values of x allow a power series to converge, you could plug in x-values one at a time, but that would be very inefficient. Instead, the Ratio Test allows you to determine whether the series converges or diverges for many values of x all at once.

Example 1. Find all values of x for which $\sum_{k=0}^{\infty} (2k+1)x^k$ converges.

Solution. Applying the Ratio Test,

$$\left|\frac{(2(k+1)+1)x^{k+1}}{(2k+1)x^k}\right| = \left|\frac{(2k+3)x}{(2k+1)}\right| = \frac{(2k+3)}{(2k+1)} \cdot |x| \longrightarrow |x|$$

The Ratio Test tells us that the series converges if this limit, $|x|$, is less than 1: $|x| < 1 \Rightarrow -1 < x < 1$. The Ratio Test also says that the series diverges if the limit is great than 1: $|x| > 1 \Rightarrow x < -1$ or $x > 1$. Finally, the Ratio Test provides no information if $|x| = 1$, so we need to check the two remaining values of x: the endpoints $x = -1$ and $x = 1$.

When $x = -1$, $\sum_{k=0}^{\infty} (2k+1)x^k = \sum_{k=0}^{\infty} (2k+1) \cdot (-1)^k$, and when $x = 1$, $\sum_{k=0}^{\infty} (2k+1)x^k = \sum_{k=0}^{\infty} (2k+1)$. Both of these series diverge (by the Test for Divergence), because in both series the terms do not approach 0.

The power series $\sum_{k=0}^{\infty} (2k+1)x^k$ converges if and only if $-1 < x < 1$. In other words, the series converges when x is in the interval $(-1, 1)$ and it diverges on the intervals $(-\infty, -1]$ and $[1, \infty)$. ◀

Example 2. Find all values of x for which $\sum_{k=0}^{\infty} \frac{x^k}{k \cdot 3^k}$ converges.

Solution. Applying the Ratio Test:

$$\left|\frac{\frac{x^{k+1}}{(k+1)\cdot 3^{k+1}}}{\frac{x^k}{k\cdot 3^k}}\right| = \left|\frac{x^{k+1}}{(k+1)\cdot 3^{k+1}} \cdot \frac{k \cdot 3^k}{x^k}\right| = \frac{k}{k+1} \cdot \frac{|x|}{3} \longrightarrow \frac{|x|}{3}$$

The Ratio Test tells us that the series converges absolutely when:

$$\frac{|x|}{3} < 1 \quad \Rightarrow \quad |x| < 3 \quad \Rightarrow \quad -3 < x < 3$$

The Ratio Test also tells us that the series diverges when:

$$\frac{|x|}{3} > 1 \quad \Rightarrow \quad |x| > 3 \quad \Rightarrow \quad x < -3 \text{ or } x > 3$$

The Ratio Test provides no information if $|x| = 3$, so we need to check the endpoints $x = -3$ and $x = 3$ separately. When $x = -3$, the series becomes:

$$\sum_{k=0}^{\infty} \frac{(-3)^k}{k \cdot 3^k} = \sum_{k=0}^{\infty} \frac{(-1)^k}{k}$$

This is the alternating harmonic series, which converges conditionally (by the Alternating Series Test). When $x = 3$, the series becomes:

$$\sum_{k=0}^{\infty} \frac{(3)^k}{k \cdot 3^k} = \sum_{k=0}^{\infty} \frac{1}{k}$$

This is the harmonic series, which diverges. In summary, the power series converges if $-3 \le x < 3$; that is, on the interval $[-3, 3)$. ◀

This power series converges *absolutely* on the interval $(-3, 3)$.

The Ratio Test is a powerful tool for determining where a power series converges. Typically, you also need to check the endpoints of an interval by replacing x with the endpoint values and then determining if the resulting numerical series converge or diverge at these values.

The Ratio Test does not help with the endpoints.

Practice 1. Find all values of x for which $\sum_{k=1}^{\infty} \frac{5^k \cdot x^k}{k}$ converges.

Interval of Convergence

In the preceding examples, the values of x for which the power series converged formed an interval. The next theorem and its corollary say that this *always* happens for *any* power series.

Interval of Convergence Theorem for Power Series

- If $\sum_{k=0}^{\infty} a_k x^k$ converges for $x = c$, it converges when $|x| < |c|$.
- If $\sum_{k=0}^{\infty} a_k x^k$ diverges for $x = d$, it diverges when $|x| > |d|$.

Proof. If $\sum_{k=0}^{\infty} a_k c^k$ converges, $\lim_{k\to\infty} a_k c^k = 0$, so there is an N such that:

$$k \geq N \quad \Rightarrow \quad \left|a_k c^k\right| < 1 \quad \Rightarrow \quad |a_k| < \frac{1}{\left|c^k\right|}$$

If $|x| < |c|$, then $\left|\frac{x}{c}\right| < 1$ and we can write:

$$\sum_{k=0}^{\infty} \left|a_k x^k\right| = \sum_{k=0}^{N-1} \left|a_k x^k\right| + \sum_{k=N}^{\infty} \left|a_k x^k\right| \leq \sum_{k=0}^{N-1} \left|a_k x^k\right| + \sum_{k=N}^{\infty} \left|\frac{x^k}{c^k}\right|$$

which is the sum of a finite number of terms and a (convergent) geometric series with ratio $\left|\frac{x}{c}\right| < 1$, hence $\sum_{k=0}^{\infty} \left|a_k x^k\right|$ converges by the Basic Comparison Test, so that $\sum_{k=0}^{\infty} a_k x^k$ converges (absolutely).

Now suppose that $\sum_{k=0}^{\infty} a_k d^k$ diverges. If $\sum_{k=0}^{\infty} a_k x^k$ were to converge with $|x| > |d|$, this would contradict the first part of the theorem. □

If $\sum_{k=0}^{\infty} a_k x^k$ for a value $x = c$, then the series also converges for all values of x closer to the origin than c. If the power series diverges for a value $x = d$, then the power series diverges for all values of x farther from the origin than d. The Interval of Convergence Theorem does *not* tell us about the convergence of the power series for values of x with $|c| < |x| < |d|$ (see margin figure).

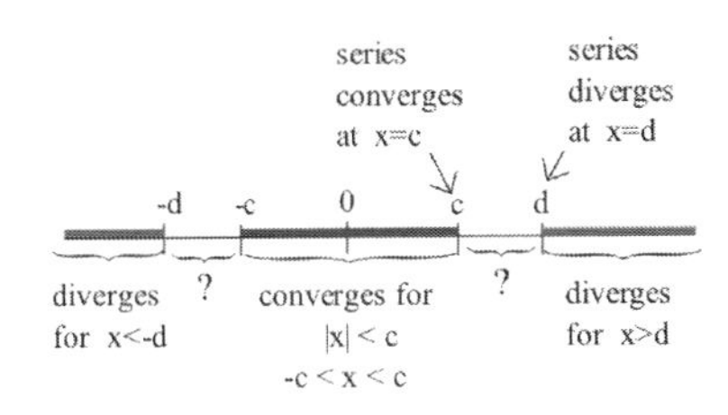

Example 3. If $\sum_{k=0}^{\infty} a_k x^k$ converges at $x = 4$ and diverges at $x = 9$, what can you conclude ("converge" or "diverge" or "no information") about the series when $x = 2, -3, -4, 5, -6, 8, -9, 10$ and -11?

Solution. We know the power series converges at $x = 4$, so with $c = 4$ in the Interval of Convergence Theorem we can conclude that the series converges for $x = 2$ and $x = -3$, because $|2| < |4|$ and $|-3| < |4|$.

We know the power series diverges at $x = 9$, so with $d = 9$ in the Interval of Convergence Theorem we can conclude that the series diverges for $x = 10$ and $x = -11$, because $|10| > |9|$ and $|-11| > |9|$. The remaining values of x ($-4, 5, -6, 8$ and -9) do not satisfy $|x| < 4$ or $|x| > 9$, so the series may converge or may diverge at those values—we don't have enough information.

The margin figure shows the regions where the Interval of Convergence Theorem guarantees convergence of this power series, guarantees divergence, and where it provides us with no information. ◀

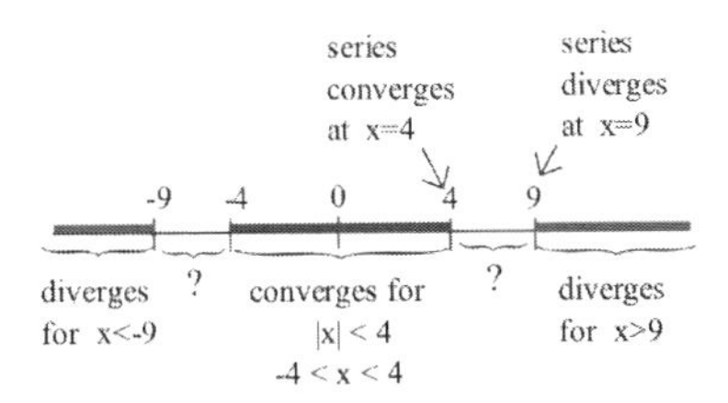

Practice 2. If a power series $\sum_{k=0}^{\infty} a_k x^k$ converges at $x = 3$ and diverges at $x = -7$, what can you say about the convergence of this series for $x = -1, 2, -3, 4, -6, 7, -8$ and 17? Sketch the regions of known convergence and known divergence.

The following corollary of the Interval of Convergence Theorem guarantees that the set of values of x where the power series converges form an interval.

> **Corollary**
>
> The values of x for which the power series $\sum_{k=0}^{\infty} a_k x^k$ converges form an interval of the form $(-R, R)$, $[-R, R)$, $(-R, R]$ or $[-R, R]$.

Idea for a Proof. The extreme cases are $R = 0$, so that the power series converges only at $x = 0$, and $R = \infty$, so that the power series converges for all values of x. Otherwise, there is some number $c > 0$ so that the series converges for $x = c$ (and, by the Interval of Convergence Theorem, for all x with $|x| < |c|$) and another number $d > c$ so that the series converges for $x = d$ (and, by the Interval of Convergence Theorem, for all x with $|x| > |d|$). Consider the set of all such values c: these are bounded above by d, so there is a *least* upper bound of all such values c. Let R be the that least upper bound.

The existence of such a least upper bound is guaranteed by the Completeness Axiom of the real numbers (discussed in more advanced mathematics courses).

The Interval of Convergence Theorem and its Corollary do *not* tell us about the convergence of the power series at the endpoints of this interval: we need to check those two points individually.

Definition:

The **interval of convergence** of a power series $\sum_{k=0}^{\infty} a_k x^k$ is the interval of values of x for which the series converges.

From Examples 1 and 2, we know the interval of convergence of the power series $\sum_{k=0}^{\infty} (2k+1)x^k$ is $(-1,1)$ and the interval of convergence of the power series $\sum_{k=0}^{\infty} \frac{x^k}{k \cdot 3^k}$ is $[-3,3)$.

Radius of Convergence

More advanced mathematics courses study power series of the form $\sum_{k=0}^{\infty} a_k z^k$, where z is allowed to be a *complex* number (such as i or $3+4i$). Instead of an interval of convergence, these power series have a disk of convergence in the complex plane. The radius of this disk is called the **radius of convergence** of the power series. We use this same terminology for power series involving a real variable x.

radius of convergence
radius of convergence
-R
0
R
diverges for x<-R
converges for |x| < R -R < x < R
diverges for x>R

Definition:

The **radius of convergence** of a power series $\sum_{k=0}^{\infty} a_k x^k$ is the number R so that the power series converges for $|x| < R$ and diverges for $|x| > R$. (The series may converge or may diverge if $|x| = R$.)

The radius of convergence of a power series is half of the length of its interval of convergence.

Example 4. What is the radius of convergence of the series from Example 1? From Example 2?

Solution. The series $\sum_{k=0}^{\infty} (2k+1)x^k$ converges if $-1 < x < 1$, so $R = 1$.

The series $\sum_{k=0}^{\infty} \frac{x^k}{k \cdot 3^k}$ converges if $-3 \le x < 3$, so $R = 3$. ◀

Practice 3. Find the radius of convergence of the series in Practice 1.

The convergence or divergence of a power series at an endpoint of the interval of convergence does not affect the value of the radius of convergence R, and the value of R does not tell you anything about the convergence of the power series at the endpoints of the interval of convergence (at $x = R$ and $x = -R$).

Summary

From the preceding discussion (and from the Idea for a Proof of the Corollary to the Interval of Convergence Theorem), we know that exactly one of three situations can occur for the radius of convergence R of a power series:

- it converges only for $x = 0$ (so that $R = 0$)
- it converges for when $|x| < R$ and diverges when $|x| > R$
- it converges for all values of x (so that $R = \infty$)

The following table displays information about the intervals and radii of convergence for several power series. Four of the series in the table have the same radius of convergence ($R = 1$) but slightly different intervals of convergence.

series	radius	interval
$\sum_{k=1}^{\infty} k! \cdot x^k$	$R = 0$	$\{0\}$
$\sum_{k=1}^{\infty} x^k$	$R = 1$	$(-1, 1)$
$\sum_{k=1}^{\infty} \frac{x^k}{k}$	$R = 1$	$[-1, 1)$
$\sum_{k=1}^{\infty} \frac{(-x)^k}{k}$	$R = 1$	$(-1, 1]$
$\sum_{k=1}^{\infty} \frac{x^k}{k^2}$	$R = 1$	$[-1, 1]$
$\sum_{k=1}^{\infty} \frac{x^k}{2^k}$	$R = 2$	$(-2, 2)$
$\sum_{k=1}^{\infty} \frac{x^k}{k!}$	$R = \infty$	$(-\infty, \infty)$

A power series looks like a very long polynomial. The domain of a regular polynomial with a finite number of terms, however, is $(-\infty, \infty)$, but a power series may converge only on a finite interval, in which case it will have a much smaller domain than any related polynomial. As we continue to work with power series we need to be alert to where the power series converges (and behaves like a finite polynomial) and where the power series diverges. We need to know the interval of convergence of the power series and, typically, we use the Ratio Test to find that interval.

10.1 Problems

In Problems 1–15, determine all values of x for which each given power series converges, then graph the interval of convergence for the series on a number line.

1. $\sum_{k=1}^{\infty} x^k$
2. $\sum_{k=1000}^{\infty} x^k$
3. $\sum_{k=1}^{\infty} 3^k \cdot x^k$
4. $\sum_{k=1}^{\infty} (5x)^k$
5. $\sum_{k=1}^{\infty} \frac{x^k}{k}$
6. $\sum_{k=1}^{\infty} \frac{x^k}{k^2}$
7. $\sum_{k=1}^{\infty} k \cdot x^k$
8. $\sum_{k=1}^{\infty} k^2 \cdot x^k$
9. $\sum_{k=1}^{\infty} k \cdot x^{2k+1}$
10. $\sum_{k=0}^{\infty} k! \cdot x^k$
11. $\sum_{k=0}^{\infty} \frac{x^k}{k!}$
12. $\sum_{k=0}^{\infty} \frac{2^k \cdot x^k}{k!}$
13. $\sum_{k=1}^{\infty} k \left(\frac{x}{4}\right)^{2k}$
14. $\sum_{k=0}^{\infty} \frac{k! \cdot x^k}{2^k}$
15. $\sum_{k=0}^{\infty} \frac{x^k}{2^k}$
16. Your friend claims that the interval of convergence for a power series of the form $\sum_{k=0}^{\infty} a_k x^k$ is the interval $(-2, 3)$. Without checking your friend's work, how can you be certain that your friend is wrong?

In Problems 17–24, find the radius of convergence for the series from the given Problem.

17. Problem 1
18. Problem 3
19. Problem 5
20. Problem 7
21. Problem 9
22. Problem 11
23. Problem 13
24. Problem 15

In Problems 25–28, use the patterns you noticed in earlier problems and Examples to build a power series with the given interval of convergence. (There are many possible correct answers—find one.)

25. $(-5, 5)$
26. $[-3, 3)$
27. $[-2, 2]$
28. $(-4, 4]$

In Problems 29–32, given the interval of convergence for a power series, find its radius of convergence.

29. $(-5, 5)$
30. $[-3, 3)$
31. $[-2, 2]$
32. $(-4, 4]$

In Problems 33–41, find the interval of convergence for each series. Then, for x in the interval of convergence, find the sum of the series as a function of x. (Hint: You already know how to find the sum of a geometric series.)

33. $\sum_{k=0}^{\infty} x^k$
34. $\sum_{k=0}^{\infty} (-x)^k$
35. $\sum_{k=0}^{\infty} (2x)^k$
36. $\sum_{k=0}^{\infty} (3x)^k$
37. $\sum_{k=1}^{\infty} x^k$
38. $\sum_{k=0}^{\infty} x^{2k}$
39. $\sum_{k=0}^{\infty} x^{3k}$
40. $\sum_{k=0}^{\infty} \left(\frac{x}{4}\right)^k$
41. $\sum_{k=0}^{\infty} (4x)^k$

10.1 Practice Answers

1. Applying the Ratio Test to $\sum_{k=1}^{\infty} \frac{5^k \cdot x^k}{k}$:

$$\left| \frac{\frac{5^{k+1} \cdot x^{k+1}}{k+1}}{\frac{5^k \cdot x^k}{k}} \right| = \left| \frac{5^{k+1} \cdot x^{k+1}}{k+1} \cdot \frac{5^k \cdot x^k}{k} \right| = 5|x| \cdot \left(\frac{k}{k+1} \right) \to 5|x|$$

For the series to converge, we need:

$$5|x| < 1 \quad \Rightarrow \quad |x| < \frac{1}{5} \quad \Rightarrow \quad -\frac{1}{5} < x < \frac{1}{5}$$

The Ratio Test also tells us that the series diverges when:

$$5|x| > 1 \quad \Rightarrow \quad |x| > \frac{1}{5} \quad \Rightarrow \quad x < -\frac{1}{5} \text{ or } x > \frac{1}{5}$$

Now we need to check the points where $|x| = \frac{1}{5}$. When $x = \frac{1}{5}$:

$$\sum_{k=1}^{\infty} \frac{5^k \cdot x^k}{k} = \sum_{k=1}^{\infty} \frac{5^k \cdot \left(\frac{1}{5}\right)^k}{k} = \sum_{k=1}^{\infty} \frac{1}{k}$$

which diverges (it's the harmonic series), and when $x = -\frac{1}{5}$:

$$\sum_{k=1}^{\infty} \frac{5^k \cdot x^k}{k} = \sum_{k=1}^{\infty} \frac{5^k \cdot \left(-\frac{1}{5}\right)^k}{k} = \sum_{k=1}^{\infty} \frac{(-1)^k}{k}$$

which converges conditionally (by the Alternating Series Test). So the series converges on the interval $\left[-\frac{1}{5}, \frac{1}{5}\right)$.

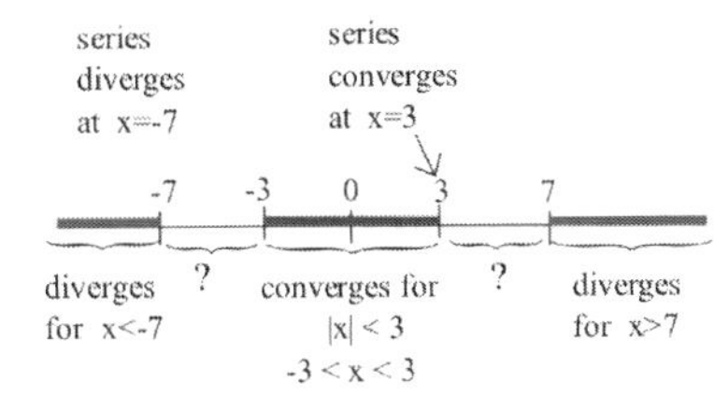

2. The series converges at $x = -1$ and $x = 2$, diverges at $x = -8$ and $x = 17$, and may converge or diverge at the other points. See the margin figure for a graph of the regions of known convergence and known divergence.

3. The radius of convergence is $R = \frac{1}{5}$ (half the length of the interval of convergence).

10.2 *Power Series:* $\sum_{k=0}^{\infty} a_k(x-c)^k$

The preceding section examined power series of the form $\sum_{k=0}^{\infty} a_k x^k$, which were guaranteed to converge at $x = 0$. As we observed, such a power series may have converged for all values of x, or for some smaller interval centered at $x = 0$. When we apply power series to approximate functions and solve problems throughout the rest of this chapter, we may need our power series to converge near some number other than $x = 0$. We can accomplish this by shifting a power series of the form $\sum_{k=0}^{\infty} a_k x^k$ to get a power series of the form $\sum_{k=0}^{\infty} a_k(x-c)^k$.

> **Definition:**
>
> A **power series** is an expression of the form:
>
> $$\sum_{k=0}^{\infty} a_k(x-c)^k = a_0 + a_1(x-c) + a_2(x-c)^2 + a_3(x-c)^3 + \cdots$$
>
> where $a_0, a_1, a_2, a_3, \ldots$ are constants, called the **coefficients** of the series, and x is a variable.

For $k = 0$ we again use the convention for power series that $(x-c)^0 = 1$ even when $x = c$.

When $x = c$, this series yields:

$$\sum_{k=0}^{\infty} a_k(c-c)^k = a_0 + a_1 \cdot 0 + a_2 \cdot 0^2 + a_3 \cdot 0^3 + \cdots = a_0$$

so it is guaranteed to converge at $x = c$, no matter the values of a_k.

Furthermore, $x = c$ is always the center of the interval of convergence for this power series (just as $x = 0$ was the center of the intervals of convergence for the series we studied in the preceding section). The radius of convergence is—like it was in Section 10.1—half the length of the interval of convergence (see margin figure). If a power series centered at $x = c$ converges for some value of x, then the series converges for all values of x closer to $x = c$; if a power series centered at $x = c$ diverges for another value of x, then the series diverges for all values of x farther from $x = c$.

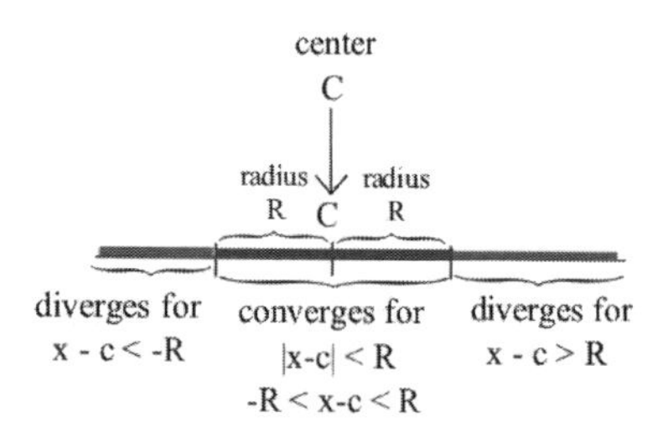

Example 1. If you know $\sum_{k=0}^{\infty} a_k(x-4)^k$ converges at $x = 6$ and diverges at $x = 0$, what can you conclude ("converge" or "diverge" or "not enough information") about the series when $x = 3, 9, -1, 2$ and 7?

Solution. We know the power series converges at $x = 6$, so we can conclude that the series converges for all values of x closer to 4 than $|6-4| = 2$ units: in particular, the series converges at $x = 3$.

We know the power series diverges at $x = 0$, so we can conclude that the series diverges for all values of x farther from 4 than $|0-4| = 4$ units: in particular, the series diverges at $x = 9$ and $x = -1$.

The remaining values of x (2 and 7) do not satisfy $|x-4| < 2$ or $|x-4| > 4$, so the series may converge or may diverge for those values of x. The margin figure shows the regions where convergence of this power series is guaranteed, where divergence is guaranteed, and where we lack enough information to make a determination. ◄

Practice 1. If you know that the power series $\sum_{k=0}^{\infty} a_k(x+5)^k$ converges at $x = -1$ and diverges at $x = 1$, what can you conclude about the series when $x = -2, -9, 0, -11$ and 3?

The Ratio Test remains our primary tool for finding an interval of convergence of a power series, even if the power series is centered at $x = c$ rather than at $x = 0$.

Example 2. Find the interval and radius of convergence of $\sum_{k=1}^{\infty} \frac{(x-5)^k}{k \cdot 2^k}$.

Solution. Applying the Ratio Test:

$$\left| \frac{\frac{(x-5)^{k+1}}{(k+1)2^{k+1}}}{\frac{(x-5)^k}{k \cdot 2^k}} \right| = \left| \frac{(x-5)^{k+1}}{(k+1)2^{k+1}} \cdot \frac{k \cdot 2^k}{(x-5)^k} \right| = \left(\frac{k}{k+1} \right) \cdot \frac{|x-5|}{2} \longrightarrow \frac{|x-5|}{2}$$

The Ratio Test guarantees absolute convergence when:

$$\frac{|x-5|}{2} < 1 \Rightarrow |x-5| < 2 \Rightarrow -2 < x-5 < 2 \Rightarrow 3 < x < 7$$

and says that the power series diverges if $x < 3$ or $x > 7$. We still need to check the endpoints of this interval of convergence: $x = 3$ and $x = 7$.

When $x = 3$, the power series becomes:

$$\sum_{k=1}^{\infty} \frac{(3-5)^k}{k \cdot 2^k} = \sum_{k=1}^{\infty} \frac{(-2)^k}{k \cdot 2^k} = \sum_{k=1}^{\infty} \frac{(-1)^k}{k}$$

(the alternating harmonic series), which converges conditionally. When $x = 7$, the power series becomes:

$$\sum_{k=1}^{\infty} \frac{(7-5)^k}{k \cdot 2^k} = \sum_{k=1}^{\infty} \frac{(2)^k}{k \cdot 2^k} = \sum_{k=1}^{\infty} \frac{1}{k}$$

(the harmonic series), which diverges. The interval of convergence is therefore $[3,7)$ and the radius of convergence is $R = \frac{1}{2}(7-3) = 2$. ◄

This power series converges *absolutely* on the interval $(3,7)$.

Practice 2. Find the interval and radius of convergence of $\sum_{k=1}^{\infty} \frac{k \cdot (x-3)^k}{5^k}$.

10.2 Problems

In Problems 1–15, determine the interval of convergence and radius of convergence for the given power series, then graph the interval of convergence on a number line.

1. $\sum_{k=0}^{\infty} (x+2)^k$
2. $\sum_{k=0}^{\infty} (x-3)^k$
3. $\sum_{k=1}^{\infty} (x+5)^k$
4. $\sum_{k=1}^{\infty} \frac{(x+3)^k}{k}$
5. $\sum_{k=1}^{\infty} \frac{(x-2)^k}{k}$
6. $\sum_{k=1}^{\infty} \frac{(x-5)^k}{k^2}$
7. $\sum_{k=1}^{\infty} \frac{(x-7)^{2k+1}}{k^2}$
8. $\sum_{k=1}^{\infty} \frac{(x+1)^{2k}}{k^3}$
9. $\sum_{k=0}^{\infty} (2x-6)^k$
10. $\sum_{k=0}^{\infty} (3x+1)^k$
11. $\sum_{k=0}^{\infty} \frac{(x-5)^k}{k!}$
12. $\sum_{k=0}^{\infty} k! \cdot (x+2)^k$
13. $\sum_{k=3}^{\infty} k! \cdot (x-7)^k$
14. $\sum_{k=3}^{\infty} (x-7)^{2k}$
15. Your friend claims the interval of convergence for a power series of the form $\sum_{k=0}^{\infty} a_k(x-4)^k$ is the interval $(1,9)$. Without checking his work, how can you be certain that your friend is wrong?
16. Determine which of the following intervals *could* be the interval of convergence for a power series of the form $\sum_{k=0}^{\infty} a_k(x-4)^k$: $(2,6)$, $(0,4)$, $\{0\}$, $[1,7]$, $(-1,9]$, $\{4\}$, $[3,5)$, $[-4,4)$, $\{3\}$.
17. Determine which of the following intervals *could* be the interval of convergence for a power series of the form $\sum_{k=0}^{\infty} a_k(x-7)^k$: $(3,10)$, $(5,9)$, $\{0\}$, $[1,13]$, $(-1,15]$, $\{4\}$, $[3,11)$, $[0,14)$, $\{7\}$.
18. Fill in each blank with a number so the resulting interval *could* be the interval of convergence for a power series of the form $\sum_{k=0}^{\infty} a_k(x-3)^k$: $(0,\_\_)$, $(\_\_,7)$, $[0,\_\_]$, $(\_\_,15]$, $[\_\_,11)$, $[0,\_\_)$, $\{\_\_\}$.
19. Fill in each blank with a number so the resulting interval *could* be the interval of convergence for a power series of the form $\sum_{k=0}^{\infty} a_k(x-1)^k$: $(0,\_\_)$, $(\_\_,7)$, $[0,\_\_]$, $(\_\_,5]$, $[\_\_,11)$, $[0,\_\_)$, $\{\_\_\}$.
20. Fill in each blank with a number so the resulting interval *could* be the interval of convergence for a power series of the form $\sum_{k=0}^{\infty} a_k(x+1)^k$: $(-2,\_\_)$, $(\_\_,7)$, $[-3,\_\_]$, $(\_\_,5]$, $[\_\_,11)$, $[-4,\_\_)$, $\{\_\_\}$.

In Problems 21–24, given the interval of convergence for a power series, find its radius of convergence.

21. $(0,6)$
22. $[0,8)$
23. $(2,8]$
24. $[3,7]$

In Problems 25–28, use the patterns you noticed in earlier problems and Examples to build a power series with the given interval of convergence. (There are many possible correct answers—find one.)

25. $(0,6)$
26. $[0,8)$
27. $(2,8]$
28. $[3,7]$

In Problems 29–34, find the interval of convergence for each series. Then, for x in the interval of convergence, find the sum of the series as a function of x. (Hint: You already know how to find the sum of a geometric series.)

29. $\sum_{k=0}^{\infty} (x-3)^k$
30. $\sum_{k=0}^{\infty} \left(\frac{x-6}{2}\right)^k$
31. $\sum_{k=0}^{\infty} \left(\frac{x-6}{5}\right)^k$
32. $\sum_{k=0}^{\infty} \left(\frac{x}{3}\right)^{2k}$
33. $\sum_{k=0}^{\infty} \left(\frac{1}{2}\sin(x)\right)^k$
34. $\sum_{k=0}^{\infty} \left(\frac{1}{2}\cos(x)\right)^k$

In 35–42, the letters a and b represent positive constants. Find the interval of convergence for each series.

35. $\sum_{k=1}^{\infty} (x-a)^k$

36. $\sum_{k=1}^{\infty} (x+b)^k$

37. $\sum_{k=1}^{\infty} \frac{(x-a)^k}{k}$

38. $\sum_{k=1}^{\infty} \frac{(x-a)^k}{k^2}$

39. $\sum_{k=1}^{\infty} (ax)^k$

40. $\sum_{k=1}^{\infty} \left(\frac{x}{a}\right)^k$

41. $\sum_{k=1}^{\infty} (ax-b)^k$

42. $\sum_{k=1}^{\infty} (ax+b)^k$

10.2 Practice Answers

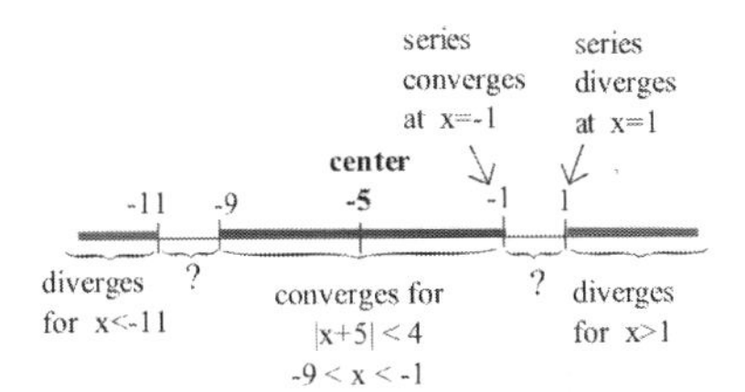

1. The center of the power series is at $x = -5$ and the series converges at $x = -1$, so it must converge at all points less than $|-1-(-5)| = 4$ units from $x = -5$, in particular at $x = -2$. The series diverges at $x = 1$, so it must diverge at all points more than $|1-(-5)| = 6$ units from $x = -5$, in particular at $x = 3$. The series may converge or diverge at the other points. See the margin figure for a graph of the regions of known convergence and known divergence.

2. Applying the Ratio Test:

$$\left|\frac{\frac{(k+1)\cdot(x-3)^{k+1}}{5^{k+1}}}{\frac{k\cdot(x-3)^k}{5^k}}\right| = \left|\frac{(k+1)\cdot(x-3)^{k+1}}{5^{k+1}} \cdot \frac{5^k}{k\cdot(x-3)^k}\right| = \left(\frac{k+1}{k}\right)\cdot\frac{|x-3|}{5} \to \frac{|x-3|}{5}$$

The Ratio Test tells us that the power series converges when:

$$\frac{|x-3|}{5} < 1 \Rightarrow |x-3| < 5 \Rightarrow -5 < x-3 < 5 \Rightarrow -2 < x < 8$$

and that the power series diverges for $x < -2$ and $x > 8$. We need to check the endpoints of this interval separately. When $x = -2$ the power series becomes:

$$\sum_{k=1}^{\infty} \frac{k\cdot(-2-3)^k}{5^k} = \sum_{k=1}^{\infty} (-1)^k \cdot k$$

which diverges (by the Test for Divergence). When $x = 8$, the power series becomes:

$$\sum_{k=1}^{\infty} \frac{k\cdot(8-3)^k}{5^k} = \sum_{k=1}^{\infty} k$$

which also diverges (by the Test for Divergence). The interval of convergence is therefore $(-2, 8)$ and the radius of convergence is $R = \frac{1}{2}\cdot 10 = 5$ (half the length of the interval of convergence).

10.3 *Representing Functions as Power Series*

We know from our work with geometric series that the function:

$$G(x) = \sum_{k=0}^{\infty} x^k = 1 + x + x^2 + x^3 + x^4 + \cdots$$

has domain $-1 < x < 1$ (that is, the geometric series converges for $|x| < 1$) and for values of x in that domain we know that:

$$G(x) = \sum_{k=0}^{\infty} x^k = \frac{1}{1-x} = (1-x)^{-1}$$

The function $\frac{1}{1-x}$ has a much larger domain, $(-\infty, 1) \cup (1, \infty)$, than the corresponding power series function $G(x)$. But on their common domain, $(-1, 1)$, these two functions agree.

Can we find power series representations for other functions? If so, on what interval does the power series converge to the same value as the function? We will investigate the answers to these questions throughout the next few sections.

In this section, we obtain power series representations for several functions related to $\frac{1}{1-x}$ using our knowledge of the geometric series. We will also examine some applications of these power series representations of functions.

Substitution in Power Series

One simple but powerful method for obtaining a power series for a function is to make a substitution into a known power series representation. If we begin with the geometric series:

$$\frac{1}{1-u} = \sum_{k=0}^{\infty} u^k = 1 + u + u^2 + u^3 + u^4 + \cdots$$

and make the substitution $u = -x$ we get:

$$\frac{1}{1-(-x)} = \sum_{k=0}^{\infty} (-x)^k = 1 + (-x) + (-x)^2 + (-x)^3 + (-x)^4 + \cdots$$

which we can rewrite as:

$$\frac{1}{1+x} = \sum_{k=0}^{\infty} (-1)^k \cdot x^k = 1 - x + x^2 - x^3 + x^4 + \cdots$$

This new power series is also a geometric series, and it converges when (and only when) $|-x| < 1 \Rightarrow |x| < 1 \Rightarrow -1 < x < 1$.

Similar substitutions (along with some straightforward algebra) lead to a variety of other power series representations.

Example 1. Find power series for $\frac{1}{1-x^2}$ and $\frac{x}{1-x}$.

Solution. For the first function, use the substitution $u = x^2$ in the geometric series formula:

$$\frac{1}{1-u} = \sum_{k=0}^{\infty} u^k = 1 + u + u^2 + u^3 + u^4 + \cdots$$
$$\Rightarrow \frac{1}{1-x^2} = \sum_{k=0}^{\infty} (x^2)^k = 1 + (x^2) + (x^2)^2 + (x^2)^3 + (x^2)^4 + \cdots$$
$$\Rightarrow \frac{1}{1-x^2} = \sum_{k=0}^{\infty} x^{2k} = 1 + x^2 + x^4 + x^6 + x^8 + \cdots$$

which converges if $x^2 < 1 \Rightarrow \sqrt{x^2} < \sqrt{1} \Rightarrow |x| < 1 \Rightarrow -1 < x < 1$. For the second series, rewrite the function as a product:

$$\frac{x}{1-x} = x \cdot \frac{1}{1-x} = x \cdot \sum_{k=0}^{\infty} x^k = x \cdot \left[1 + x + x^2 + x^3 + x^4 + \cdots\right]$$
$$= \sum_{k=0}^{\infty} x^{k+1} = x + x^2 + x^3 + x^4 + x^5 + \cdots$$

This is also a geometric series with ratio x, so it converges when $|x| < 1 \Rightarrow -1 < x < 1$. ◀

Practice 1. Find power series for $\frac{1}{1-x^3}$, $\frac{1}{1+x^2}$ and $\frac{5x}{1+x}$.

Differentiation and Integration of Power Series

One feature of polynomials that makes them very easy to differentiate and integrate is that we can differentiate and integrate them term-by-term. The same result holds true for power series.

Term-by-Term Differentiation of Power Series

If $f(x)$ is defined by a power series:

$$f(x) = \sum_{k=0}^{\infty} a_k x^k = a_0 + a_1 x + a_2 x^2 + a_3 x^3 + \cdots$$

that converges for $-R < x < R$, then:

$$f'(x) = \sum_{k=1}^{\infty} k \cdot a_k x^{k-1} = a_1 + 2 \cdot a_2 x + 3 \cdot a_3 x^2 + 4 \cdot a_4 x^3 + \cdots$$

and this new power series also converges for $-R < x < R$.

The power series for f and the power series for f' may differ in whether they converge or diverge at the endpoints of the interval of convergence, but they both converge for $-R < x < R$.

The proof of this statement is rather long and highly technical, so we will omit it, but this result allows us to find power series of even more functions based on the geometric series.

Example 2. Find a power series for $\frac{1}{(1-x)^2}$.

Solution. Because $(1-x)^{-2}$ is the derivative of $(1-x)^{-1}$, we can write:

$$\frac{1}{(1-x)^2} = \frac{d}{dx}\left[\frac{1}{1-x}\right] = \frac{d}{dx}\left[\sum_{k=0}^{\infty} x^k\right] = \sum_{k=0}^{\infty} \frac{d}{dx}\left(x^k\right)$$
$$= \sum_{k=0}^{\infty} k \cdot x^{k-1} = \sum_{k=1}^{\infty} k \cdot x^{k-1} = 1 + 2x + 3x^2 + 4x^3 + 5x^4 + \cdots$$

See Problem 41.

You can use the Ratio Test on the new power series to verify that its interval of convergence is $(-1, 1)$, the same as the original series. ◀

Term-by-Term Integration of Power Series

If $f(x)$ is defined by a power series:

$$f(x) = \sum_{k=0}^{\infty} a_k x^k = a_0 + a_1 x + a_2 x^2 + a_3 x^3 + \cdots$$

that converges for $-R < x < R$, then:

$$\int f(x)\,dx = C + \sum_{k=0}^{\infty} \frac{a_k}{k+1} x^{k+1} = C + a_0 x + \frac{a_1}{2} x^2 + \frac{a_2}{3} x^3 + \cdots$$

and this new power series also converges for $-R < x < R$.

The power series for f and its antiderivative may differ in whether they converge or diverge at the endpoints of the interval of convergence, but they both converge for $-R < x < R$.

Proof. Let $F(x) = \sum_{k=0}^{\infty} \frac{a_k}{k+1} x^{k+1}$ so that, using term-by-term differentiation:

$$F'(x) = \sum_{k=0}^{\infty} (k+1) \cdot \frac{a_k}{k+1} x^k = \sum_{k=0}^{\infty} a_k x^k = f(x)$$

Here we use the result of the previous theorem, which we did not prove.

Therefore $\int f(x)\,dx = F(x) + C$. To find the interval of convergence for $F(x)$, note that:

$$k \geq 0 \Rightarrow k+1 \geq 1 \Rightarrow \frac{1}{k+1} \leq 1 \Rightarrow \frac{|a_k|}{k+1}|x|^k \leq |a_k| \cdot |x^k|$$

Because $\sum_{k=0}^{\infty} a_k x^k$ converges for $|x| < R$, it converges absolutely on this open interval, hence so does $\sum_{k=0}^{\infty} \frac{a_k}{k+1} x^k$ (by the Basic Comparison Test), and so does:

$$x \cdot \sum_{k=0}^{\infty} \frac{a_k}{k+1} x^k = \sum_{k=0}^{\infty} \frac{a_k}{k+1} x^{k+1}$$

(Convergence at the endpoints $x = -R$ and $x = R$ must be determined on a case-by-case basis.) □

Example 3. Find power series for $\ln(1-x)$ and $\arctan(x)$

Solution. We need to recognize that these functions are integrals of functions whose power series we already know. For the first function:

$$\begin{aligned}\ln(1-x) &= \int \frac{-1}{1-x}\,dx = -\int \left[\sum_{k=0}^{\infty} x^k\right] dx = C - \sum_{k=0}^{\infty} \frac{1}{k+1}\cdot x^{k+1} \\ &= C - x - \frac{1}{2}x^2 - \frac{1}{3}x^3 - \frac{1}{4}x^4 - \cdots\end{aligned}$$

Substituting $x = 0$ into both sides of this equation yields:

$$0 = \ln(1) = \ln(1-0) = C - 0 - \frac{1}{2}\cdot 0^2 - \frac{1}{3}\cdot 0^3 - \cdots = C \Rightarrow C = 0$$

so that:

$$\ln(1-x) = -x - \frac{1}{2}x^2 - \frac{1}{3}x^3 - \frac{1}{4}x^4 - \cdots = -\sum_{k=0}^{\infty}\frac{1}{k+1}\cdot x^{k+1}$$

You should be able to check that the interval of convergence for this new power series is $-1 \le x < 1$, which agrees with the interval of convergence of the original series (except at the left endpoint).

See Problem 37.

For the second series, we can use the second result from Practice 1:

$$\frac{1}{1+x^2} = \sum_{k=0}^{\infty}(-x^2)^k = \sum_{k=0}^{\infty}(-1)^k\cdot x^{2k} = 1 - x^2 + x^4 - x^6 + x^8 - \cdots$$

and then apply term-by-term integration to this power series to get:

$$\begin{aligned}\arctan(x) &= \int \frac{1}{1+x^2}\,dx = \int \left[\sum_{k=0}^{\infty}(-1)^k\cdot x^{2k}\right] dx \\ &= C + \sum_{k=0}^{\infty}\frac{(-1)^k}{2k+1}x^{2k+1} = C + x - \frac{1}{3}x^3 + \frac{1}{5}x^5 - \frac{1}{7}x^7 + \cdots\end{aligned}$$

To determine the value of C, substitute $x = 0$ to get:

$$0 = \arctan(0) = C + 0 - \frac{1}{3}\cdot 0^3 + \frac{1}{5}\cdot 0^5 - \frac{1}{7}\cdot 0^7 + \cdots = C \Rightarrow C = 0$$

so that:

$$\arctan(x) = \sum_{k=0}^{\infty}\frac{(-1)^k}{2k+1}x^{2k+1} = x - \frac{1}{3}x^3 + \frac{1}{5}x^5 - \frac{1}{7}x^7 + \cdots$$

You should be able to check that the interval of convergence for this new power series is $-1 \le x \le 1$, which agrees with the interval of convergence of the original series (except at the endpoints). ◀

See Problem 39.

Practice 2. Find a power series for $\ln(1+x)$.

Applications of Power Series

An important application of power series is the use of these "infinite polynomials" in place of a complicated integrand to help evaluate difficult integrals.

Example 4. Express the definite integral $\int_0^{\frac{1}{2}} \arctan\left(x^2\right)\, dx$ as a numerical series. Then approximate the value of the integral by calculating the sum of the first four terms of that numerical series.

Solution. From Example 3, we know that:

$$\arctan(u) = \sum_{k=0}^{\infty} \frac{(-1)^k}{2k+1} u^{2k+1} = u - \frac{1}{3}u^3 + \frac{1}{5}u^5 - \frac{1}{7}u^7 + \cdots$$

so substituting $u = x^2$ into this power series give us:

$$\arctan(x^2) = \sum_{k=0}^{\infty} \frac{(-1)^k}{2k+1} x^{4k+2} = x^2 - \frac{1}{3}x^6 + \frac{1}{5}x^{10} - \frac{1}{7}x^{14} + \cdots$$

Term-by-term integration of this power series yields:

$$\int_0^{\frac{1}{2}} \left[\sum_{k=0}^{\infty} \frac{(-1)^k}{2k+1} x^{4k+2}\right] dx = \int_0^{\frac{1}{2}} \left[x^2 - \frac{1}{3}x^6 + \frac{1}{5}x^{10} - \frac{1}{7}x^{14} + \cdots\right] dx$$

$$= \sum_{k=0}^{\infty} \frac{(-1)^k}{2k+1} \left[\frac{x^{4k+3}}{4k+3}\right]_0^{\frac{1}{2}} = \left[\frac{1}{3}x^3 - \frac{1}{21}x^7 + \frac{1}{55}x^{11} - \frac{1}{105}x^{15} + \cdots\right]_0^{\frac{1}{2}}$$

$$= \sum_{k=0}^{\infty} \frac{(-1)^k}{(2k+1)(4k+3)\cdot 2^{4k+3}} = \frac{1}{3\cdot 2^3} - \frac{1}{21\cdot 2^7} + \frac{1}{55\cdot 2^{11}} - \frac{1}{105\cdot 2^{15}} + \cdots$$

Adding up the first four terms of this series gives the approximation:

$$\int_0^{\frac{1}{2}} \arctan(x^2)\, dx \approx 0.04130323$$

Because this numerical series is an alternating series, the Alternating Series Estimation Bound tells us that the difference between this approximation and the actual value of the integral is smaller than the absolute value of the next term in the series:

$$\left|\int_0^{\frac{1}{2}} \arctan(x^2)\, dx - 0.04130323\right| \le \frac{1}{9\cdot 19\cdot 2^{19}} \approx 0.0000000112$$

We can therefore say that the value of the definite integral is between 0.0413032188 and 0.0413032415. ◀

Practice 3. Express the definite integral $\int_0^{0.2} x^2 \ln(1+x)\, dx$ as a numerical series. Then approximate the value of the integral by calculating the sum of the first four terms of the numerical series.

Wrap-Up

We obtained all of the power series used in this section from the geometric series via substitution, differentiation and integration. Many important functions, however, are not related to a geometric series, so future sections will discuss methods for representing more general functions using power series. The following table collects some of the power series representations we have obtained in this section.

$$\frac{1}{1-x} = \sum_{k=0}^{\infty} x^k = 1 + x + x^2 + x^3 + x^4 + \cdots$$

$$\frac{1}{1+x} = \sum_{k=0}^{\infty} (-1)^k \cdot x^k = 1 - x + x^2 - x^3 + x^4 - \cdots$$

$$\frac{1}{1-x^2} = \sum_{k=0}^{\infty} x^{2k} = 1 + x^2 + x^4 + x^6 + x^8 + \cdots$$

$$\frac{1}{1+x^2} = \sum_{k=0}^{\infty} (-1)^k \cdot x^{2k} = 1 - x^2 + x^4 - x^6 + x^8 - \cdots$$

$$\ln(1-x) = -\sum_{k=1}^{\infty} \frac{x^k}{k} = -x - \frac{1}{2}x^2 - \frac{1}{3}x^3 - \frac{1}{4}x^4 - \cdots$$

$$\ln(1+x) = \sum_{k=1}^{\infty} (-1)^{k+1} \cdot \frac{x^k}{k} = x - \frac{1}{2}x^2 + \frac{1}{3}x^3 - \frac{1}{4}x^4 + \cdots$$

$$\arctan(x) = \sum_{k=0}^{\infty} (-1)^k \cdot \frac{x^{2k+1}}{2k+1} = x - \frac{1}{3}x^3 + \frac{1}{5}x^5 - \frac{1}{7}x^7 + \cdots$$

$$\frac{1}{(1-x)^2} = \sum_{k=1}^{\infty} k \cdot x^{k-1} = 1 + 2x + 3x^2 + 4x^3 + 5x^4 + \cdots$$

10.3 Problems

In Problems 1–14, use substitution and a known power series to find a power series for the function.

1. $\dfrac{1}{1-x^4}$
2. $\dfrac{1}{1-x^5}$
3. $\dfrac{1}{1+x^4}$
4. $\dfrac{1}{1+x^5}$
5. $\dfrac{1}{5+x}$
6. $\dfrac{1}{3-x}$
7. $\dfrac{x^2}{1+x^3}$
8. $\dfrac{x}{1+x^4}$
9. $\ln\left(1+x^2\right)$
10. $\ln\left(1+x^3\right)$
11. $x\arctan(x^2)$
12. $\arctan(x^3)$
13. $\dfrac{1}{(1-x^2)^2}$
14. $\dfrac{1}{(1+x^2)^2}$
15. $\dfrac{1}{(1-x)^3}$
16. $\dfrac{1}{(1-x^2)^3}$
17. $\dfrac{1}{(1+x^2)^3}$
18. $\dfrac{1}{(1-x)^4}$

In 19–26, represent each integral as a series, then calculate the sum of the first three terms.

19. $\displaystyle\int_0^{\frac{1}{2}} \frac{1}{1-x^3}\,dx$
20. $\displaystyle\int_0^{\frac{1}{2}} \frac{1}{1+x^3}\,dx$
21. $\displaystyle\int_0^{\frac{3}{5}} \ln(1+x)\,dx$
22. $\displaystyle\int_0^{\frac{1}{2}} \ln(1+x^2)\,dx$
23. $\displaystyle\int_0^{\frac{1}{2}} x^2\arctan(x)\,dx$
24. $\displaystyle\int_0^{\frac{1}{2}} \arctan(x^3)\,dx$
25. $\displaystyle\int_0^{0.3} \frac{1}{(1-x)^2}\,dx$
26. $\displaystyle\int_0^{0.7} \frac{x^3}{(1-x)^2}\,dx$

In 27–32, represent each numerator as a power series, then use the power series to help find the limit.

27. $\lim_{x \to 0} \frac{\arctan(x)}{x}$

28. $\lim_{x \to 0} \frac{\ln(1-x)}{2x}$

29. $\lim_{x \to 0} \frac{\ln(1+x)}{2x}$

30. $\lim_{x \to 0} \frac{\arctan\left(x^2\right)}{x}$

31. $\lim_{x \to 0} \frac{\arctan\left(x^2\right)}{x^2}$

32. $\lim_{x \to 0} \frac{\arctan(x) - x}{x^3}$

33. $\lim_{x \to 0} \frac{\ln\left(1-x^2\right)}{3x}$

34. $\lim_{x \to 0} \frac{\ln\left(1+x^2\right)}{3x}$

In Problems 35–42, determine a power series for each function and then determine the interval of convergence of each power series.

35. $\frac{1}{1+x}$

36. $\frac{1}{1-x^2}$

37. $\ln(1-x)$

38. $\ln(1+x)$

39. $\arctan(x)$

40. $\arctan(x^2)$

41. $\frac{1}{(1-x)^2}$

42. $\frac{1}{(1+x)^2}$

10.3 Practice Answers

1. For the first function, use the substitution $u = x^3$ in the geometric series formula:

$$\begin{aligned} \frac{1}{1-u} &= \sum_{k=0}^{\infty} u^k = 1 + u + u^2 + u^3 + u^4 + \cdots \\ \Rightarrow \frac{1}{1-x^3} &= \sum_{k=0}^{\infty} (x^3)^k = 1 + (x^3) + (x^3)^2 + (x^3)^3 + (x^3)^4 + \cdots \\ \Rightarrow \frac{1}{1-x^3} &= \sum_{k=0}^{\infty} x^{3k} = 1 + x^3 + x^6 + x^9 + x^{12} + \cdots \end{aligned}$$

For the second function, use the substitution $u = -x^2$:

$$\begin{aligned} \frac{1}{1-u} &= \sum_{k=0}^{\infty} u^k = 1 + u + u^2 + u^3 + u^4 + \cdots \\ \Rightarrow \frac{1}{1-(-x^2)} &= \sum_{k=0}^{\infty} (-x^2)^k = 1 + (-x^2) + (-x^2)^2 + (-x^2)^3 + \cdots \\ \Rightarrow \frac{1}{1+x^2} &= \sum_{k=0}^{\infty} (-1)^k \cdot x^{2k} = 1 - x^2 + x^4 - x^6 - \cdots \end{aligned}$$

For the third function, use the substitution $u = -x$:

$$\begin{aligned} \frac{1}{1-u} &= \sum_{k=0}^{\infty} u^k = 1 + u + u^2 + u^3 + u^4 + \cdots \\ \Rightarrow \frac{1}{1-(-x)} &= \sum_{k=0}^{\infty} (-x)^k = 1 + (-x) + (-x)^2 + (-x)^3 + \cdots \\ \Rightarrow \frac{1}{1+x} &= \sum_{k=0}^{\infty} (-1)^k \cdot x^k = 1 - x + x^2 - x^3 - \cdots \end{aligned}$$

and then multiply both sides of this last equation by $5x$:

$$\frac{5x}{1+x} = 5x \cdot \sum_{k=0}^{\infty} (-1)^k \cdot x^k = 5x\left[1 - x + x^2 - x^3 - \cdots\right]$$
$$= \sum_{k=0}^{\infty} 5(-1)^k \cdot x^{k+1} = 5x - 5x^2 + 5x^3 - 5x^4 + \cdots$$

2. Use the first result from Example 3:

$$\ln(1-u) = -u - \frac{1}{2}u^2 - \frac{1}{3}u^3 - \frac{1}{4}u^4 - \cdots = -\sum_{k=0}^{\infty} \frac{1}{k+1} \cdot u^{k+1}$$

and substitute $u = -x$:

$$\ln(1-(-x)) = x - \frac{1}{2}x^2 + \frac{1}{3}x^3 - \frac{1}{4}x^4 + \cdots = \sum_{k=0}^{\infty} \frac{(-1)^k}{k+1} \cdot x^{k+1}$$

3. Multiply the result of Practice 2 by x^2:

$$x^2 \cdot \ln(1+x) = x^3 - \frac{1}{2}x^4 + \frac{1}{3}x^5 - \frac{1}{4}x^6 + \cdots = \sum_{k=0}^{\infty} \frac{(-1)^k}{k+1} \cdot x^{k+3}$$

and then apply term-by-term integration to get:

$$\int_0^{0.2} x^2 \cdot \ln(1+x)\, dx = \left[\sum_{k=0}^{\infty} \frac{(-1)^k}{(k+1)(k+4)} \cdot x^{k+4}\right]_0^{0.2}$$
$$= \left[\frac{1}{4}x^4 - \frac{1}{10}x^5 + \frac{1}{18}x^6 - \frac{1}{28}x^7 + \cdots\right]_0^{0.2}$$
$$= \frac{1}{4}(0.2)^4 - \frac{1}{10}(0.2)^5 + \frac{1}{18}(0.2)^6 - \frac{1}{28}(0.2)^7 + \cdots$$
$$\approx 0.000371$$

10.4 MacLaurin and Taylor Series

Having found several power series (all variations of the geometric series) that converge to familiar functions such as $\ln(1+x)$ and $\arctan(x)$, we turn our attention to more general functions, asking:

- Does the function have a power series expansion?
- Where does this power series converge?
- Where does this power series converge to the original function?

Once we determine a power series for a new function, we can use it to approximate function values, compute integrals and evaluate limits.

MacLaurin Series

Our first result tells us that *if* a function has a power series expansion, the coefficients of that power series must follow a familiar pattern.

Theorem:

If a function $f(x)$ has a power series representation $f(x) = \sum_{k=0}^{\infty} a_k x^k$ valid for $|x| < R$

then the coefficients of the power series must be:

$$a_k = \frac{f^{(k)}(0)}{k!}$$

Proof. Suppose that:

$$f(x) = \sum_{k=0}^{\infty} a_k x^k = a_0 + a_1 x + a_2 x^2 + a_3 x^3 + \cdots + a_n x^n + \cdots$$

Putting $x = 0$ into this equation yields:

$$f(0) = a_0 + a_1 \cdot 0 + a_2 \cdot 0^2 + a_3 \cdot 0^3 + \cdots + a_n \cdot 0^n + \cdots = a_0$$

so we know that $a_0 = f(0) = \dfrac{f^{(0)}(0)}{0!}$, proving the coefficient formula for $k = 0$. Differentiating the equation in the hypothesis yields:

We use the conventions that $0! = 1$ and that $f^{(0)}(x) = f(x)$.

$$f'(x) = a_1 + 2a_2 x + 3a_3 x^2 + \cdots + n \cdot a_n x^{n-1} + \cdots$$
$$\Rightarrow f'(0) = a_1 + 2a_2 \cdot 0 + 3a_3 \cdot 0^2 + \cdots + n \cdot a_n \cdot 0^{n-1} + \cdots = a_1$$

so that $a_1 = f'(0) = \dfrac{f'(0)}{1!}$, proving the coefficient formula for $k = 1$. Differentiating again yields:

$$f''(x) = 2a_2 + 3 \cdot 2a_3 x + \cdots + n(n-1) \cdot a_n x^{n-2} + \cdots$$
$$\Rightarrow \quad f''(0) = 2a_2 + 3 \cdot 2a_3 \cdot 0 + \cdots + n(n-1) \cdot a_n \cdot 0^{n-2} + \cdots = 2a_2$$

so that $a_2 = \dfrac{f''(0)}{2} = \dfrac{f''(0)}{2!}$, proving the coefficient formula for $k = 2$.
Differentiating yet again:

$$f'''(x) = 3 \cdot 2 \cdot 1 a_3 + \cdots + n(n-1)(n-1) \cdot a_n x^{n-3} + \cdots$$
$$\Rightarrow f'''(0) = 3 \cdot 2 \cdot 1 a_3 + \cdots + n(n-1)(n-2) \cdot a_n \cdot 0^{n-3} + \cdots = 3 \cdot 2 \cdot 1 a_3$$

so that $a_3 = \dfrac{f'''(0)}{3 \cdot 2 \cdot 1} = \dfrac{f'''(0)}{3!}$, proving the coefficient formula for $k = 3$.
In general:

$$f^{(n)}(x) = n(n-1)(n-2) \cdots 3 \cdot 2 \cdot 1 a_n + [\text{terms containing powers of } x]$$
$$\Rightarrow \quad f^{(n)}(0) = n(n-1)(n-2) \cdots 3 \cdot 2 \cdot 1 a_n + [0] = n! \cdot a_n$$

so that $a_n = \dfrac{f^{(n)}(0)}{n!}$. □

You may recognize this coefficient pattern from Section 8.7, where we called the polynomial:

$$P(x) = f(0) + \frac{f'(0)}{1!}x + \frac{f''(0)}{2!}x^2 + \frac{f'''(0)}{3!}x^3 + \cdots + \frac{f^{(n)}(0)}{n!}$$

the **MacLaurin polynomial** for $f(x)$. If we continue to add terms (forever) to this polynomial, we get the **MacLaurin series** for $f(x)$.

The **MacLaurin series** for $f(x)$ is:

$$\sum_{k=0}^{\infty} \frac{f^{(k)}(0)}{k!}x^k = f(0) + \frac{f'(0)}{1!}x + \frac{f''(0)}{2!}x^2 + \frac{f'''(0)}{3!}x^3 + \cdots + \frac{f^{(n)}(0)}{n!} + \cdots$$

This definition of a MacLaurin series and the preceding result about the form of its coefficients do *not* say that every function can be written as a power series. But *if* a function can be written as a power series, its coefficients must follow the above pattern. Fortunately, many important functions (such as $\sin(x)$ and e^x) can be written as power series.

The preceding proof also does *not* tell us where a MacLaurin series converges: we will need to apply techniques from Chapter 9 (typically the Ratio Test) to determine the interval of convergence for a MacLaurin series. Nor does the proof tell us that the series actually converges to the original function at any point (other than $x = 0$): to show that the series actually converges to the original function on its interval of convergence, we will need a result to be proved in Section 10.5.

Example 1. Find the MacLaurin series for $f(x) = \sin(x)$ and determine the radius of convergence of the series.

Solution. $f(x) = \sin(x) \Rightarrow f(0) = \sin(0) = 0 \Rightarrow a_0 = f(0) = 0.$ Computing the derivatives of $\sin(x)$:

$$f'(x) = \cos(x) \Rightarrow f'(0) = \cos(0) = 1 \Rightarrow a_1 = \frac{f'(0)}{1!} = \frac{1}{1} = 1$$
$$f''(x) = -\sin(x) \Rightarrow f''(0) = -\sin(0) = 0 \Rightarrow a_2 = \frac{f''(0)}{2!} = \frac{0}{2} = 0$$
$$f'''(x) = -\cos(x) \Rightarrow f'''(0) = -\cos(0) = -1 \Rightarrow a_3 = \frac{f'''(0)}{3!} = \frac{-1}{6}$$
$$f^{(4)}(x) = \sin(x) \Rightarrow f^{(4)}(0) = \sin(0) = 0 \Rightarrow a_4 = \frac{f^{(4)}(0)}{4!} = \frac{0}{24} = 0$$

This derivative pattern repeats, cycling through the values $1, 0, -1$ and 0 so that $a_5 = \frac{1}{5!}$, $a_6 = \frac{0}{6!} = 0$, $a_7 = \frac{-1}{7!}$, $a_8 = 0$, $a_9 = \frac{1}{9!}$, and so on:

Notice that the MacLaurin series for $\sin(x)$, an odd function, contains only odd powers of x. Also notice that it alternates between positive and negative coefficients.

$$\sin(x) = x - \frac{1}{3!}x^3 + \frac{1}{5!}x^5 - \frac{1}{7!}x^7 + \frac{1}{9!}x^9 - \cdots = \sum_{k=0}^{\infty} \frac{(-1)^k}{(2k+1)!}x^{2k+1}$$

To find the radius of convergence, apply the Ratio Test:

$$\left| \frac{\frac{(-1)^{k+1}}{(2(k+1)+1)!}x^{2(k+1)+1}}{\frac{(-1)^k}{(2k+1)!}x^{2k+1}} \right| = \left| \frac{x^{2k+3}}{(2k+3)!} \cdot \frac{(2k+1)!}{x^{2k+1}} \right| = \frac{(2k+1)!}{(2k+3)!}x^2$$
$$= \frac{x^2}{(2k+3)(2k+2)} \longrightarrow 0 < 1$$

for any value of x, so the interval of convergence is $(-\infty, \infty)$, hence the radius of convergence is $R = \infty$. ◀

You have two options here: proceed as in Example 1, or differentiate the result of Example 1.

Practice 1. Find the MacLaurin series for $f(x) = \cos(x)$ and determine the radius of convergence of the series.

Example 2. Find the MacLaurin series for $f(x) = e^x$ and determine the radius of convergence of the series.

Solution. With $f(x) = e^x$, $f'(x) = e^x \Rightarrow f''(x) = e^x \Rightarrow f'''(x) = e^x$ and in fact $f^{(k)}(x) = e^x$ for any integer $k \geq 0$, so $f^{(k)}(0) = e^0 = 1$ for all such k. Therefore the coefficients of the MacLaurin series for $f(x) = e^x$ all have the form $a_k = \frac{1}{k!}$, so we can write:

$$e^x = \sum_{k=0}^{\infty} \frac{1}{k!}x^k = 1 + x + \frac{1}{2!}x^2 + \frac{1}{3!}x^3 + \frac{1}{4!}x^4 + \cdots + \frac{1}{n!}x^n + \cdots$$

To find the radius of convergence for this series, apply the Ratio Test:

$$\left| \frac{\frac{1}{(k+1)!}x^{k+1}}{\frac{1}{k!}x^k} \right| = \frac{k!}{(k+1)!}|x| = \frac{|x|}{k+1} \longrightarrow 0$$

for any value of x, so the interval of convergence this MacLaurin series is $(-\infty, \infty)$ and its radius of convergence is $R = \infty$. ◀

Approximation Using MacLaurin Series

The MacLaurin series for $\sin(x)$ converges for every value of x (although we have not yet shown that it actually converges to $\sin(x)$ anywhere other than $x = 0$). The margin figure shows the graphs of $\sin(x)$ and the first few MacLaurin polynomials x, $x - \frac{1}{6}x^3$ and $x - \frac{1}{6}x^3 + \frac{1}{120}x^5$ for $-\pi \leq x \leq \pi$. While these low-degree MacLaurin polynomials appear to approximate $\sin(x)$ well near $x = 0$, the farther x gets from 0, the worse the approximation.

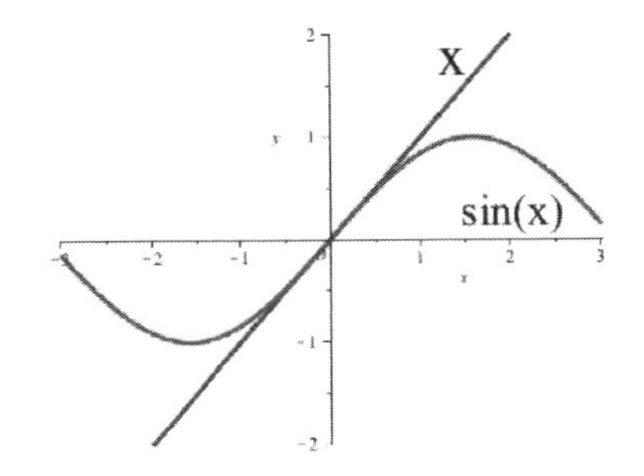

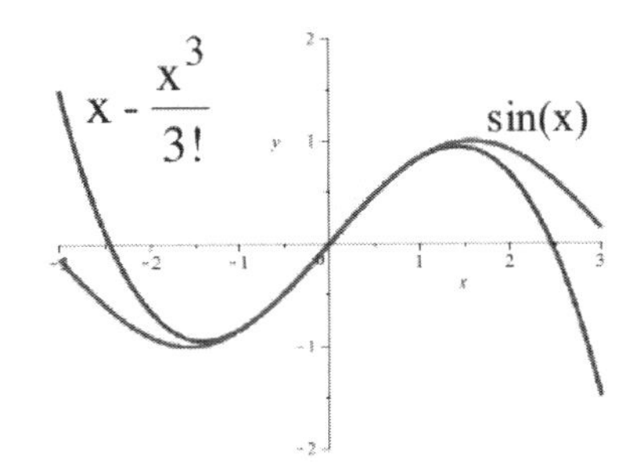

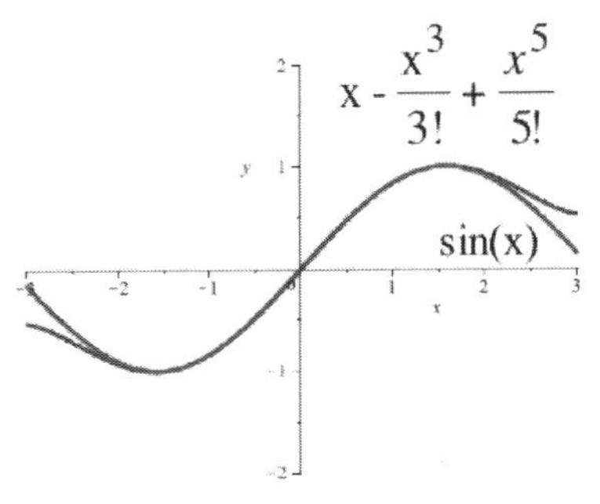

Example 3. Use a MacLaurin series to represent $\sin(0.5)$ as a numerical series. Approximate the value of $\sin(0.5)$ by computing the partial sum of the first three non-zero terms of this series and give a bound on the "error" between this approximation and the exact value of $\sin(0.5)$.

Solution. Putting $x = 0.5$ into the MacLaurin Series for $\sin(x)$:

$$\sin(0.5) = (0.5) - \frac{1}{3!}(0.5)^3 + \frac{1}{5!}(0.5)^5 - \frac{1}{7!}(0.5)^7 + \frac{1}{9!}(0.5)^9 - \cdots$$

so that:

$$\begin{aligned}\sin(0.5) &\approx (0.5) - \frac{1}{3!}(0.5)^3 + \frac{1}{5!}(0.5)^5 \\ &= \frac{1}{2} - \frac{1}{48} + \frac{1}{3840} \approx 0.479427083333\end{aligned}$$

Because the series in question is an alternating series, the difference between the approximation of $\sin(0.5)$ and the exact value of $\sin(0.5)$ is less than the absolute value of the next term in the alternating series:

$$\text{"error"} < \frac{1}{7!}(0.5)^7 = \frac{1}{645120} \approx 0.00000155$$

If you use the first four nonzero terms to approximate $\sin(0.5)$:

$$\sin(0.5) \approx (0.5) - \frac{1}{3!}(0.5)^3 + \frac{1}{5!}(0.5)^5 - \frac{1}{7!}(0.5)^7 \approx 0.47942553323$$

then the "error" is less than $\frac{1}{9!}(0.5)^9 = \frac{1}{185794560} \approx 5.4 \times 10^{-9}$. ◀

We were able to obtain a bound for the error in the approximation of $\sin(0.5)$ because the series in question was an alternating series, a type of series for which we have an error bound.

Many power series, however, are not alternating series. In Section 10.5 we will develop a general error bound for MacLaurin series.

Practice 2. Use the sum of the first two nonzero terms of the MacLaurin series for $\cos(x)$ to approximate the value of $\cos(0.2)$. Give a bound on the "error" between this approximation and the exact value of $\cos(0.2)$.

Practice 3. Evaluate the partial sums of the first six terms of the numerical series for $e = e^1$ and $\frac{1}{\sqrt{e}} = e^{-\frac{1}{2}}$. Compare these partial sums with the values your calculator gives.

The numerical series for e is not an alternating series, so we do not have a bound for the approximation yet. We will in the next section.

Calculator Note: When you press the buttons on your calculator to evaluate $\sin(0.5)$ or $\cos(0.2)$, the calculator does not look up the answer in a table. Instead, it has been programmed with series representations for sine, cosine and other functions, and it calculates a partial sum of an appropriate series to obtain a numerical answer. It adds enough terms so that the eight or nine digits shown on the display are (usually) correct. In Section 10.5 we examine these methods in more detail and consider how to determine the number of terms needed in the partial sum to achieve the desired number of accurate digits in the answer.

Substitution in MacLaurin Series

Now that we know MacLaurin series for $\sin(x)$, $\cos(x)$ and e^x, we can use techniques from Section 10.3 to quickly determine MacLaurin series representations of more complicated fucntions.

Example 4. Represent $\sin\left(x^3\right)$ and $\int \sin\left(x^3\right)\,dx$ as power series. Use the first three non-zero terms of the second series to approximate $\int_0^1 \sin\left(x^3\right)\,dx$ and obtain a bound for the "error."

Solution. Starting with the MacLaurin series for $\sin(u)$:

$$\sin(u) = u - \frac{1}{3!}u^3 + \frac{1}{5!}u^5 - \frac{1}{7!}u^7 + \cdots$$

put $u = x^3$ to get:

$$\begin{aligned}\sin\left(x^3\right) &= x^3 - \frac{1}{3!}\left(x^3\right)^3 + \frac{1}{5!}\left(x^3\right)^5 - \frac{1}{7!}\left(x^3\right)^7 + \cdots \\ &= x^3 - \frac{1}{3!}x^9 + \frac{1}{5!}x^{15} - \frac{1}{7!}x^{21} + \cdots\end{aligned}$$

Integrating this result term by term yields:

$$\begin{aligned}\int \sin\left(x^3\right)\,dx &= \int \left[x^3 - \frac{1}{6}x^9 + \frac{1}{120}x^{15} - \frac{1}{5040}x^{21} + \cdots\right]dx \\ &= C + \frac{1}{4}x^4 - \frac{1}{60}x^{10} + \frac{1}{1920}x^{16} - \frac{1}{110880}x^{22} + \cdots\end{aligned}$$

Approximating the definite integral:

$$\int_0^1 \sin\left(x^3\right)\,dx \approx \frac{1}{4} - \frac{1}{60} + \frac{1}{1920} \approx 0.2338542$$

A bound for the "error" between this approximation and the exact value of the definite integral is $\frac{1}{22 \cdot 7!} = \frac{1}{110880} \approx 0.0000090$. Using just one more term:

$$\int_0^1 \sin\left(x^3\right)\,dx \approx \frac{1}{4} - \frac{1}{60} + \frac{1}{1920} - \frac{1}{110880} \approx 0.233845515$$

gives an estimate within $\frac{1}{28 \cdot 9!} \approx 0.000000098$ of the exact value. ◀

Practice 4. Represent $x \cdot \cos\left(x^3\right)$ and $\int x \cdot \cos\left(x^3\right)\,dx$ as MacLaurin series. Use the first two nonzero terms of the second series to approximate $\int_0^{\frac{1}{2}} x \cdot \cos\left(x^3\right)\,dx$ and obtain a bound for the "error."

Taylor Series

The coefficients for a MacLaurin series (or polynomial) for a function $f(x)$ depend only on the values $f(0)$ and $f^{(k)}(0)$. As a consequence, the MacLaurin polynomials for $f(x)$ typically do a very good job of approximating the values of the original function near $x = 0$, as you can observe in this graph of $\sin(x)$ and its first few MacLaurin polynomials:

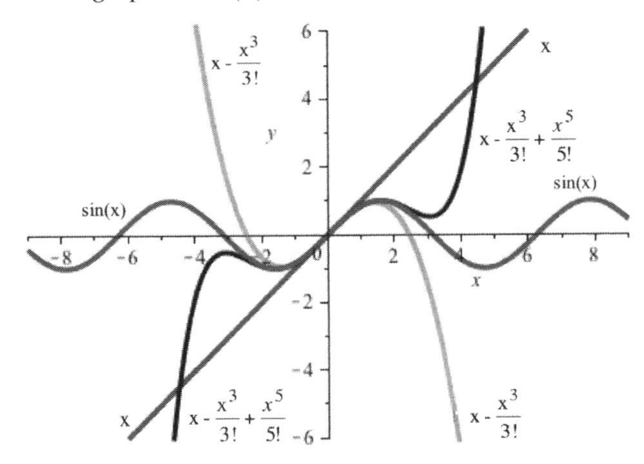

The figure above also demonstrates, however, that for values of x not close to 0, the values of the MacLaurin polynomials for $f(x)$ can be quite far from the values of the original function $f(x)$. Even though we know that the MacLaurin series for $\sin(x)$ converges for any value of x, for values of x far away from 0 we might need to add up hundreds of terms in order to achieve a good approximation of $\sin(x)$. For example, the first two nonzero terms of the MacLaurin series for $\sin(x)$ approximate $\sin(0.1)$ correctly to six decimal places, but you need 11 terms to approximate $\sin(5)$ with the same accuracy.

If you need to approximate a function with polynomials near a value away from $x = 0$, you can either use significantly more terms in a MacLaurin polynomial for that function, or you can "shift" the power series to center it at another value $x = c$. We call these "shifted" power series **Taylor series** and their partial sums **Taylor polynomials**. Typically the Taylor polynomials of a function centered at $x = c$ provide good approximations to $f(x)$ when x is close to c.

Theorem:

If a function $f(x)$ has a power series representation

$$f(x) = \sum_{k=0}^{\infty} a_k (x-c)^k \text{ valid for } |x-c| < R$$

then the coefficients of the power series must be:

$$a_k = \frac{f^{(k)}(c)}{k!}$$

This theorem generalizes the corresponding result about MacLaurin series from the beginning of this section. To prove the new theorem, merely replace 0 with c in the original proof.

The **Taylor series** for $f(x)$ centered at $x = c$ is:

$$\sum_{k=0}^{\infty} \frac{f^{(k)}(c)}{k!}(x-c)^k = f(c) + \frac{f'(c)}{1!}(x-c) + \frac{f''(c)}{2!}(x-c)^2 + \cdots$$

Taylor series and MacLaurin series were developed by the Scottish mathematician and astronomer James Gregory (1638–1675), but the results were not published until after his death. The English mathematician Brook Taylor (1685–1731) independently rediscovered these results and included them in a 1715 book. The Scottish mathematician and engineer Colin MacLaurin (1698–1746) quoted Taylor's work in his widely read 1742 *Treatise on Fluxions*, with the result that Taylor series centered at $c = 0$ became known as MacLaurin series.

A Maclaurin series is merely a Taylor series centered at $c = 0$, hence a MacLaurin series is a special case of Taylor series.

You should notice that the first term of the Taylor series for $f(x)$ is simply the value of the function f at the point $x = c$: it provides the best constant-function approximation of $f(x)$ near $x = c$. The sum of the first two terms of the Taylor series for a function $f(x)$:

$$f(c) + f'(c) \cdot (x-c)$$

resembles our usual formula in an equation of the line tangent to the graph of $f(x)$ at $x = c$ and gives the linear approximation of $f(x)$ near $x = c$ that we first examined in Chapter 2. The Taylor series formula extends these approximations to higher-degree polynomials, and the partial sums of the Taylor series provide higher-degree polynomial approximations of $f(x)$ near $x = c$.

Multiplying Power Series

We can add and subtract power series term by term, and you have already multiplied a power series by monomials such as x and x^2 to create new power series. Occasionally, you may find it useful to multiply a power series by another power series. The method for multiplying series is the same method used to multiply a polynomial by another polynomial, but it becomes very tedious to obtain more than the first few terms of the resulting product.

Example 5. Find the first five nonzero terms of the MacLaurin series for $\frac{1}{1-x} \cdot \sin(x)$.

Solution. Starting with the MacLaurin series for $\frac{1}{1-x}$ and $\sin(x)$:

$$\frac{1}{1-x} = 1 + x + x^2 + x^3 + x^4 + \cdots$$

$$\sin(x) = x - \frac{1}{6}x^3 + \frac{1}{120}x^5 - \cdots$$

multiply each term in the first series by each term in the second series:

$$\begin{aligned}\frac{1}{1-x}\cdot\sin(x) &= \left[1+x+x^2+x^3+\cdots\right]\cdot\left[x-\frac{1}{6}x^3+\frac{1}{120}x^5-\cdots\right]\\ &= 1\cdot\left[x-\frac{1}{6}x^3+\frac{1}{120}x^5-\cdots\right]+x\left[x-\frac{1}{6}x^3+\frac{1}{120}x^5-\cdots\right]+x^2\left[x-\frac{1}{6}x^3+\frac{1}{120}x^5-\cdots\right]\\ &\quad+x^3\left[x-\frac{1}{6}x^3+\frac{1}{120}x^5-\cdots\right]+x^4\left[x-\frac{1}{6}x^3+\frac{1}{120}x^5-\cdots\right]+\cdots\\ &= \left[x-\frac{1}{6}x^3+\frac{1}{120}x^5-\cdots\right]+\left[x^2-\frac{1}{6}x^4+\cdots\right]+\left[x^3-\frac{1}{6}x^5+\cdots\right]+\left[x^4\cdots\right]+\left[x^5\cdots\right]\\ &= x+x^2+\frac{5}{6}x^3+\frac{5}{6}x^4+\frac{101}{120}x^5+\cdots\end{aligned}$$

We know that *if* the function $\frac{1}{1-x}\cdot\sin(x)$ has a MacLaurin series, these must be the first five non-zero terms of that series. In order to show that this power series actually converges to $\frac{1}{1-x}\cdot\sin(x)$ on its interval of convergence, we need a theorem due to Abel (proved in more advanced courses) that says the product of two convergent power series also converges on their common interval of convergence. ◄

Practice 5. Find the first three nonzero terms of the MacLaurin series for $e^x\cdot\sin(x)$.

It is also possible to divide one power series by another power series using a procedure similar to "long division" of a polynomial by a polynomial, but we will not discuss that (quite tedious) process here.

Wrap-Up

The table below collects information about several important MacLaurin series developed in this section and the previous one.

$$e^x=\sum_{k=0}^{\infty}\frac{1}{k!}x^k=1+x+\frac{1}{2!}x^2+\frac{1}{3!}x^3+\cdots \qquad \text{valid on } (-\infty,\infty)$$

$$\sin(x)=\sum_{k=0}^{\infty}\frac{(-1)^k}{(2k+1)!}x^{2k+1}=x-\frac{1}{3!}x^3+\frac{1}{5!}x^5-\frac{1}{7!}x^7+\frac{1}{9!}x^9-\cdots \qquad \text{valid on } (-\infty,\infty)$$

$$\cos(x)=\sum_{k=0}^{\infty}\frac{(-1)^k}{(2k)!}x^{2k}=1-\frac{1}{2!}x^2+\frac{1}{4!}x^4-\frac{1}{6!}x^6+\frac{1}{8!}x^8-\cdots \qquad \text{valid on } (-\infty,\infty)$$

$$\frac{1}{1-x}=\sum_{k=0}^{\infty}x^k=1+x+x^2+x^3+x^4+\cdots \qquad \text{valid on } (-1,1)$$

$$\ln(1+x)=\sum_{k=1}^{\infty}(-1)^{k+1}\cdot\frac{x^k}{k}=x-\frac{1}{2}x^2+\frac{1}{3}x^3-\frac{1}{4}x^4+\cdots \qquad \text{valid on } (-1,1]$$

$$\arctan(x)=\sum_{k=0}^{\infty}(-1)^k\cdot\frac{x^{2k+1}}{2k+1}=x-\frac{1}{3}x^3+\frac{1}{5}x^5-\frac{1}{7}x^7+\cdots \qquad \text{valid on } [-1,1]$$

10.4 Problems

In Problems 1–14, use the MacLaurin series coefficient formula to find the first several terms of the MacLaurin series for the given function, then compare the result with the series representation found in Section 10.3.

1. $\ln(1+x)$
2. $\ln(1+x)$
3. $\arctan(x)$
4. $\dfrac{1}{1-x}$

In Problems 5–8, find the first several terms of the MacLaurin series for the given function.

5. $\cos(x)$ to the x^6 term
6. $\tan(x)$ to the x^5 term
7. $\sec(x)$ to the x^4 term
8. e^{3x} to the x^4 term

In Problems 9–13, find the first several terms of the Taylor series for the given function centered at the given point c.

9. $\ln(x)$ for $c = 1$
10. $\sin(x)$ for $c = \pi$
11. $\sin(x)$ for $c = \dfrac{\pi}{2}$
12. $\sqrt{x}$ for $c = 1$
13. $\sqrt{x}$ for $c = 9$

In Problems 14–17, use the first three nonzero terms of a MacLaurin series to approximate the given numerical values. Then compare the approximation with the value your calculator provides.

14. $\sin(0.1)$, $\sin(0.2)$, $\sin(0.5)$, $\sin(1)$ and $\sin(2)$
15. $\cos(0.1)$, $\cos(0.2)$, $\cos(0.5)$, $\cos(1)$ and $\cos(2)$
16. $\ln(1.1)$, $\ln(1.2)$, $\ln(1.3)$, $\ln(2)$ and $\ln(3)$
17. $\arctan(0.1)$, $\arctan(0.2)$, $\arctan(0.5)$, $\arctan(1)$, $\arctan(2)$

In Problems 18–23, find the first three nonzero terms of a power series for the integral.

18. $\displaystyle\int \cos\left(x^2\right)\,dx$
19. $\displaystyle\int \sin\left(x^2\right)\,dx$
20. $\displaystyle\int \cos\left(x^3\right)\,dx$
21. $\displaystyle\int \sin\left(x^3\right)\,dx$
22. $\displaystyle\int e^{x^2}\,dx$
23. $\displaystyle\int e^{-x^2}\,dx$
24. $\displaystyle\int e^{x^3}\,dx$
25. $\displaystyle\int e^{-x^3}\,dx$
26. $\displaystyle\int \ln(x)\,dx$
27. $\displaystyle\int x\sin(x)\,dx$
28. $\displaystyle\int x\ln(x)\,dx$
29. $\displaystyle\int x^2\sin(x)\,dx$

In Problems 30–37, use a series representation to help compute the limit.

30. $\displaystyle\lim_{x\to 0}\frac{1-\cos(x)}{x}$
31. $\displaystyle\lim_{x\to 0}\frac{1-\cos(x)}{x^2}$
32. $\displaystyle\lim_{x\to 0}\frac{\ln(x)}{x-1}$
33. $\displaystyle\lim_{x\to 0}\frac{1-e^x}{x}$
34. $\displaystyle\lim_{x\to 0}\frac{1+x-e^x}{x^2}$
35. $\displaystyle\lim_{x\to 0}\frac{\sin(x)}{x}$
36. $\displaystyle\lim_{x\to 0}\frac{x-\sin(x)}{x^3}$
37. $\displaystyle\lim_{x\to 0}\frac{x-\frac{1}{6}x^3-\sin(x)}{x^5}$
38. Use MacLaurin series for e^x and e^{-x} to find a series representation for $\cosh(x) = \dfrac{e^x+e^{-x}}{2}$.
39. Use MacLaurin series for e^x and e^{-x} to find a series representation for $\sinh(x) = \dfrac{e^x-e^{-x}}{2}$.
40. Use results from the previous two problems to show that $\mathbf{D}(\cosh(x)) = \sinh(x)$.
41. Use results from previous problems to show that $\mathbf{D}(\sinh(x)) = \cosh(x)$.

Euler's Formula: So far we have only discussed series involving real numbers, but sometimes it is useful to replace the variable in a power series with a complex number. Problems 42–44 ask you to make such a substitution and then to obtain and use one of the most famous formulas in mathematics: Euler's formula. Recall that $i = \sqrt{-1}$ is called the **complex unit** and that its powers follow the pattern $i^2 = -1$, $i^3 = (i^2)(i) = -i$, $i^4 = (i^2)^2 = (-1)^2 = 1$, $i^5 = (i^4)(i) = i$, and so on.

42. (a) Substitute $x = i\theta$ into the MacLaurin series for e^x to obtain a series for $e^{i\theta}$.

    (b) Simplify each power of i to rewrite the series for $e^{i\theta}$.

(c) Sort the terms in the simplified series into those terms that do not contain i and those terms that do contain i. Then rewrite the series for $e^{i\theta}$ in the form:

$$e^{i\theta} = [\text{terms that do not contain } i] + i \cdot [\text{terms that do contain } i]$$

(d) You should recognize the sum in each bracket as the MacLaurin series for an elementary function. Rewrite the series for $e^{i\theta}$ as:

$$e^{i\theta} = [\text{function of } \theta] + i \cdot [\text{other function of } \theta]$$

43. In Problem 42 you should have obtained the result:

$$e^{i\theta} = \cos(\theta) + i \cdot \sin(\theta)$$

Use Euler's formula to compute the values of $e^{i\left(\frac{\pi}{2}\right)}$ and $e^{\pi i}$.

44. Use Euler's formula to show that $e^{\pi i} + 1 = 0$. This is one of the most remarkable and beautiful formulas in mathematics because it connects five of the most fundamental constants: the additive identity 0, the multiplicative identity 1, the complex unit i and the two most commonly used transcendental numbers (π and e) in a simple yet non-obvious way.

The Binomial Theorem

You have probably seen the pattern for expanding $(1+x)^n$ where n is a non-negative integer:

$$\begin{aligned}
(1+x)^0 &= 1 \\
(1+x)^1 &= 1 + x \\
(1+x)^2 &= 1 + 2x + x^2 \\
(1+x)^3 &= 1 + 3x + 3x^2 + x^3 \\
(1+x)^4 &= 1 + 4x + 6x^2 + 4x^3 + x^4 \\
(1+x)^5 &= 1 + 5x + 10x^2 + 10x^3 + 5x^4 + x^5
\end{aligned}$$

using either Pascal's triangle (see margin) or binomial coefficients:

$$\binom{n}{k} = \frac{n(n-1)(n-2)\cdots(n-k+1)}{k!} = \frac{n!}{k! \cdot (n-k)!}$$

for any positive integers n and k with $k \le n$, defining $\binom{n}{0} = 1$. Binomial coefficients allow us to write the expansion of $(1+x)^n$ for non-negative integer powers n in a very compact way:

$$(1+x)^n = \sum_{k=0}^{n} \binom{n}{k} x^k$$

1
1 1
1 2 1
1 3 3 1
1 4 6 4 1

Notice that each entry in the interior of **Pascal's triangle** is the sum of the two numbers immediately above it.

When n is a positive integer, $(1+x)^n$ expands into a polynomial of degree n, but what happens when n is a negative integer? Or a non-integer? Newton himself investigated this question, leading him to a general pattern that allowed him to quickly write a MacLaurin series expansion for $(1+x)^m$ when m is any real number:

Binomial Series Theorem:

If m is any real number and $|x| < 1$

then $(1+x)^m = \sum_{k=0}^{\infty} \binom{m}{k} x^k$ where:

$$\binom{m}{k} = \frac{m(m-1)(m-2)\cdots(m-k+1)}{k!}$$

As before, we define: $\binom{m}{0} = 1$

The remaining problems guide you through an investigation and (the idea behind a) proof of this theorem.

45. Calculate $\binom{3}{0}$, $\binom{3}{1}$, $\binom{3}{2}$ and $\binom{3}{3}$, then verify that:
    (a) they agree with the entries in the third row of Pascal's triangle.
    (b) they agree with the coefficients in the expansion of $(1+x)^3$.
46. Calculate $\binom{4}{0}$, $\binom{4}{1}$, $\binom{4}{2}$, $\binom{4}{3}$ and $\binom{4}{4}$, then verify that:
    (a) they agree with the entries in the fourth row of Pascal's triangle.
    (b) they agree with the coefficients in the expansion of $(1+x)^4$.
47. Determine the first five terms of the MacLaurin series for $(1+x)^{\frac{5}{2}}$.
48. Determine the first five terms of the MacLaurin series for $(1+x)^{-\frac{3}{2}}$.
49. Determine the first five terms of the MacLaurin series for $(1+x)^{-\frac{1}{2}}$ and use this result to find the first five non-zero terms in the MacLaurin series for $\arcsin(x)$.
50. Use the first result from the preceding problem to approximate $\sqrt{2}$.
51. Determine the first four terms of the MacLaurin series for $(1+x)^m$. (This is the beginning of a proof of the Binomial Series Theorem.)

10.4 Practice Answers

1. Differentiating the MacLaurin series for $\sin(x)$ yields:

$$\begin{aligned}\cos(x) &= \mathbf{D}\left(\sin(x)\right) \\ &= \mathbf{D}\left(x - \frac{1}{3!}x^3 + \frac{1}{5!}x^5 - \frac{1}{7!}x^7 + \frac{1}{9!}x^9 - \frac{1}{11!}x^{11} + \cdots\right) \\ &= 1 - \frac{3}{3!}x^2 + \frac{5}{5!}x^4 - \frac{7}{7!}x^6 + \frac{9}{9!}x^8 - \frac{11}{11!}x^{10} + \cdots \\ &= 1 - \frac{1}{2!}x^2 + \frac{1}{4!}x^4 - \frac{1}{6!}x^6 + \frac{1}{8!}x^8 - \frac{1}{10!}x^{10} + \cdots \\ &= \sum_{k=0}^{\infty} \frac{(-1)^k}{(2k)!}x^{2k}\end{aligned}$$

2. Putting $x = 0.2$ into the MacLaurin series obtained in Practice 1:

$$\cos(0.2) \approx 1 - \frac{1}{2!}(0.2)^2 = 1 - \frac{0.04}{2} = 0.98$$

Because the full MacLaurin series:

$$\cos(0.2) = 1 - \frac{1}{2!}(0.2)^2 + \frac{1}{4!}(0.2)^4 - \frac{1}{6!}(0.2)^6 + \cdots$$

is a convergent alternating series, the "error" when approximating $\cos(0.2)$ by 0.98 is no bigger than the absolute value of the next term in the series, which is:

$$\frac{1}{4!}(0.2)^4 = \frac{0.0016}{24} \approx 0.000067$$

so that $|\cos(0.2) - 0.98| < 0.000067$.

In fact, $\cos(0.2) \approx 0.9800665778$.

3. Starting with the MacLaurin series:

$$e^x = 1 + x + \frac{1}{2!}x^2 + \frac{1}{3!}x^3 + \frac{1}{4!}x^4 + \frac{1}{5!}x^5 + \cdots$$

and using the first six terms with $x = 1$:

$$e = e^1 \approx 1 + 1 + \frac{1}{2!} + \frac{1}{3!} + \frac{1}{4!} + \frac{1}{5!} \approx 2.71666666666$$

Your calculator should report the approximation: $e^1 \approx 2.718281828$

To approximate $\frac{1}{\sqrt{e}}$, substitute $x = -\frac{1}{2}$:

$$\begin{aligned}e^{-\frac{1}{2}} &\approx 1 - \frac{1}{2} + \frac{1}{2!}\left(-\frac{1}{2}\right)^2 + \frac{1}{3!}\left(-\frac{1}{2}\right)^3 + \frac{1}{4!}\left(-\frac{1}{2}\right)^4 + \frac{1}{5!}\left(-\frac{1}{2}\right)^5 \\ &\approx 0.6065104167\end{aligned}$$

Your calculator should report the approximation: $e^{-\frac{1}{2}} \approx 0.6065306597$

4. Substitute $u = x^3$ into the MacLaurin series for $\cos(u)$ and multiply the result by x:

$\cos(u) = 1 - \frac{1}{2!}u^2 + \frac{1}{4!}u^4 - \frac{1}{6!}u^6 + \frac{1}{8!}u^8 - \cdots$

$$x \cdot \cos\left(x^3\right) = x \cdot \left[1 - \frac{1}{2!}(x^3)^2 + \frac{1}{4!}(x^3)^4 - \frac{1}{6!}(x^3)^6 + \frac{1}{8!}(x^3)^8 - \cdots\right]$$
$$= x - \frac{1}{2!}x^7 + \frac{1}{4!}x^{13} - \frac{1}{6!}x^{19} + \frac{1}{8!}x^{25} - \cdots$$

then integrate term by term:

$$\int x \cdot \cos\left(x^3\right)\, dx = C + \frac{1}{2}x^2 - \frac{1}{8 \cdot 2!}x^8 + \frac{1}{14 \cdot 4!}x^{14} - \frac{1}{20 \cdot 6!}x^{20} + \cdots$$

and use the first two nonzero terms of this antiderivative to estimate the value of the definite integral:

$$\int_0^{\frac{1}{2}} x \cdot \cos\left(x^3\right)\, dx \approx \left[\frac{1}{2}x^2 - \frac{1}{8 \cdot 2!}x^8\right]_0^{\frac{1}{2}} = \frac{1}{8} - \frac{1}{8 \cdot 2 \cdot 2^8} = \frac{511}{4096}$$

or about 0.124755859375. Because the series for the exact value of the integral is an alternating series, the "error" is no bigger than:

$$\frac{1}{14 \cdot 4!}\left(\frac{1}{2}\right)^{14} \approx 0.000000182$$

5. Multiply the two MacLaurin series:

$$e^x = 1 + x + \frac{1}{2}x^2 + \frac{1}{6}x^3 + \cdots$$
$$\sin(x) = x - \frac{1}{6}x^3 + \frac{120^5}{x} + \cdots$$

to get a MacLaurin series for $e^x \cdot \sin(x)$:

$$\left[1 + x + \frac{1}{2}x^2 + \frac{1}{6}x^3 + \cdots\right] \cdot \left[x - \frac{1}{6}x^3 + \frac{120^5}{x} + \cdots\right]$$
$$= 1 \cdot \left[x - \frac{1}{6}x^3 + \frac{1}{120}x^5 + \cdots\right] + x \cdot \left[x - \frac{1}{6}x^3 + \frac{120^5}{x} + \cdots\right]$$
$$+ \frac{1}{2}x^2 \cdot \left[x - \frac{1}{6}x^3 + \frac{120^5}{x} + \cdots\right]$$
$$+ \frac{1}{6}x^3 \cdot \left[x - \frac{1}{6}x^3 + \frac{120^5}{x} + \cdots\right] \cdots$$
$$= \left[x - \frac{1}{6}x^3 + \frac{1}{120}x^5 + \cdots\right] + \left[x^2 - \frac{1}{6}x^4 + \cdots\right]$$
$$+ \left[\frac{1}{2}x^3 - \frac{1}{12}x^5 + \cdots\right] + \left[\frac{1}{6}x^4 + \cdots\right] + \cdots$$
$$= x + x^2 + \frac{1}{3}x^3 + 0x^4 - \frac{3}{40}x^5 + \cdots$$

so the sum of the first three nonzero terms is $x + x^2 + \frac{1}{3}x^3$.

10.5 *Approximation Using Taylor Polynomials*

If a function has a power series representation, we now have a formula to determine the coefficients of that power series. Using techniques from Chapter 9, we can then find the interval of convergence for the power series. And *if* evaluating a power series at a point results in an alternating numerical series, we can even use the Estimation Bound for Alternating Series to get a bound on the "error" between a partial sum approximation and the exact value of the series:

$$|(\text{exact value}) - (\text{approximation})| < |\text{next term in the series}|$$

If evaluating the power series at a point does not result in an alternating numerical series, we do not yet have a bound on the size of the error of the approximation.

In this section we obtain a bound on the error when approximating a Taylor series for any $f(x)$ with a corresponding Taylor polynomial:

$$\text{"error"} = |(\text{exact value of } f(x)) - (\text{approximation of } f(x))|$$

The bound we get is valid even if the Taylor series is not an alternating series, and the pattern for the error bound looks very much like the next term in the series (the first unused term in the partial sum of the Taylor series). In computer and calculator applications, this error bound can help us work efficiently by allowing us to use only the number of terms we really need.

As a very important bonus, this error bound will allow us to show that a function is equal to its Taylor series on its interval of convergence (for most "well-behaved" functions).

Taylor Polynomials

For a function $f(x)$, the n-th degree **Taylor Polynomial** (centered at $x = c$) is:

$$\begin{aligned} P_n(x) &= \sum_{k=0}^{n} \frac{f^{(k)}(c)}{k!}(x-c)^k \\ &= f(c) + f'(c)\cdot(x-c) + \frac{f''(c)}{2!}(x-c)^2 + \cdots + \frac{f^{(n)}(c)}{n!}(x-c)^n \end{aligned}$$

Example 1. Write the first four Taylor Polynomials, $P_0(x)$ through $P_3(x)$, centered at $x = 0$ for e^x, then graph them for $-1 < x < 1$.

Solution. With $c = 0$, a Taylor polynomial is a MacLaurin polynomial and the MacLaurin series for e^x is:

$$e^x = 1 + x + \frac{1}{2!}x^2 + \frac{1}{3!}x^3 + \frac{1}{4!}x^4 + \cdots = \sum_{k=0}^{\infty} \frac{1}{k!}x^k$$

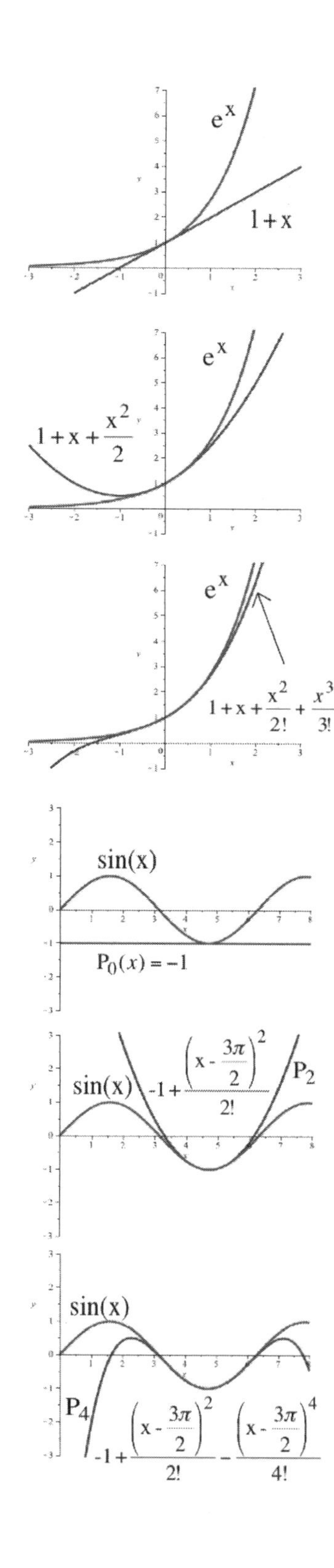

so $P_0(x) = 1$, $P_1(x) = 1 + x$, $P_2(x) = 1 + x + \frac{1}{2!}x^2$ and:

$$P_3(x) = 1 + x + \frac{1}{2!}x^2 + \frac{1}{3!}x^3$$

Graphs of e^x, $P_1(x)$, $P_2(x)$ and $P_3(x)$ appear in the margin. ◀

Practice 1. Write the Taylor polynomials $P_0(x)$, $P_2(x)$ and $P_4(x)$ centered at $x = 0$ for $\cos(x)$, then graph them for $-\pi < x < \pi$. (What are the Taylor polynomials $P_1(x)$ and $P_3(x)$?)

When we center a Taylor polynomial at $x = c \neq 0$, the Taylor polynomial approximates the function well for values of x near c.

Example 2. Write the Taylor polynomials $P_0(x)$, $P_2(x)$ and $P_4(x)$ centered at $x = \frac{3\pi}{2}$ for $\sin(x)$, then graph them for $0 < x < 8$.

Solution. The Taylor series centered at $x = \frac{3\pi}{2}$ for $\sin(x)$ is:

$$\sin(x) = -1 + \frac{1}{2!}\left(x - \frac{3\pi}{2}\right)^2 - \frac{1}{4!}\left(x - \frac{3\pi}{2}\right)^4 + \frac{1}{6!}\left(x - \frac{3\pi}{2}\right)^6 - \cdots$$

so $P_0(x) = -1$, $P_2(x) = -1 + \frac{1}{2}\left(x - \frac{3\pi}{2}\right)^2$ and:

$$P_4(x) = -1 + \frac{1}{2}\left(x - \frac{3\pi}{2}\right)^2 - \frac{1}{24}\left(x - \frac{3\pi}{2}\right)^4$$

Graphs of $\sin(x)$, $P_0(x)$, $P_2(x)$ and $P_4(x)$ appear in the margin. ◀

Practice 2. Write the Taylor polynomials $P_0(x)$, $P_1(x)$ and $P_3(x)$ centered at $x = \frac{\pi}{2}$ for $\cos(x)$, then graph them for $-1 < x < 4$.

Looking at the graphs from the preceding Examples, you should notice that how well a function can be approximated by its Taylor polynomial appears to depend on:

- The number of terms, n, in the Taylor polynomial.
- How close the point of approximation, x, is to the the center of the Taylor series, c.

The Remainder

Approximation of functions using Taylor polynomials is a useful tool, but in many applied situations you want to know how good the approximation is, or how many terms of a series you require to obtain a needed level of accuracy. If two terms of a series provide the desired accuracy for your application, it is a waste of time, resources and money to use 100 terms. On the other hand, sometimes even 100 terms may not provide the accuracy you need. Fortunately, it is possible to obtain a guarantee on how close a particular Taylor polynomial approximates an exact value. Then you can work efficiently and use the smallest possible number of terms in a Taylor polynomial. The next theorem gives a pattern for the size of the "error" in a Taylor polynomial approximation.

Taylor's Formula with Remainder

If f has $n+1$ derivatives in an interval I containing c, and x is in I,

then there is a number z, strictly between c and x, so that $f(x) = P_n(x) + R_n(x)$ where:

$$R_n(x) = \frac{f^{(n+1)}(z)}{(n+1)!} \cdot (x-c)^{n+1}$$

This result says that $f(x)$ is equal to its n-th degree Taylor polynomial plus a **remainder**, and the remainder $R_n(x)$ has the form given in the theorem. Notice that the formula for $R_n(x)$ looks very much like the pattern for the $(n+1)$-st term of the Taylor series for $f(x)$, except that it involves $f^{(n+1)}(z)$ instead of $f^{(n+1)}(c)$.

This particular formula for $R_n(x)$ is called the **Lagrange form** of the remainder, named for the French-Italian mathematician and astronomer Joseph-Louis Lagrange (1736–1813).

The main idea of the proof of the Taylor's Formula with Remainder is straightforward, but the details are somewhat technical, so we will set aside the proof for the moment.

Notice that the formula for the remainder, $\frac{f^{(n+1)}(z)}{(n+1)!} \cdot (x-c)^{n+1}$, involves three pieces: $(n+1)!$, $(x-c)^{n+1}$ and $f^{(n+1)}(z)$ for some z (strictly) between x and c. Notice also that:

- When you make the number of terms, n, in the Taylor polynomial bigger, the $(n+1)!$ in the denominator of the formula becomes bigger, making the remainder smaller.

Usually. Both $f^{(n+1)}(z)$ and $(x-c)^{n+1}$ also depend on n and affect the size of the remainder.

- When you make the point of approximation, x, closer to the the center of the Taylor series, c, the factor $(x-c)^{n+1}$ becomes smaller, making the remainder smaller.

- With $n = 0$, Taylor's Formula becomes the Mean Value Theorem:

$$f(x) = f(c) + f'(z) \cdot (x-c) \Rightarrow f'(z) = \frac{f(x) - f(c)}{x-c}$$

We used Rolle's Theorem to prove the Mean Value Theorem, so we will rely on repeated applications of Rolle's Theorem to prove Taylor's Formula.

This last fact allows us to think of Taylor's Formula with Remainder as a generalization of the Mean Value Theorem.

Applying the Taylor Remainder Formula

In practice, you will typically use Taylor's Formula with Remainder in one of two ways:

- You know the Taylor polynomial $P_n(x)$ for $f(x)$, so you know x, c and n, allowing you to evaluate $(n+1)!$ and $(x-c)^{n+1}$ exactly. That leaves the piece $f^{(n+1)}(z)$ for some z between x and c. If you can find a bound for the value of $\left|f^{(n+1)}(z)\right|$ for all z between x and c, then you can use this bound with the known values of $(n+1)!$ and $(x-c)^{n+1}$ to obtain a *bound* for the remainder term $R_n(x)$.
- Someone tells you the amount of acceptable "error," so you know the values of x, c and $R_n(x)$. You then need to find a value of n that guarantees the required accuracy.

Corollary: A Bound for the Remainder $R_n(x)$

If f has $n+1$ derivatives in an interval I containing c, x is in I, and $\left|f^{(n+1)}(z)\right| \le M$ for all z between x and c,

then "error" $= |f(x) - P_n(x)| = |R_n(x)| \le \dfrac{M}{(n+1)!} \cdot |x-c|^{n+1}$

Example 3. You need to approximate the values of e^x using the MacLaurin polynomial $P_3(x) = 1 + x + \frac{1}{2}x^2 + \frac{1}{6}x^3$. Find a bound for the "error" of the approximation, $R_3(x)$, if x is in the interval:

(a) $[-1,1]$ (b) $[-3,2]$ (c) $[-0.2, 0.3]$

Solution. You know that $f(x) = e^x$, $c = 0$ (corresponding to a Maclaurin series), $n = 3 \Rightarrow (n+1)! = 4! = 24$ and $f^{(n+1)}(x) = f^{(4)}(x) = e^x$.

(a) For x in the interval $[-1,1]$:

$$\left|(x-c)^{n+1}\right| = \left|x^4\right| \le |1|^4 = 1 \quad \text{and} \quad \left|f^{(n+1)}(x)\right| = |e^x| \le e^1$$

because e^x is increasing on $[-1,1]$. A "crude" but "easy to use" bound for e^1 is $e^1 < 3^1 = 3 = M$. Then:

$$|R_3(x)| < \frac{M}{(n+1)!} \cdot |x-c|^{n+1} < \frac{3}{24} \cdot 1 = 0.125$$

When $-1 < x < 1$, $P_3(x) = 1 + x + \frac{1}{2}x^2 + \frac{1}{6}x^3$ is within 0.125 of e^x.

For a more precise approximation of e, you can use:

$$e^1 < 2.72^1 = 2.72$$

resulting in the bound:

$$|R_3(x)| < \frac{2.72}{24} = \frac{17}{150} \approx 0.1133$$

(b) For x in the interval $[-3,2]$:

$$\left|(x-c)^{n+1}\right| = \left|x^4\right| \le \left|(-3)^4\right| = 81 \quad \text{and} \quad \left|f^{(n+1)}(x)\right| = |e^x| \le e^2$$

because e^x is increasing on $[-3,2]$. A "crude" but "easy to use" bound for e^2 is $e^2 < 3^2 = 9 = M$. Then:

$$|R_3(x)| < \frac{M}{(n+1)!} \cdot |x-c|^{n+1} < \frac{9}{24} \cdot 81 = 30.375$$

Obviously you cannot have much confidence when using $P_3(x)$ to approximate e^x on the interval $[-3,2]$.

For a more precise bound, you can use:

$$e^2 < 2.72^2 < 7.4$$

resulting in:

$$|R_3(x)| < \frac{(2.72)^2}{24} \cdot 81 < 24.9696$$

(c) For x in the interval $[-0.2, 0.3]$:

$$\left|(x-c)^{n+1}\right| = \left|x^4\right| \leq \left|(0.3)^4\right| = 0.0081 \text{ and } \left|f^{(n+1)}(x)\right| = |e^x| \leq e^{0.3}$$

because e^x is increasing on $[-0.2, 0.3]$. A good bound for $e^{0.3}$ is $e^{0.3} < 2.72^{0.3} < 1.4 = M$. Then:

$$|R_3(x)| < \frac{M}{(n+1)!} \cdot |x-c|^{n+1} < \frac{1.4}{24} \cdot 0.0081 = 0.0004725$$

When $-0.2 < x < 0.3$, $P_3(x)$ is within 0.0004725 of e^x.

When the interval is small, you can be confident that $P_3(x)$ provides a good approximation of e^x, but as the interval grows, so does your bound on the remainder. ◀

To guarantee a good approximation on a larger interval, you typically need $(n+1)!$ to be larger, so you need to use a higher-degree Taylor Polynomial $P_n(x)$.

Practice 3. Find a value of n to guarantee that $P_n(x)$ is within 0.001 of e^x for x in the interval $[-3,2]$.

Example 4. You need to approximate the values of $f(x) = \sin(x)$ on the interval $\left[-\frac{\pi}{2}, \frac{\pi}{2}\right]$ with an error less that 10^{-10}. How many terms of the MacLaurin series for $\sin(x)$ do you need?

Solution. For every value of n, $\left|f^{(n+1)}(x)\right|$ is either equal to $|\sin(x)|$ or $|\cos(x)|$, so $M = 1$ works as a bound for $\left|f^{(n+1)}(z)\right|$. Hence:

$$\text{"error"} = |R_n(x)| < \frac{1}{(n+1)!} \cdot |x-0|^{n+1} \leq \frac{\left(\frac{\pi}{2}\right)^{n+1}}{(n+1)!}$$

so we need to find a value of n such that:

$$\frac{\left(\frac{\pi}{2}\right)^{n+1}}{(n+1)!} < 10^{-10}$$

holds. A bit of numerical experimentation on a calculator (see margin) shows that $n+1 = 16$ works, so we can take $n = 15$ and use:

$$\frac{\left(\frac{\pi}{2}\right)^{14}}{14!} \approx 6.39 \times 10^{-9}$$

$$\frac{\left(\frac{\pi}{2}\right)^{15}}{15!} \approx 6.69 \times 10^{-10}$$

$$\frac{\left(\frac{\pi}{2}\right)^{16}}{16!} \approx 6.57 \times 10^{-11}$$

$$P_{15}(x) = x - \frac{1}{3!}x^3 + \frac{1}{5!}x^5 - \frac{1}{7!}x^7 + \frac{1}{9!}x^9 - \frac{1}{11!}x^{11} + \frac{1}{13!}x^{13} - \frac{1}{15!}x^{15}$$

to approximate $\sin(x)$. If $-\frac{\pi}{2} < x < \frac{\pi}{2}$, then $|P_{15}(x) - \sin(x)| < 10^{-10}$ will hold. ◀

Practice 4. How many terms of the MacLaurin series for e^x do you need in order to approximate e^x to within 10^{-10} for $0 \le x \le 1$?

Proving Taylor's Formula

As observed previously, Taylor's Formula generalizes the Mean Value Theorem, so we will apply Rolle's Theorem repeatedly to prove it. Given values of x, n and c, we need to find a z strictly between x and c so that:

$$f(x) - P_n(x) = R_n(x) = \frac{f^{(n+1)}(z)}{(n+1)!} \cdot (x-c)^{n+1}$$

Proof. Given fixed values of x, n and c, define α so that:

$$\alpha = \frac{R_n(x)}{(x-c)^{n+1}} \quad \Rightarrow \quad R_n(x) = \alpha \cdot (x-c)^{n+1}$$

We now need to show that $\alpha = \dfrac{f^{(n+1)}(z)}{(n+1)!}$ for an appropriate value of z. Next, define a new function $g(t)$ as:

$$g(t) = f(t) - P_n(t) - \alpha \cdot (t-c)^{n+1}$$

Because $f(c) = P_n(c)$, we know that:

$$g(c) = f(c) - P_n(c) - (c-c)^{n+1} = 0$$

Recall that we have constructed $P_n(x)$ so the values of it (and its first n derivatives) agree with the values of $f(x)$ (and its first n derivatives) at $x = c$.

and because $f^{(k)}(c) = P_n^{(k)}(c)$ for $1 \le k \le n$, we also know that:

$$g^{(k)}(c) = f^{(k)}(c) - P_n^{(k)}(c) - (n+1)(n)\cdots(n-k+2)(0)^{n-k+1} = 0$$

for $1 \le k \le n$. Furthermore, we know that:

$$g(x) = f(x) - P_n(x) - \alpha \cdot (x-c)^{n+1} = f(x) - P_n(x) - R_n(x) = 0$$

Refer to Section 3.2 to review the statement of Rolle's Theorem.

Now apply Rolle's Theorem to $g(t)$: we know that $g(c) = 0$ and $g(x) = 0$, so there is a number z_1 between c and x such that $g'(z_1) = 0$.

Next apply Rolle's Theorem to $g'(t)$: we know that $g'(c) = 0$ and $g'(z_1) = 0$, so there is a z_2 between c and z_1 with $g''(z_2) = 0$.

Keeping track of these z_k's, we know that: $c < z_{n+1} < z_n < \cdots < z_3 < z_2 < z_1 < x$

Continue applying Rolle's Theorem to $g''(t)$, $g'''(t)$, ..., $g^{(n)}(t)$ to get similar numbers $z_3, z_4, \ldots, z_{n+1}$. Let $z = z_{n+1}$. We know that z is strictly between c and x, and that $g^{(n+1)}(z) = 0$. But:

$$g^{(n+1)}(t) = f^{(n+1)}(t) - 0 - (n+1)! \cdot \alpha$$

so using the fact that $g^{(n+1)}(z) = 0$ we have:

$$f^{(n+1)}(z) = (n+1)! \cdot \alpha \Rightarrow \alpha = \frac{f^{(n+1)}(z)}{(n+1)!}$$

as required. □

Showing a Taylor Series Converges to a Function

We know that if $f(x) = e^x$ has a MacLaurin series, the only possibility for that MacLaurin series is $\sum_{k=0}^{\infty} \frac{1}{k!} x^k$, and we know that this series converges for all values of x. But does this power series converge to e^x? And if so, where? The Remainder Bound Corollary can help us answer these questions.

Example 5. Show that $\sum_{k=0}^{\infty} \frac{1}{k!} x^k$ converges to e^x for all values of x.

Solution. For any fixed value of x, we need to show that:

$$\sum_{k=0}^{\infty} \frac{1}{k!} x^k = \lim_{n \to \infty} \sum_{k=0}^{n} \frac{1}{k!} x^k = e^x$$

With $f(x) = e^x$ and $P_n(x) = \sum_{k=0}^{n} \frac{1}{k!} x^k$, this is equivalent to showing:

$$\lim_{n \to \infty} R_n(x) = \lim_{n \to \infty} [f(x) - P_n(x)] = 0$$

For $f(x) = e^x$, we know that $f^{(n+1)}(x) = e^x$ for any n. If $x > 0$, then $f^{(n+1)}(z) = e^z < e^x$ for any z between x and 0; and if $x < 0$, then $f^{(n+1)}(z) = e^z < e^0 = 1$ for any z between x and 0. Define M to be the larger of e^x and 1, so that $\left| f^{(n+1)}(z) \right| \leq M$ for any such z. Then the Remainder Bound Corollary guarantees that:

$$|R_n(x)| \leq \frac{M}{(n+1)!} \cdot |x - 0|^{n+1} = M \cdot \frac{|x|^{n+1}}{(n+1)!}$$

Because we already know the series $\sum_{k=0}^{\infty} \frac{1}{k!} x^k$ converges, we know (by the Corollary to the Test for Divergence) that $\lim_{k \to \infty} \frac{x^k}{k!} = 0$. Hence:

$$\lim_{n \to \infty} |R_n(x)| \leq M \cdot \lim_{n \to \infty} \frac{|x|^{n+1}}{(n+1)!} = 0$$

We now know that $\sum_{k=0}^{\infty} \frac{1}{k!} x^k$ converges to e^x for any value of x. ◀

Practice 5. Show that $\sum_{k=0}^{\infty} \frac{(-1)^k}{(2k+1)!} x^{2k+1}$ converges to $\sin(x)$ for all x.

Using the Remainder Bound Corollary, you can show that most functions for which you can compute infinitely many derivatives at $x = c$ are equal to their Taylor series centered at $x = c$ everywhere that the series converges. Problems 27–28 provide an example of a function $f(x)$ with a MacLaurin series that converges everywhere, but which converges to $f(x)$ only at $x = 0$.

10.5 Problems

In Problems 1–10, calculate the Taylor polynomials P_0, P_1, P_2, P_3 and P_4 for the given function centered at the given value of c. Then graph the function and the Taylor polynomials on the given interval.

1. $f(x) = \sin(x)$, $c = 0$, $[-2, 4]$
2. $f(x) = \cos(x)$, $c = 0$, $[-3, 3]$
3. $f(x) = \ln(x)$, $c = 1$, $[0.1, 3]$
4. $f(x) = \arctan(x)$, $c = 0$, $[-3, 3]$
5. $f(x) = x$, $c = 1$, $[0, 3]$
6. $f(x) = x$, $c = 9$, $[0, 20]$
7. $f(x) = (1+x)^{-\frac{1}{2}}$, $c = 0$, $[-2, 3]$
8. $f(x) = e^{2x}$, $c = 0$, $[-2, 4]$
9. $f(x) = \sin(x)$, $c = \frac{\pi}{2}$, $[-1, 5]$
10. $f(x) = \sin(x)$, $c = \pi$, $[-1, 5]$

In Problems 11–18, use the given function $f(x)$ and the given value of n to determine a formula for $R_n(x)$ and find a bound for $|R_n(x)|$ on the given interval. This bound for $|R_n(x)|$ is our "guaranteed accuracy" for $P_n(x)$ to approximate $f(x)$ on the given interval. Use $c = 0$ (so each $P_n(x)$ will be a MacLaurin polynomial).

11. $f(x) = \sin(x)$, $n = 5$, $\left[-\frac{\pi}{2}, \frac{\pi}{2}\right]$
12. $f(x) = \sin(x)$, $n = 9$, $\left[-\frac{\pi}{2}, \frac{\pi}{2}\right]$
13. $f(x) = \sin(x)$, $n = 5$, $[-\pi, \pi]$
14. $f(x) = \sin(x)$, $n = 9$, $[-\pi, \pi]$
15. $f(x) = \cos(x)$, $n = 10$, $[-1, 2]$
16. $f(x) = \cos(x)$, $n = 10$, $[-1, 5]$
17. $f(x) = e^x$, $n = 6$, $[-1, 2]$
18. $f(x) = e^x$, $n = 10$, $[-1, 3]$

In Problems 19–24, determine the number of terms of the Taylor series for $f(x)$ you need to use in order to approximate $f(x)$ to within the specified error on the given interval. (For each function, use $c = 0$.)

19. $f(x) = \sin(x)$ within 0.001 on $[-1, 1]$
20. $f(x) = \sin(x)$ within 0.001 on $[-3, 3]$
21. $f(x) = \sin(x)$ within 0.00001 on $[-1.6, 1.6]$
22. $f(x) = \cos(x)$ within 0.001 on $[-2, 2]$
23. $f(x) = e^x$ within 0.001 on $[0, 2]$
24. $f(x) = e^x$ within 0.001 on $[-1, 4]$
25. Show that the MacLaurin series for $\cos(x)$ converges to $\cos(x)$ for all values of x.
26. Show that the Taylor series for e^x centered at $x = 3$ converges to e^x for all values of x.
27. Define the function $f(x)$ as:
$$f(x) = \begin{cases} e^{-x^{-2}} & \text{if } x \neq 0 \\ 0 & \text{if } x = 0 \end{cases}$$
   (a) Show that: $f'(0) = \lim_{h \to 0} \frac{e^{-h^{-2}}}{h}$
   (b) Use the change of variable $y = \frac{1}{h}$ along with L'Hôpital's Rule to show that $f'(0) = 0$.
28. Define $f(x)$ as in Problem 27.
   (a) Show that $f^{(k)}(0) = 0$ for all $k \geq 0$.
   (b) If $P_n(x)$ is a MacLaurin polynomial for $f(x)$, show that $P_n(x) = 0$ for all n and all x.
   (c) On what interval does the MacLaurin series for $f(x)$ converge?
   (d) On what interval is the MacLaurin series for $f(x)$ equal to $f(x)$?

Series Approximations of π

The following problems illustrate some of the ways series have been used to obtain very precise approximations of π. Several of these methods use the MacLaurin series for $\arctan(x)$:

$$\arctan(x) = x - \frac{x^3}{3} + \frac{x^5}{5} - \frac{x^7}{7} + \cdots = \sum_{k=0}^{\infty} \frac{(-1)^k}{2k+1} x^{2k+1}$$

which converges rapidly if $|x|$ is close to 0.

Method I: $\tan\left(\frac{\pi}{4}\right) = 1$, so:

$$\frac{\pi}{4} = \arctan(1) = 1 - \frac{1}{3} + \frac{1}{5} - \frac{1}{7} + \frac{1}{9} - \cdots = \sum_{k=0}^{\infty} \frac{(-1)^k}{2k+1}$$

$$\Rightarrow \pi = 4\arctan(1) = \left[1 - \frac{1}{3} + \frac{1}{5} - \frac{1}{7} + \frac{1}{9} - \cdots\right]$$

29. (a) Approximate π as $4\arctan(1) = 4 - \frac{4}{3} + \frac{4}{5} - \frac{4}{7} + \frac{4}{9}$ and compare this result with the value your calculator gives for π.

    (b) The series for $\arctan(1)$ is an alternating series, so we have an "easy" error bound. Use the error bound for an alternating series to find a bound for the error if you were to use 50 terms of the series for $\arctan(1)$ (instead of five).

    (c) Using the error bound for an alternating series, how many terms of the $4\arctan(1)$ series do you need in order to guarantee that the series approximation of π is within 0.0001 of the exact value of π? (The $4\arctan(1)$ series converges so slowly that it is not used to approximate π.)

Method II: $\tan(\alpha + \beta) = \dfrac{\tan(\alpha) + \tan(\beta)}{1 - \tan(\alpha)\tan(\beta)}$, so:

$$\tan\left(\arctan\left(\frac{1}{2}\right) + \arctan\left(\frac{1}{3}\right)\right) = \frac{\frac{1}{2} + \frac{1}{3}}{1 - \frac{1}{2} \cdot \frac{1}{3}} = 1$$

$$\Rightarrow \frac{\pi}{4} = \arctan(1) = \arctan\left(\frac{1}{2}\right) + \arctan\left(\frac{1}{3}\right)$$

Because the series for $\arctan\left(\frac{1}{2}\right)$ and $\arctan\left(\frac{1}{3}\right)$ converge much more rapidly than the series for $\arctan(1)$, this approximation method leads to a more efficient estimate of π.

30. (a) Approximate π as using the first four terms of the series for $\arctan\left(\frac{1}{2}\right)$ and $\arctan\left(\frac{1}{3}\right)$ and compare this result with the value your calculator gives for π.

    (b) The series for $\arctan\left(\frac{1}{2}\right)$ and $\arctan\left(\frac{1}{3}\right)$ are each alternating series. Use the error bound for an alternating series to find a bound for the error if you use 10 terms of each series.

    (c) How many terms of each series do you need in order to guarantee that the series approximation of π is within 0.0001 of the exact value of π?

Method III: Putting $\beta = \alpha$ in the angle addition formula for $\tan(x)$ used in Method II and letting $\tan(\alpha) = \frac{1}{5}$ yields:

$$\begin{aligned} \tan(2\alpha) &= \frac{2\tan(\alpha)}{1-\tan^2(\alpha)} = \frac{\frac{2}{5}}{\frac{24}{25}} = \frac{5}{12} \\ \Rightarrow \quad \tan(4\alpha) &= \frac{2\tan(2\alpha)}{1-\tan^2(2\alpha)} = \frac{\frac{5}{6}}{\frac{119}{144}} = \frac{120}{119} \\ \Rightarrow \quad \tan\left(4\alpha - \frac{\pi}{4}\right) &= \frac{\tan(4\alpha)-1}{1+\tan(4\alpha)\cdot 1} = \frac{\frac{1}{119}}{1+\frac{120}{119}} = \frac{1}{239} \\ \Rightarrow \quad 4\alpha - \frac{\pi}{4} &= \arctan\left(\frac{1}{239}\right) \\ \Rightarrow \quad \frac{\pi}{4} &= 4\arctan\left(\frac{1}{5}\right) - \arctan\left(\frac{1}{239}\right) \end{aligned}$$

Mathematician and astronomer John Machin (1686–1751) first obtained this result around 1706. He used it to approximate π to 100 decimal places.

31. (a) Approximate π using the first three terms of the series for $\arctan\left(\frac{1}{5}\right)$ and $\arctan\left(\frac{1}{239}\right)$ and compare this result with the value your calculator gives for π.

    (b) Explain why Method III yields a series that converges more rapidly (requiring fewer terms for a "good" approximation of π) than Methods I and II.

Method IV: Carl Friedrich Gauss (1777–1855) worked out many such formulas involving arctan, including this one with three terms:

By 1958, the advent of computers allowed mathematicians working a century after Gauss' death to approximate π accurate to more than 10,000 decimal places.

$$\frac{\pi}{4} = 12\arctan\left(\frac{1}{18}\right) + 8\arctan\left(\frac{1}{57}\right) - 5\arctan\left(\frac{1}{239}\right)$$

32. (a) Approximate π using the first three terms of of the series for $\arctan\left(\frac{1}{18}\right)$, $\arctan\left(\frac{1}{57}\right)$ and $\arctan\left(\frac{1}{239}\right)$ and compare this result with the value your calculator gives for π.

    (b) Explain why Method IV yields a series that converges more rapidly (requiring fewer terms for a "good" approximation of π) than Methods I, II and III.

Calculator Notes

Imagine that you are in charge of designing or selecting an algorithm for a calculator to employ when its user pushes the **sin** button. You know that if the value of θ is relatively close to 0, then using a "few" terms of the MacLaurin series for $\sin(x)$ will approximate the value of $\sin(\theta)$ accurate to 10 digits (the size of the display of the calculator). If $-\frac{\pi}{2} \le \theta \le \frac{\pi}{2}$, then:

$$\theta - \frac{\theta^3}{3!} + \frac{\theta^5}{5!} - \frac{\theta^7}{7!} + \frac{\theta^9}{9!} - \frac{\theta^{11}}{11!}$$

will give the value of $\sin(\theta)$ with an "error" less than:

$$\frac{1}{13!} \cdot \left(\frac{\pi}{2}\right)^1 3 < \frac{(0.76)^{13}}{13!} < 5 \times 10^{-12}$$

You could rewrite the polynomial above as:

$$\theta\left(1-\frac{\theta^2}{2\cdot 3}\left(1-\frac{\theta^2}{4\cdot 5}\left(1-\frac{\theta^2}{6\cdot 7}\left(1-\frac{\theta^2}{8\cdot 9}\left(1-\frac{\theta^2}{10\cdot 11}\right)\right)\right)\right)\right)$$

This new pattern may look more complicated, but it uses fewer multiplications and avoids very large values such as 11! and θ^{11}.

If $|\theta| > 1$, θ^{11} will be very large; if $|\theta| < 1$, it will be very small.

This algorithm should work well for values of θ near 0, but you also want your algorithm to provide the same accuracy when θ is larger, say 10 or 101.7. Rather than computing many more terms of the Maclaurin series for $\sin(x)$, some algorithms simply shift the problem closer to 0. First, you can use the fact that $\sin(x) = \sin(x - 2\pi)$ to keep shifting the problem until the argument resides in the interval $[0, 2\pi]$:

$$\begin{aligned} \sin(10) &= \sin(10 - 2\pi) \approx \sin(3.71681469) \\ \sin(101.7) &= \sin(101.7 - 2\pi) = \sin(101.7 - 4\pi) = \cdots \\ &= \sin(101.7 - 32\pi) \approx \sin(1.169035085) \end{aligned}$$

Once the argument is between 0 and 2π, you can use additional trigonometric facts. If the $\theta > \pi$, use $\sin(x) = -\sin(x - \pi)$ to replace θ with $\theta - \pi$ (and keep track of the change in sign of the answer). Finally, you can shift the problem into the interval $\left[0, \frac{\pi}{2}\right]$: if the new value of θ is larger than $\frac{\pi}{2}$, use $\sin(x) = \sin(\pi - x)$ to replace θ with $\pi - \theta$.

Calculators encounter other major problems, however, when evaluating the sine or exponential function of a very large number. Because calculators only store the leading finite number of digits of a number (usually 10 or 12 digits), the calculator cannot distinguish between large numbers that differ only past that leading number of stored digits: one particular calculator correctly says that $(10^{12} + 1) - 10^{12} = 1$, but it incorrectly reports that $(10^{13} + 1) - 10^{13} = 0$. Because it calculates that "$10^{13} + 1 = 10^{13}$," it also would falsely report the same values for $\sin(10^{13} + 1)$ and $\sin(10^{13})$.

In fact, the people who programmed this particular type of calculator recognized that problem, so the calculator produces an error message if it is asked to calculate $\sin(10^{11})$. This calculator reports that $e^{230} \approx 7.7 \times 10^{99}$ but yields an error message when asked to compute e^{231} because the largest number it can display is 9.9×10^{99} and e^{231} exceeds that value. What happens on your calculator?

10.5 Practice Answers

1. The MacLaurin series for $\cos(x)$ is:

$$1 - \frac{x^2}{2!} + \frac{x^4}{4!} - \frac{x^6}{6!} + \frac{x^8}{8!} - \frac{x^{10}}{10!} + \cdots = \sum_{k-0}^{\infty} \frac{(-1)^K}{(2k)!} x^{2k}$$

so $P_0(x) = 1$, $P_2(x) = 1 - \frac{x^2}{2}$ and $P_4(x) = 1 - \frac{x^2}{2} + \frac{x^4}{24}$.

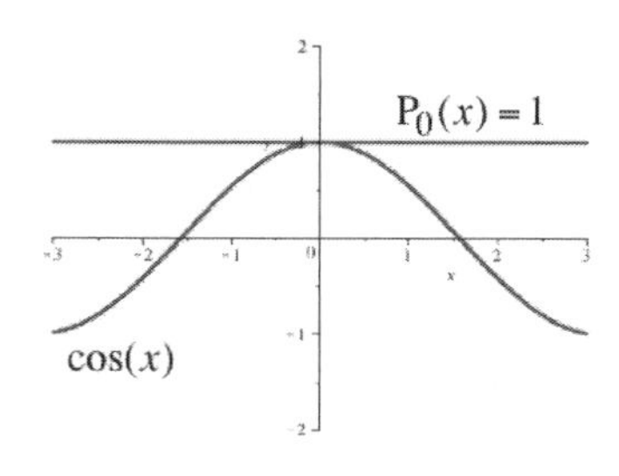

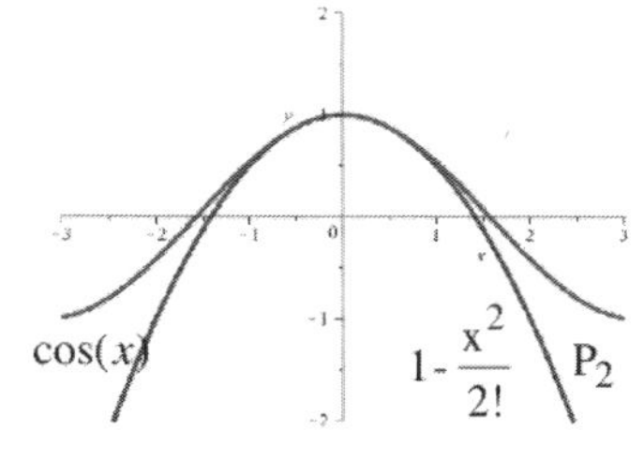

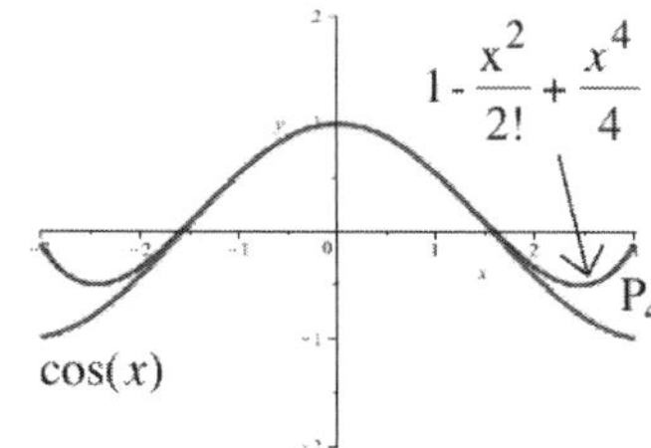

See margin for graphs. Because all odd-indexed coefficients are 0:

$$P_1(x) = P_0(x) = 1$$

$$P_3(x) = P_2(x) = 1 - \frac{x^2}{2}$$

$$P_5(x) = P_4(x) = 1 - \frac{x^2}{2} + \frac{x^4}{24}$$

2. With $f(x) = \cos(x)$ we know that $f\left(\frac{\pi}{2}\right) = 0$ and:

$$f'(x) = -\sin(x) \Rightarrow f'\left(\frac{\pi}{2}\right) = -1$$

$$f''(x) = -\cos(x) \Rightarrow f''\left(\frac{\pi}{2}\right) = 0$$

$$f'''(x) = \sin(x) \Rightarrow f'''\left(\frac{\pi}{2}\right) = 1$$

so that $P_0(x) = 0$, $P_1(x) = -\left(x - \frac{\pi}{2}\right)$ and:

$$P_3(x) = -\left(x - \frac{\pi}{2}\right) + \frac{1}{6}\left(x - \frac{\pi}{2}\right)^3$$

See lower margin figure for graphs.

3. Using the result of Example 3(b), we know that:

$$|R_n(x)| < \frac{9}{(n+1)!} \cdot 3^{n+1} = \frac{3^{n+3}}{(n+1)!}$$

so we need:

$$\frac{3^{n+3}}{(n+1)!} < 0.001 \Rightarrow \frac{(n+1)!}{3^{n+3}} > 1000$$

Experimenting with a calculator reveals that $n = 13$ works.

4. Because $0 \le x \le 1$ and $e^z < e^1 = e < 2.72$ for $0 < z < 1$:

$$|R_n(x)| = \frac{\left|f^{(n+1)}(z)\right|}{(n+1)!}|x - 0|^{n+1} < \frac{2.72}{(n+1)!}$$

hence we need:

$$\frac{2.72}{(n+1)!} < 10^{-10} \Rightarrow (n+1)! > 2.72 \times 10^{10} \Rightarrow n \ge 13$$

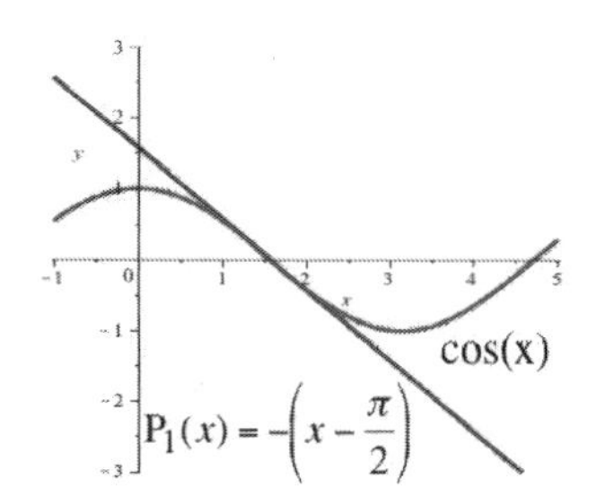

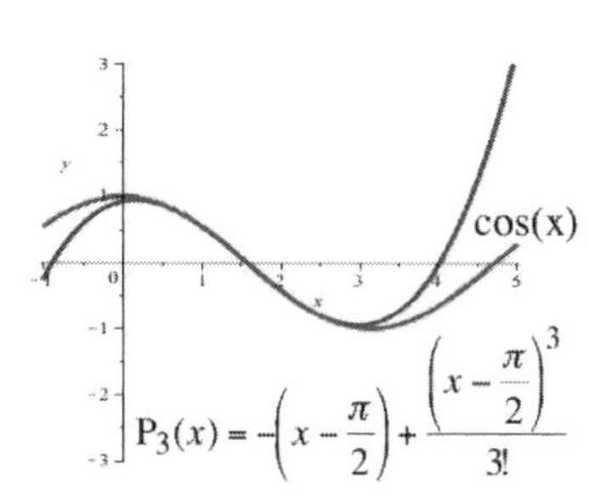

5. Any derivative of $f(x) = \sin(x)$ equals $\pm\sin(x)$ or $\pm\cos(x)$, so $\left|f^{(n+1)}(z)\right| \le 1$ for any z. Hence:

$$|R_n(x)| = \frac{\left|f^{(n+1)}(z)\right|}{(n+1)!}|x - 0|^{n+1} \le \frac{|x|^{n+1}}{(n+1)!}$$

As noted in the solution to Example 5, this expression approaches 0 as $n \to \infty$ (no matter the value of x).

11
Polar and Parametric Curves

The rectangular coordinate system, while immensely useful, is not the only way to assign an address to a point in the plane — and sometimes it is not the most useful way to describe the location of a point or the shape of curve. This chapter examines two additional ways to plot points and describe curves in a plane: polar coordinates and parametric coordinates. We then extend calculus techniques you have already learned to compute arclengths, areas and rates of change for curves, regions and functions described using these new coordinate systems.

11.1 Polar Coordinates

In many experimental situations, your location is fixed and you — or your instruments, such as radar — take readings in different directions (see margin). You can record this information in a table (below left) and graph it using rectangular coordinates with the angle on the horizontal axis and the measurement on the vertical axis (below right):

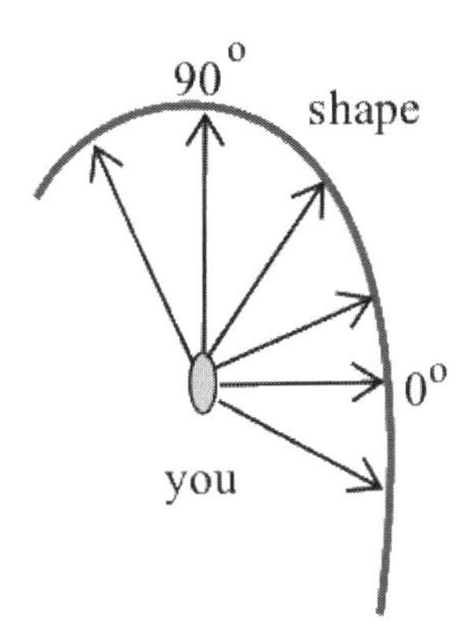

Taking Measurements

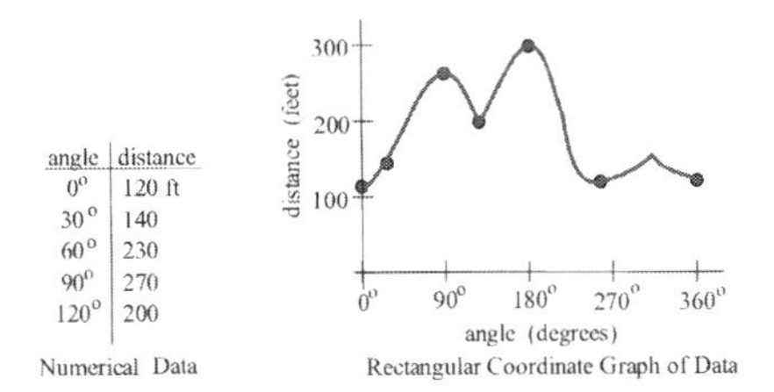

angle	distance
0°	120 ft
30°	140
60°	230
90°	270
120°	200

Numerical Data

Rectangular Coordinate Graph of Data

Sometimes, however, you will find it more useful to plot the information in a manner similar to the way in which it was collected: as magnitudes along radial lines (see margin) using the **polar coordinate system**.

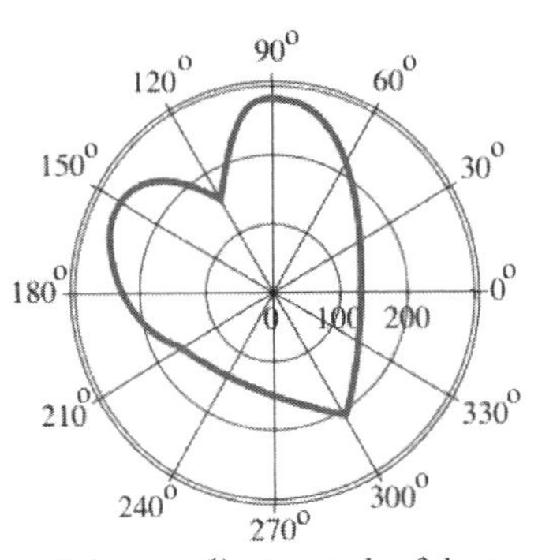

Polar coordinate graph of data

In this section we introduce polar coordinates and examine some of their uses. We graph points and functions in polar coordinates, consider how to change back and forth between the rectangular and polar coordinate systems, investigate slopes of lines tangent to polar graphs, and tackle some of the many applications in which polar coordinates arise: they provide a "natural" and easy way to represent certain types of information.

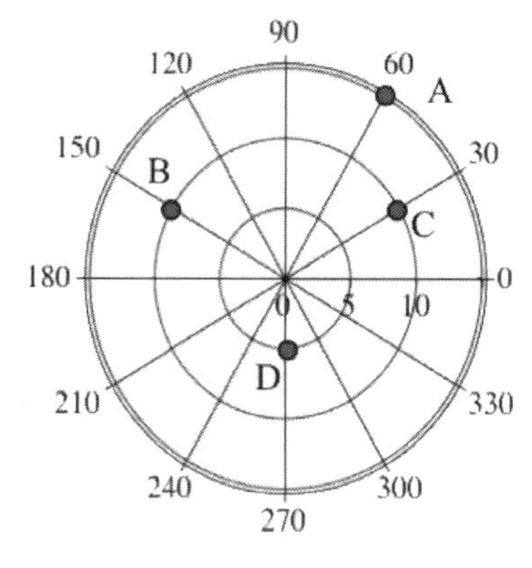

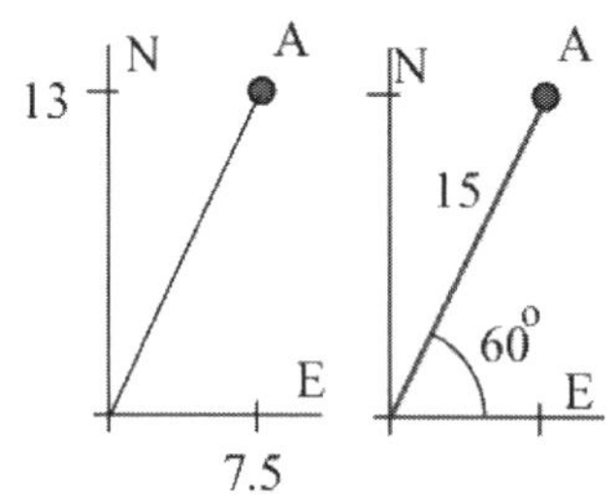

Example 1. SOS! You've just received a distress signal from a ship located at position A on your radar screen (see margin). Describe its location to your captain so your vessel can speed to the rescue.

Solution. You could convert the relative location of the other ship to rectangular coordinates and then tell your captain to sail due east for 7.5 miles and north for 13 miles, but that certainly is not the quickest way to reach the other ship. It would be better to tell the captain to sail for 15 miles in the direction of $60°$. If the distressed ship was at position B on the radar screen, your vessel should sail for 10 miles in the direction $150°$. ◀

Actual radar screens have $0°$ at the top of the screen, but the convention in mathematics is to put $0°$ in the direction of the positive x-axis and to measure positive angles counterclockwise from there. (And a real sailor uses the terms "bearing" and "range" instead of "direction" and "magnitude.")

Practice 1. Describe the locations of the ships at positions C and D in the top margin figure by determining a distance and a direction to those ships from your current position at the center of the radar screen.

Points in Polar Coordinates

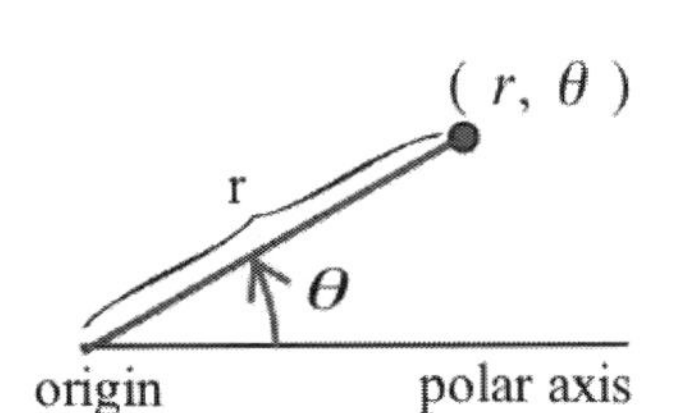

To construct a polar coordinate system we need a starting point (called the **origin** or **pole**) for the magnitude measurements and a starting direction (called the **polar axis**) for the angle measurements (see margin). A **polar coordinate pair** for a point P in the plane is an ordered pair (r, θ) where r is the directed distance along a radial line from O to P and θ is the angle formed by the polar axis and the segment OP (see margin). The angle θ is positive when the angle of the radial line OP is measured counterclockwise from the polar axis; θ is negative when measured clockwise from the polar axis.

You can use either degree or radian measure for the angle in the polar coordinate system, but when we differentiate and integrate trigonometric functions of θ we will need angles to be given in radians. You should assume that all angles are in radian measure unless the you see the "°" symbol indicating "degrees."

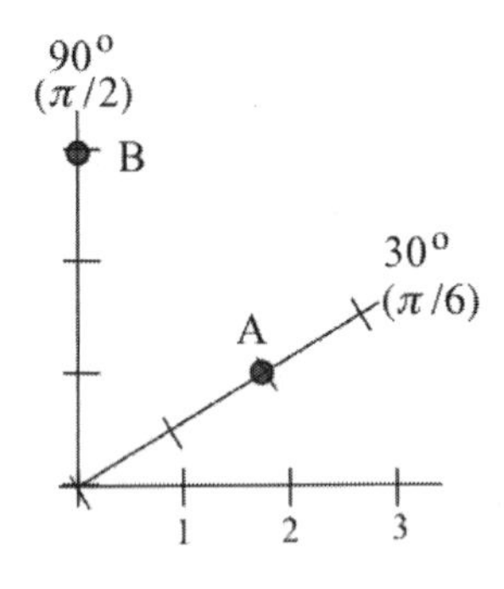

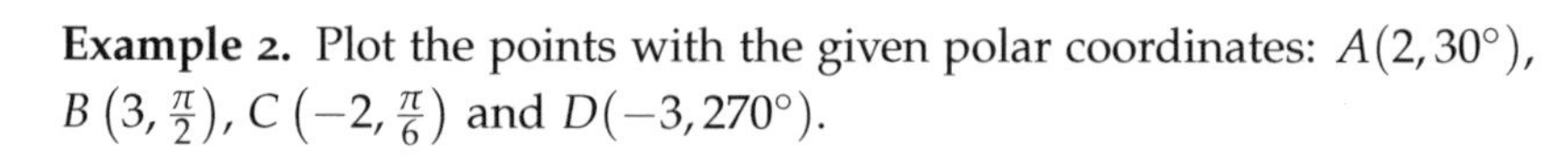
Example 2. Plot the points with the given polar coordinates: $A(2, 30°)$, $B\left(3, \frac{\pi}{2}\right)$, $C\left(-2, \frac{\pi}{6}\right)$ and $D(-3, 270°)$.

Solution. To find the location of A, we look along the ray that makes an angle of $30°$ with the polar axis, then take two steps in that direction (assuming one step corresponds to one unit on the graph). The locations of A and B appear in the margin.

To find the location of C, look along the ray that makes an angle of $\frac{\pi}{6}$ with the polar axis, then we take two steps *backwards* (because $r = -2$ is negative). The locations of C and D appear in the margin. ◀

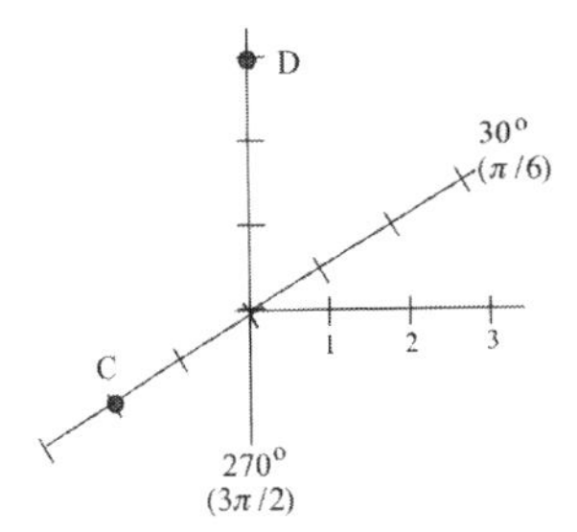

Notice that B and D have different addresses, $B\left(3, \frac{\pi}{2}\right)$ and $D(-3, 270°)$, but the same location.

Practice 2. Plot the points with polar coordinates $A\left(2, \frac{\pi}{2}\right)$, $B(2, -120°)$, $C\left(-2, \frac{\pi}{3}\right)$, $D(-2, -135°)$ and $E(2, 135°)$. Which two points coincide?

Each polar coordinate pair (r, θ) gives the location of one point, but each location has many different addresses in the polar coordinate system: the polar coordinates of a point are not unique. This non-uniqueness of addresses comes about in two ways. First, the angles $\theta, \theta \pm 360°, \theta \pm 2 \cdot 360°, \ldots$ all describe the same radial line (see below left), so the polar coordinates $(r, \theta), (r, \theta \pm 360°), (r, \theta \pm 2 \cdot 360°), \ldots$ all locate the same point.

In the rectangular coordinate system we use (x, y) and $y = f(x)$, listing the independent variable first and the dependent variable second. In the polar coordinate system we use (r, θ) and $r = f(\theta)$, listing the dependent variable first and the independent variable second, a reversal from rectangular coordinate usage.

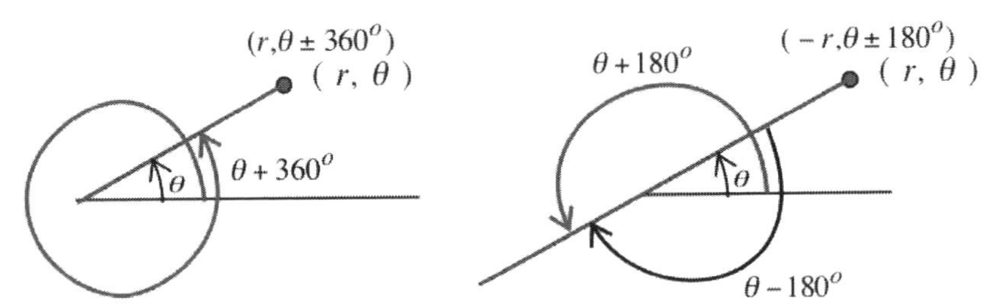

Secondly, the angle $\theta \pm 180°$ describes the radial line pointing in exactly the opposite direction from the radial line described by the angle θ (see above right), so the polar coordinates (r, θ) and $(-r, \theta \pm 180°)$ locate the same point. A polar coordinate pair gives the location of exactly one point, but the location of one point can be described by (infinitely) many different polar coordinate pairs.

Practice 3. The margin table contains measurements to the edge of a plateau taken by a remote sensor that crashed on the plateau. The figure below shows the data plotted in rectangular coordinates. Plot the data in polar coordinates and determine the shape of the plateau.

angle	distance
0°	28 feet
20°	30
40°	36
60°	27
80°	24
100°	24
130°	30
150°	22
230°	13
210°	21
180°	18
270°	10
340°	30
330°	18

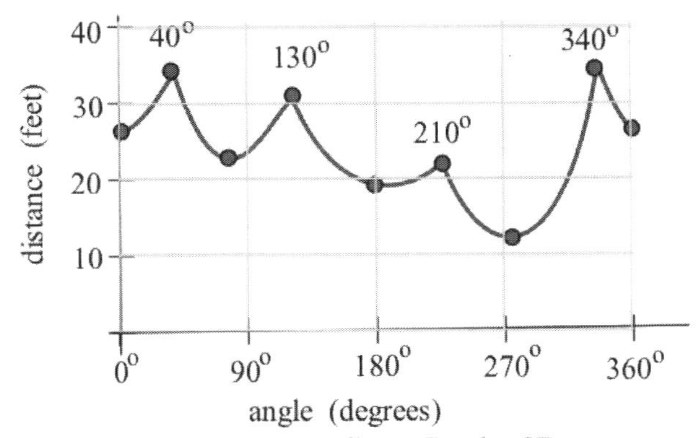

Rectangular Coordinate Graph of Data

Graphing Functions in the Polar Coordinate System

In the rectangular coordinate system, you have worked with functions given by tables of data, by graphs and by formulas. You can represent functions in the same ways using polar coordinates.

- If a table of data gives you values of a function, you can graph the function in polar coordinates by plotting individual points in a polar coordinate system and connecting the plotted points to see the shape of the graph. By hand, this is a tedious process; by calculator or computer, it is quick and easy.
- If you have a rectangular coordinate graph of magnitude as a function of angle, you can read coordinates of points on the rectangular graph and replot them in polar coodinates. In essence, as you go from the rectangular coordinate graph to the polar coordinate graph you "wrap" the rectangular graph around the "pole" at the origin of the polar coordinate system (see margin).

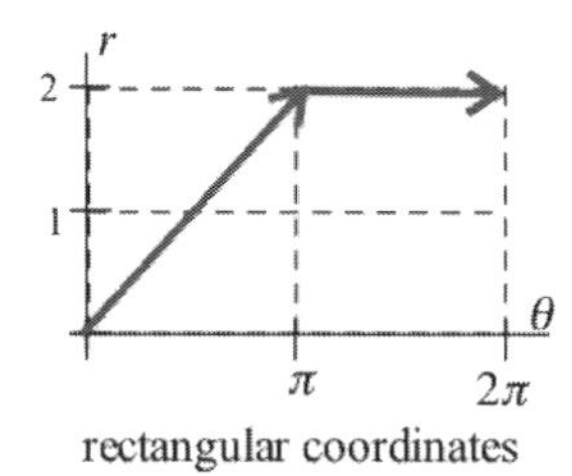

rectangular coordinates

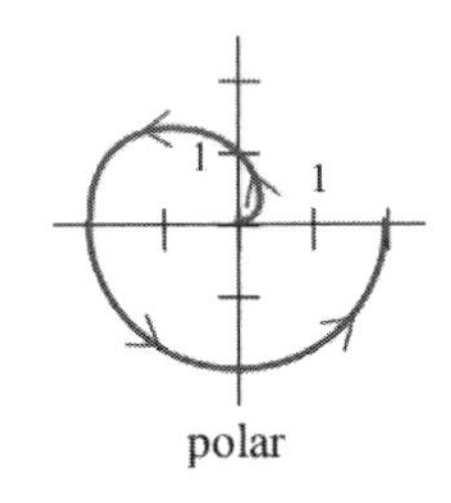

polar

- If you have a formula for a function, you (or your calculator) can graph the function to help obtain information about its behavior. Typically, you (or a calculator) creates a graph by evaluating the function at many points and then plotting the points in the polar coordinate system. Some of the following examples illustrate that functions given by simple formulas may have rather exotic graphs in the polar coordinate system.

If you already have a polar coordinate graph of a function, you can use the graph to answer questions about the behavior of the function. It is usually easy to locate the maximum value(s) of r on a polar coordinate graph and, by moving counterclockwise around the graph, you can observe where r is increasing, constant or decreasing.

Example 3. Graph $r = 2$ and $r = \pi - \theta$ in the polar coordinate system for $0 \le \theta \le 2\pi$.

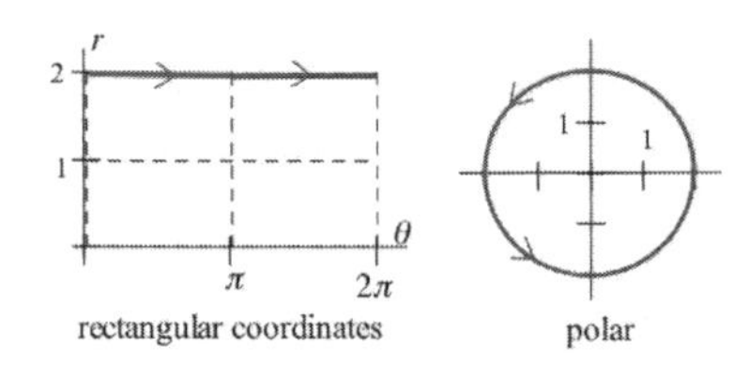

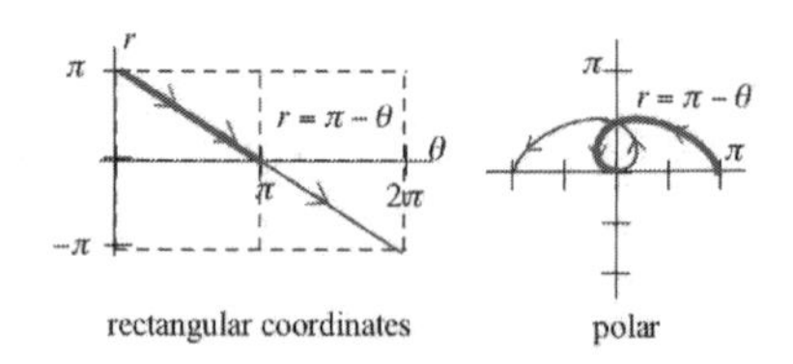

Solution. First consider $r = 2$: In every direction θ, we simply move 2 units along the radial line and plot a point. The resulting polar graph (see margin) is a circle centered at the origin with a radius of 2. In the rectangular coordinate system, the graph of a constant $y = k$ is a horizontal line; in the polar coordinate system, the graph of a constant $r = k$ is a circle with radius $|k|$.

Next consider $r = \pi - \theta$: The rectangular=coordinate graph appears in the margin. Reading the values of r and θ from the rectangular coordinate graph and plotting them in polar coordinates results in the shape in the lower margin figure. The different line thicknesses used in the figures help you see which values from the rectangular graph become which parts of the loop in the polar graph. ◀

Practice 4. Graph $r = -2$ and $r = \cos(\theta)$ in polar coordinates.

Example 4. Graph $r = \theta$ and $r = 1 + \sin(\theta)$ in polar coordinates.

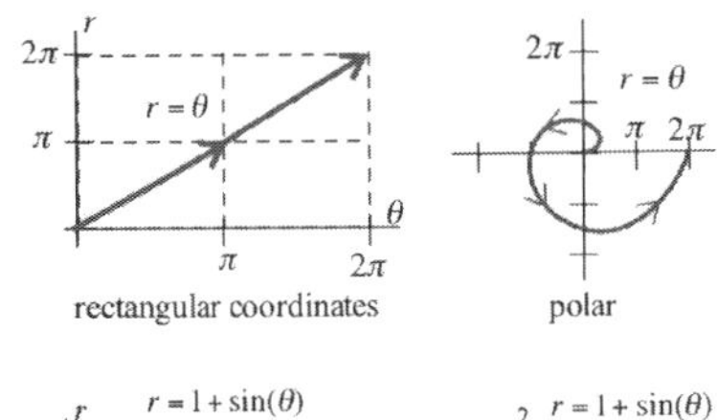

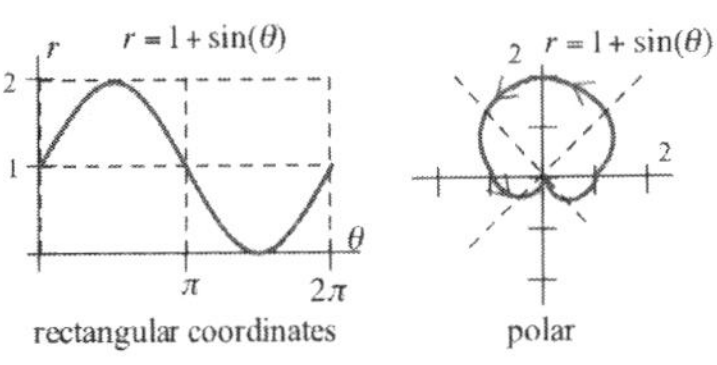

Solution. The rectangular coordinate graph of $r = \theta$ is a straight line (see top margin figure). Reading the values of r and θ from the rectangular coordinate graph and plotting them in polar coordinates results in a spiral, called an **Archimedean spiral**.

In the rectangular coordinate graph of $r = 1 + \sin(\theta)$ (see margin) the graph of the sine curve is shifted up 1 unit; in polar coordinates, the result of adding 1 to the sine function is much less obvious. ◀

Practice 5. Plot the points in the margin table in polar coordinates and connect them with a smooth curve. Describe the shape in words.

angle	distance
0	3.0 m
$\frac{\pi}{6}$	1.6
$\frac{\pi}{4}$	1.7
$\frac{\pi}{3}$	1.9
$\frac{\pi}{2}$	2.0

The graphs below show the effects of adding various constants to the rectangular and polar graphs of $r = \sin(\theta)$:

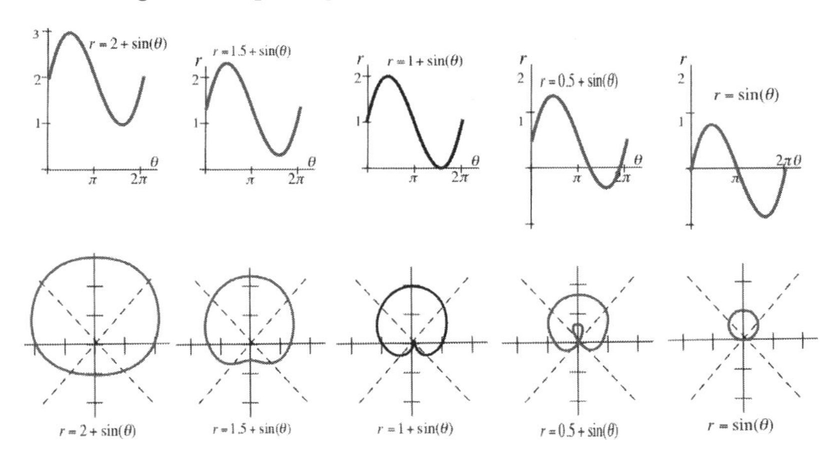

In rectangular coordinates, the result is a graph shifted up or down by k units; in polar coordinates, the result may be a graph with an entirely different shape.

The next set of graphs show the effects of adding a constant to the independent variable in rectangular and polar coordinates:

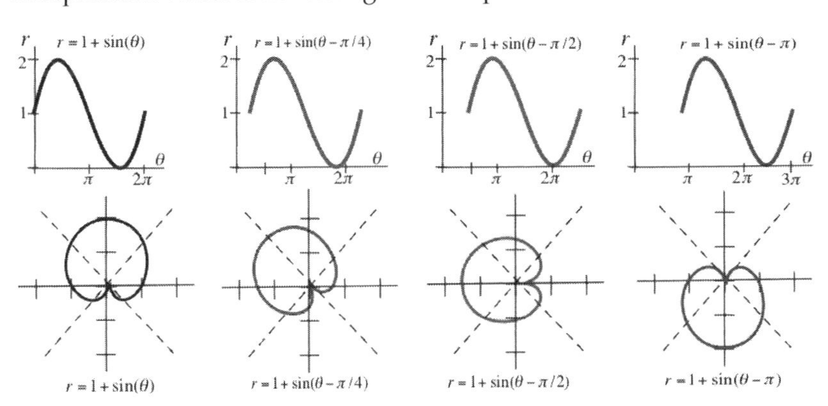

The result in rectangular coordinates is a horizontal shift of the original graph; the result in polar coordinates is a **rotation** of the original. Finding formulas for rotated figures in rectangular coordinates can be quite difficult, but rotations are easy in polar coordinates.

The formulas and names of several functions with exotic shapes in polar coordinates arise in the Problems. Many of them are difficult to graph "by hand," but by using a graphing calculator or computer you can appreciate the shapes and easily examine the effects of changing some of the constants in their formulas.

Converting Between Coordinate Systems

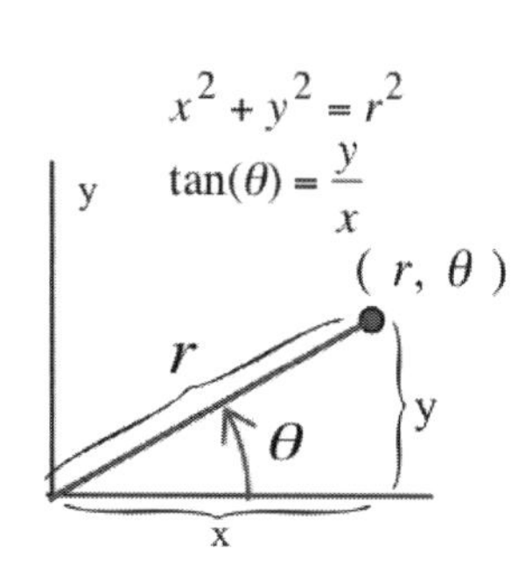

Sometimes you need both rectangular and polar coordinates in the same application, so it becomes necessary to change back and forth between the systems. If you place the two origins together and align the polar axis with the positive x-axis, the conversions involve straightforward applications of trigonometry and right triangles (see margin).

Polar to Rectangular: $x = r \cdot \cos(\theta)$, $y = r \cdot \sin(\theta)$
Rectangular to Polar: $r^2 = x^2 + y^2$, $\tan(\theta) = \frac{y}{x}$ (if $x \neq 0$)

Example 5. Convert (a) the polar coordinate point $P(7, 0.4)$ to rectangular coordinates and (b) the rectangular coordinate point $R(12, 5)$ to polar coordinates.

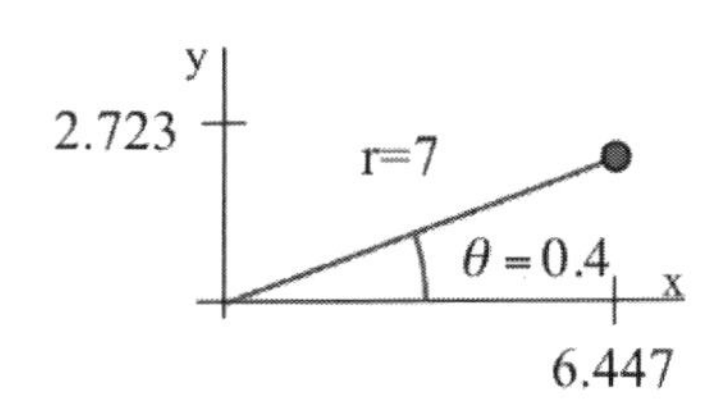

Solution. (a) $r = 7$ and $\theta = 0.4$ so $x = 7 \cdot \cos(0.4) \approx 7(0.921) = 6.447$ and $y = 7 \cdot \sin(0.4) \approx 7(0.389) = 2.723$. (b) $x = 12$ and $y = 5$ so $r^2 = x^2 + y^2 = 144 + 25 = 169$, and $\tan(\theta) = \frac{y}{x} = \frac{5}{12}$; we can take $r = 13$ and $\theta = \arctan\left(\frac{5}{12}\right) \approx 0.395$. The polar coordinate addresses $(13, 0.395 \pm n \cdot 2\pi)$ and $(-13, 0.395 \pm (2n+1) \cdot \pi)$ give the location of the same point for any integer n. ◄

You can also use these conversion formulas to convert equations from one system to the other.

Example 6. Convert the linear equation $y = 3x + 5$ (see margin) from rectangular coordinates to polar coordinates.

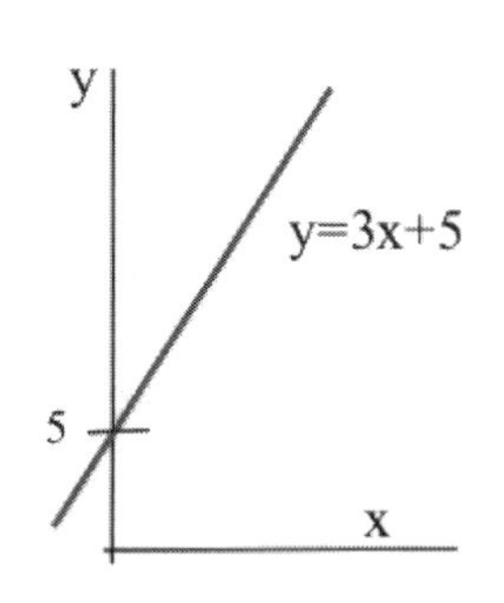

Solution. Replacing x with $r \cdot \cos(\theta)$ and y with $r \cdot \sin(\theta)$:

$$y = 3x + 5 \Rightarrow r \cdot \sin(\theta) = 3r \cdot \cos(\theta) + 5$$
$$\Rightarrow r \cdot [\sin(\theta) - 3\cos(\theta)] = 5 \Rightarrow r = \frac{5}{\sin(\theta) - 3\cos(\theta)}$$

This final representation is valid only when $\sin(\theta) - 3\cos(\theta) \neq 0$. ◄

Practice 6. Convert the polar coordinate equation $r^2 = 4r \cdot \sin(\theta)$ to a rectangular coordinate equation.

Example 7. A robotic arm has a hand at the end of a 12-inch forearm connected to an 18-inch upper arm (see margin). Determine the position of the hand, relative to the shoulder, if $\theta = 45° = \frac{\pi}{4}$ and $\varphi = 30° = \frac{\pi}{6}$.

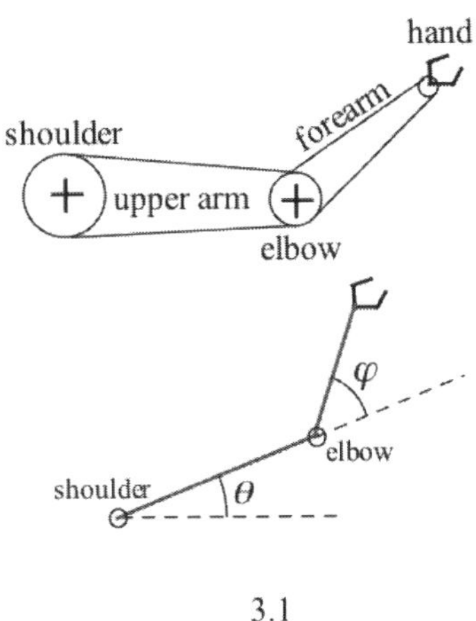

Solution. The hand is $12 \cdot \cos\left(\frac{\pi}{4} + \frac{\pi}{6}\right) \approx 3.1$ inches to the right of the elbow and $12 \cdot \sin\left(\frac{\pi}{4} + \frac{\pi}{6}\right) \approx 11.6$ inches above the elbow. Similarly, the elbow is $18 \cdot \cos\left(\frac{\pi}{4}\right) \approx 12.7$ inches to the right of the shoulder and $18 \cdot \sin\left(\frac{\pi}{4}\right) \approx 12.7$ inches above the shoulder. Finally, the hand is approximately $3.1 + 12.7 = 15.8$ inches to the right of the shoulder and approximately $11.6 + 12.7 = 24.3$ inches above the shoulder. In polar coordinates, the hand is approximately 29 inches from the shoulder, at an angle of about 57° (about 0.994 radians) above the horizontal. ◀

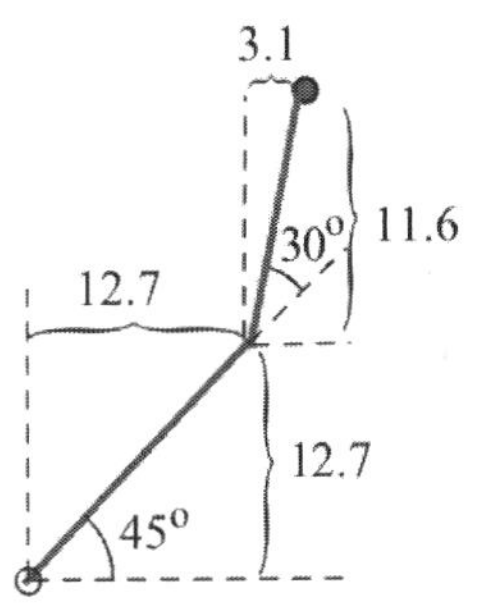

Practice 7. Determine the position of the hand, relative to the shoulder, when $\theta = 30°$ and $\varphi = 45°$.

Which Coordinate System Should You Use?

There are no rigid rules. Use whichever coordinate system is easier or more "natural" for the problem or data you have. Sometimes it is unclear which system to use until you have graphed the data both ways. Some problems are easier if you switch back and forth between the systems. Generally, the polar coordinate system is easier if:

- the data consists of measurements in various directions (radar)
- your problem involves locations in relatively featureless locations (deserts, oceans, sky)
- rotations are involved

Typically, the rectangular coordinate system is easier if:

- the data consists of measurements given as functions of time or location (temperature, height)
- your problem involves locations in situations with an established grid (a city, a chess board)
- translations are involved

11.1 Problems

1. Give the locations in polar coordinates (using radians) of the points labeled A, B and C below.

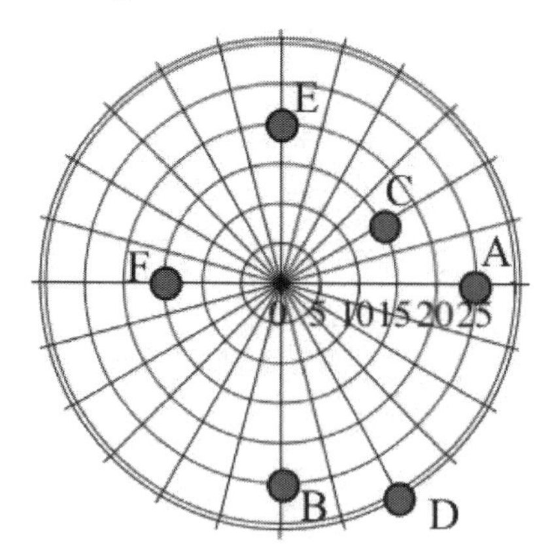

2. Give the locations in polar coordinates of the points labeled D, E and F above.
3. Give the locations in polar coordinates of the points labeled A, B and C below.

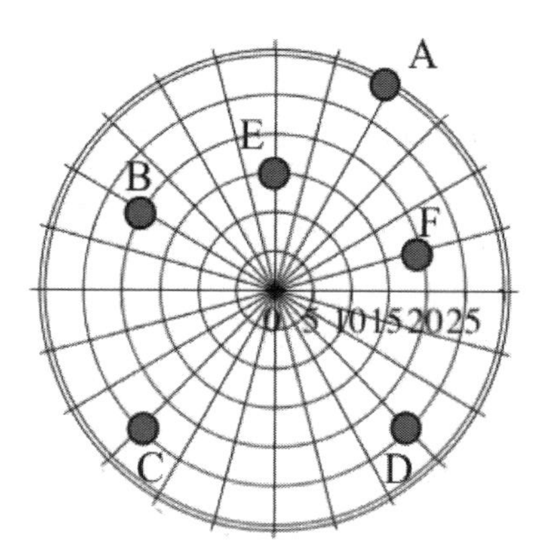

4. Give the locations in polar coordinates of the points labeled D, E and F above.

In Problems 5–8, plot the points A–D in polar coordinates, connect the dots in order (A to B to C to D to A) using line segments, and name the approximate shape of the resulting figure.

5. $A(3,0^\circ)$, $B(2,120^\circ)$, $C(2,200^\circ)$, $D(2.8,315^\circ)$
6. $A(3,30^\circ)$, $B(2,130^\circ)$, $C(3,150^\circ)$, $D(2,280^\circ)$
7. $A(2,0.175)$, $B(3,2.269)$, $C(2,2.618)$, $D(3,4.887)$
8. $A(3,0.524)$, $B(2,2.269)$, $C(3,2.618)$, $D(2,4.887)$

In Problems 9–14, use the given rectangular coordinate graph of the function $r = f(\theta)$ to sketch the polar coordinate graph of $r = f(\theta)$.

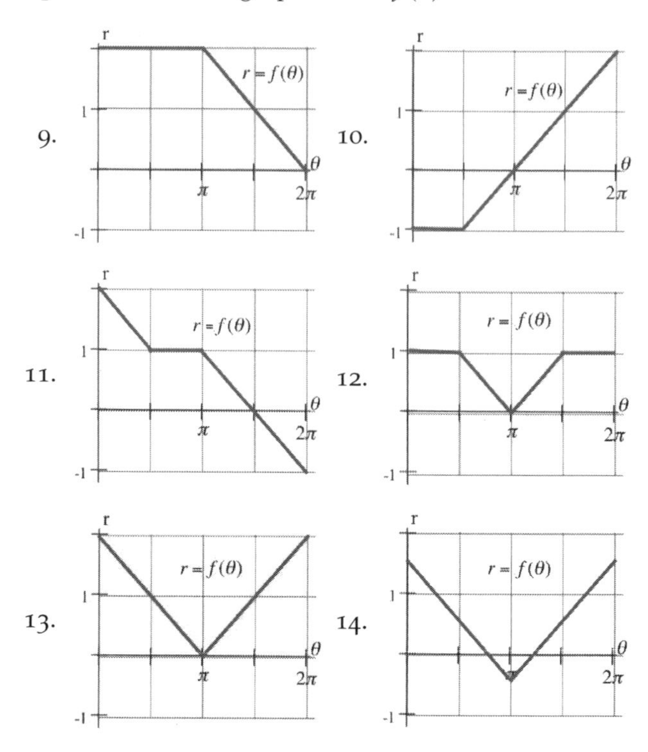

15. The rectangular coordinate graph of $r = f(\theta)$ appears below left.
    (a) Sketch the rectangular coordinate graphs of $r = 1 + f(\theta)$, $r = 2 + f(\theta)$ and $r = -1 + f(\theta)$.
    (b) Sketch the polar coordinate graphs of $r = 1 + f(\theta)$, $r = 2 + f(\theta)$ and $r = -1 + f(\theta)$.

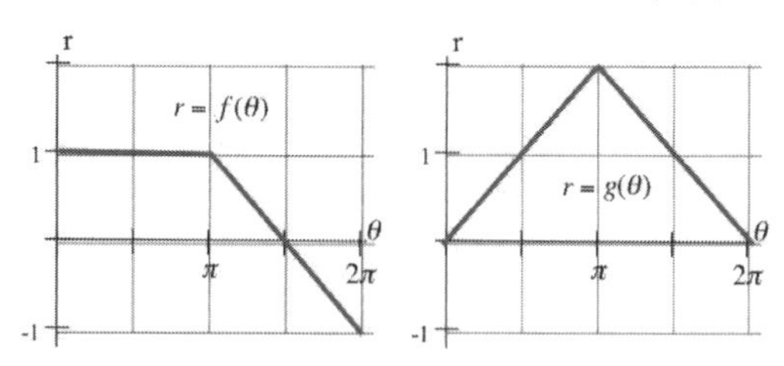

16. The rectangular coordinate graph of $r = g(\theta)$ appears above right.
    (a) Sketch the rectangular coordinate graphs of $r = 1 + g(\theta)$, $r = 2 + g(\theta)$ and $r = -1 + g(\theta)$.
    (b) Sketch the polar coordinate graphs of $r = 1 + g(\theta)$, $r = 2 + g(\theta)$ and $r = -1 + g(\theta)$.

17. The rectangular coordinate graph of $r = f(\theta)$ appears below left.
    (a) Sketch the rectangular coordinate graphs of $r = 1 + f(\theta)$, $r = 2 + f(\theta)$ and $r = -1 + f(\theta)$.
    (b) Sketch the polar coordinate graphs of $r = 1 + f(\theta)$, $r = 2 + f(\theta)$ and $r = -1 + f(\theta)$.

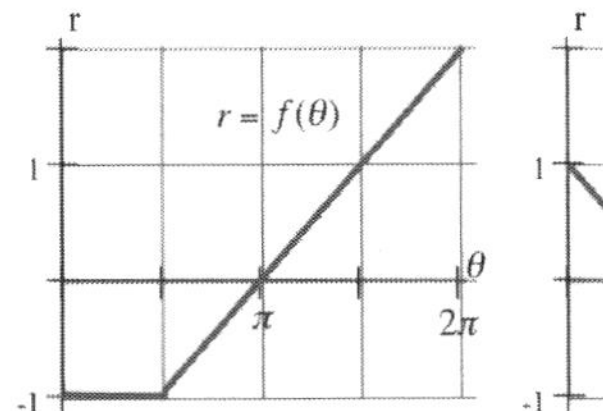

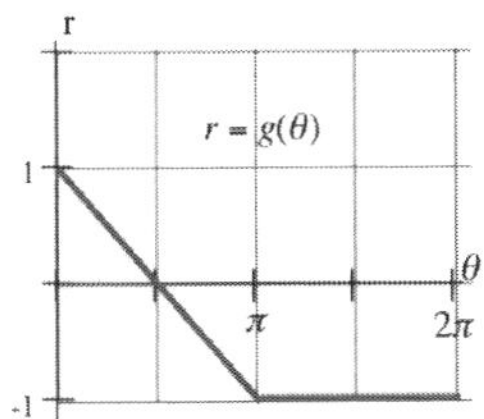

18. The rectangular coordinate graph of $r = g(\theta)$ appears above right.
    (a) Sketch the rectangular coordinate graphs of $r = 1 + g(\theta)$, $r = 2 + g(\theta)$ and $r = -1 + g(\theta)$.
    (b) Sketch the polar coordinate graphs of $r = 1 + g(\theta)$, $r = 2 + g(\theta)$ and $r = -1 + g(\theta)$.

19. If the rectangular coordinate graph of $r = f(\theta)$ has a horizontal asymptote of $r = 3$ as θ grows arbitrarily large, what does that tell you about the polar coordinate graph of $r = f(\theta)$ for large values of θ?

20. If $\lim_{\theta \to \frac{\pi}{6}} f(\theta) = \infty$ so that the rectangular coordinate graph of $r = f(\theta)$ has a vertical asymptote at $\theta = \frac{\pi}{6}$, what does that tell you about the polar coordinate graph of $r = f(\theta)$ for θ near $\frac{\pi}{6}$?

In Problems 21–40, graph the functions in polar coordinates for $0 \le \theta \le 2\pi$.

21. $r = -3$
22. $r = 5$
23. $\theta = \frac{\pi}{6}$
24. $\theta = \frac{5\pi}{3}$
25. $r = 4 \cdot \sin(\theta)$
26. $r = -2 \cdot \cos(\theta)$
27. $r = 2 + \sin(\theta)$
28. $r = -2 + \sin(\theta)$
29. $r = 2 + 3 \cdot \sin(\theta)$
30. $r = \sin(2\theta)$
31. $r = \tan(\theta)$
32. $r = 1 + \tan(\theta)$
33. $r = 3\sec(\theta)$
34. $r = 3\csc(\theta)$
35. $r = \dfrac{1}{\sin(\theta) + \cos(\theta)}$
36. $r = \dfrac{\theta}{2}$
37. $r = 2\theta$
38. $r = \theta^2$
39. $r = \dfrac{1}{\theta}$
40. $r = \sin(2\theta)\cos(3\theta)$

41. $r = \sin(m\theta) \cdot \cos(n\theta)$ produces lovely graphs for various small integer values of m and n. Use a calculator or computer to find values of m and n that result in shapes you find interesting.

42. Graph $r = \dfrac{1}{1 + 0.5 \cdot \cos(\theta + \alpha)}$ for $0 \le \theta \le 2\pi$ and for $\alpha = 0$, $\frac{\pi}{6}$, $\frac{\pi}{4}$ and $\frac{\pi}{2}$. Describe how the graphs are related.

43. Graph $r = \dfrac{1}{1 + 0.5 \cdot \cos(\theta - \alpha)}$ for $0 \le \theta \le 2\pi$ and for $\alpha = 0$, $\frac{\pi}{6}$, $\frac{\pi}{4}$ and $\frac{\pi}{2}$. Describe how the graphs are related.

44. Graph $r = \cos(n\theta)$ for $0 \le \theta \le 2\pi$ and for $n = 1$, 2, 3 and 4. Count the number of "petals" on each graph. Predict the number of "petals" for the graphs of $r = \sin(n\theta)$ for $n = 5$, 6 and 7, then test your prediction by creating those graphs.

45. Repeat the steps in Problem 44 using $r = \cos(n\theta)$.

In Problems 46–49, convert the rectangular coordinate locations to polar coordinates.

46. $(0,3)$, $(5,0)$, $(1,2)$
47. $(-2,3)$, $(2,-3)$, $(0,-4)$
48. $(0,-2)$, $(4,4)$, $(3,-3)$
49. $(3,4)$, $(-1,-3)$, $(-7,12)$

In Problems 50–53, convert the polar coordinate locations to rectangular coordinates.

50. $(3,0)$, $(5,90^\circ)$ and $(1,\pi)$
51. $(-2,3)$, $(2,-3)$ and $(0,-4)$
52. $(0,3)$, $(5,0)$ and $(1,2)$
53. $(2,3)$, $(-2,-3)$ and $(0,4)$

For 54–60, refer to the robotic arm shown below.

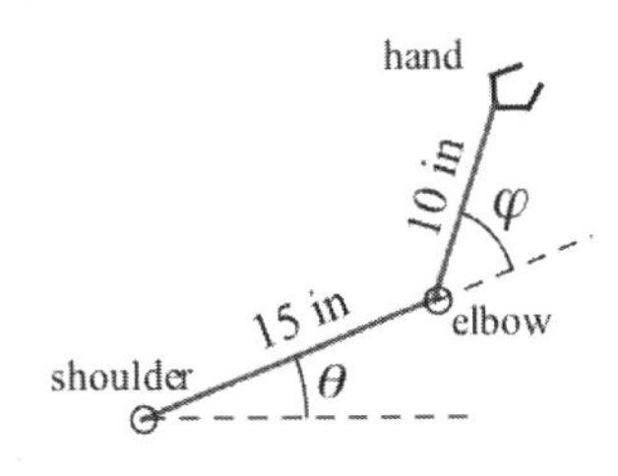

54. Determine the position of the hand, relative to the shoulder, when $\theta = 60°$ and $\varphi = -45°$.

55. Determine the position of the hand, relative to the shoulder, when $\theta = -30°$ and $\varphi = 30°$.

56. Determine the position of the hand, relative to the shoulder, when $\theta = 0.6$ and $\varphi = 1.2$.

57. Determine the position of the hand, relative to the shoulder, when $\theta = -0.9$ and $\varphi = 0.4$.

58. If the robot's shoulder pivots so $-\frac{\pi}{2} \le \theta \le \frac{\pi}{2}$, but the elbow is broken and φ is always 0, sketch the points the hand can reach.

59. If the robot's shoulder pivots so $-\frac{\pi}{2} \le \theta \le \frac{\pi}{2}$, and the elbow pivots so $-\frac{\pi}{2} \le \varphi \le \frac{\pi}{2}$, sketch the points the hand can reach.

60. If the robot's shoulder pivots so $-\frac{\pi}{2} \le \theta \le \frac{\pi}{2}$, and the elbow pivots completely so $-\pi \le \varphi \le \pi$, sketch the points the hand can reach.

61. Graph $r = \dfrac{1}{1 + a \cdot \cos(\theta)}$ for $0 \le \theta \le 2\pi$ and $a = 0.5,\ 0.8,\ 1,\ 1.5$ and 2. What shapes do the various values of a produce?

62. Repeat Problem 61 with $r = \dfrac{1}{1 + a \cdot \sin(\theta)}$.

63. Show that the polar form of the linear equation $Ax + By + C = 0$ is:

$$r \cdot (A \cdot \cos(\theta) + B \cdot \sin(\theta)) + C = 0$$

64. Show that the equation of the line through the polar coordinate points (r_1, θ_1) and (r_2, θ_2) is:

$$r\left[r_1 \sin(\theta - \theta_1) + r_2 \sin(\theta_2 - \theta)\right] = r_1 r_2 \sin(\theta_2 - \theta_1)$$

65. Show that the graph of $r = a \cdot \sin(\theta) + b \cdot \cos(\theta)$ is a circle through the origin with center $\left(\frac{b}{2}, \frac{a}{2}\right)$ and radius $\frac{1}{2}\sqrt{a^2 + b^2}$.

Some Exotic Curves (and Names)

An inexpensive resource for these shapes and names is *A Catalog Of Special Plane Curves* by J. Dennis Lawrence, Dover Publications, 1972; the page numbers given below refer to that book.

Many of the following curves were discovered and named even before polar coordinates came about. In most cases the curve describes the path of a point moving on or around some object. You may enjoy using your calculator or a computer to graph some of these curves, or you can invent your own exotic shapes.

Some classics:

- Cissoid ("like ivy") of Diocles (about 200 B.C.): $r = a\sin(\theta) \cdot \tan(\theta)$
- Right Strophoid ("twisting") of Barrow (1670): $r = a\left[\sec(\theta) - 2\cos(\theta)\right]$
- Trisectrix of MacLaurin (1742): $r = a\sec(\theta) - 4a\cos(\theta)$
- Lemniscate ("ribbon") of Bernoulli (1694): $r^2 = a^2\cos(2\theta)$
- Conchoid ("shell") of Nicomedes (225 B.C.): $r = a + b\sec(\theta)$
- Hippopede ("horse fetter") of Proclus (about 75 B.C.):

$$r^2 = 4b\left[a - b\sin^2(\theta)\right] \quad \text{for} b = 3,\ a = 1,\ 2,\ 3,\ 4$$

- Devil's Curve of Cramer (1750):

$$r^2\left[\sin^2(\theta) - \cos^2(\theta)\right] = a^2\sin^2(\theta) - b^2\cos^2(\theta) \quad \text{for } a = 2,\ b = 3$$

- Nephroid ("kidney") of Freeth: $r = a\left[1 + 2\sin(\theta^2)\right]$ for $a = 3$

Some of our own:

Based on their names, what shapes do you expect for the following curves?

- Piscatoid of Pat (1992): $r = \sec(\theta) - 3\cos(\theta)$ for $-1.1 \le \theta \le 1.1$, with window $-2 \le x \le 1$ and $-1 \le y \le 1$
- Kermitoid of Kelcey (1992):

  $$r = 2.5\sin(2\theta)\left[\theta - 4.71\right] \cdot \text{INT}\left(\frac{\theta}{\pi}\right) + \left[5\sin^3(\theta) - 3\sin^9(\theta)\right] \cdot \left[1 - \text{INT}\left(\frac{\theta}{\pi}\right)\right]$$

  for $0 \le \theta \le 2\pi$ with window $-3 \le x \le 3$ and $-1 \le y \le 4$
- Bovine Oculoid: $r = 1 + \text{INT}\left(\frac{\theta}{2\pi}\right)$ for $0 \le \theta \le 6\pi$ with window $-5 \le x \le 5$ and $-4 \le y \le 4$

11.1 Practice Answers

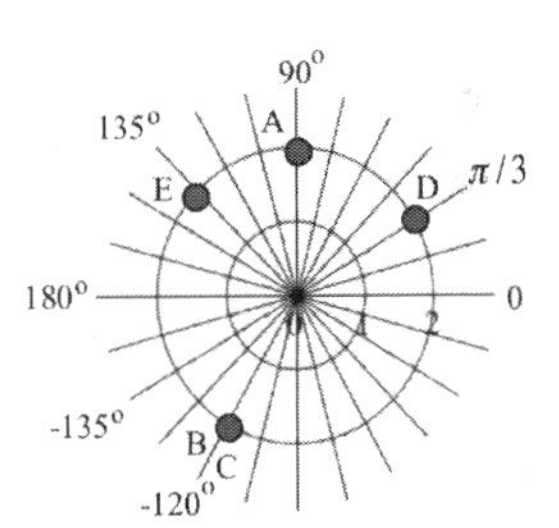

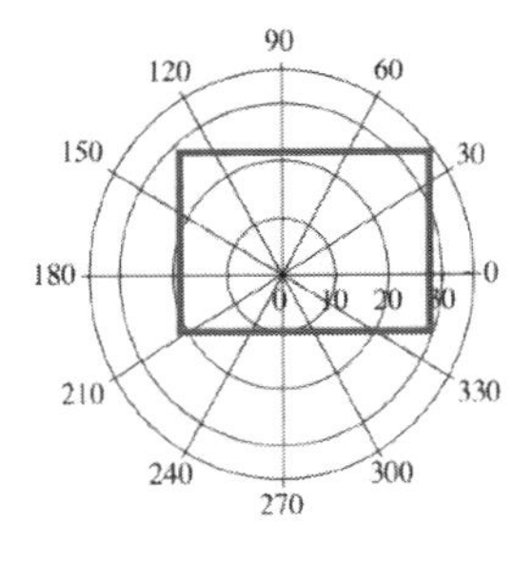

1. Point C is at a distance of 10 miles in the direction $30°$; D is 5 miles away at $270°$.
2. See first margin figure.
3. See second margin figure. The plateau is roughly rectangular.
4. The graphs for $r = -2$ appear below:

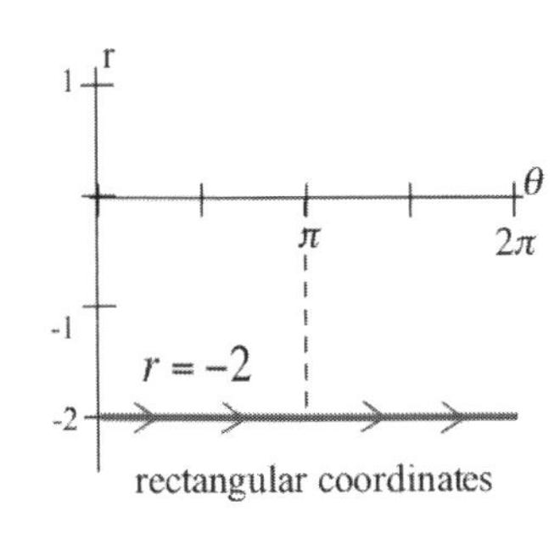

rectangular coordinates

polar coordinates

The graphs for $r = \cos(\theta)$ appear below. Note that in polar coordinates $r = \cos(\theta)$ traces out a circle *twice*: once as θ goes from 0 to π, and a second time as θ goes from π to 2π.

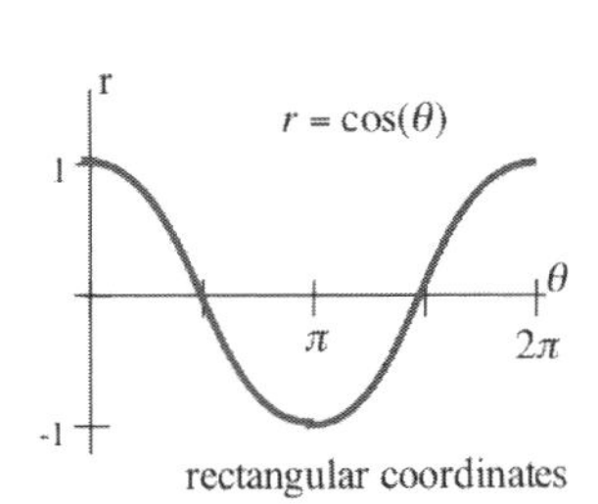

rectangular coordinates

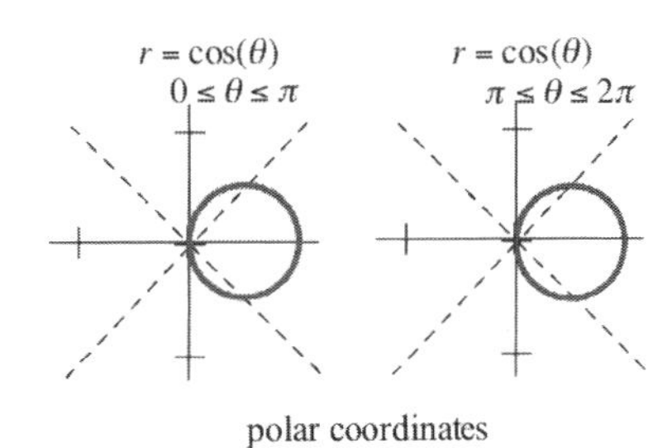

polar coordinates

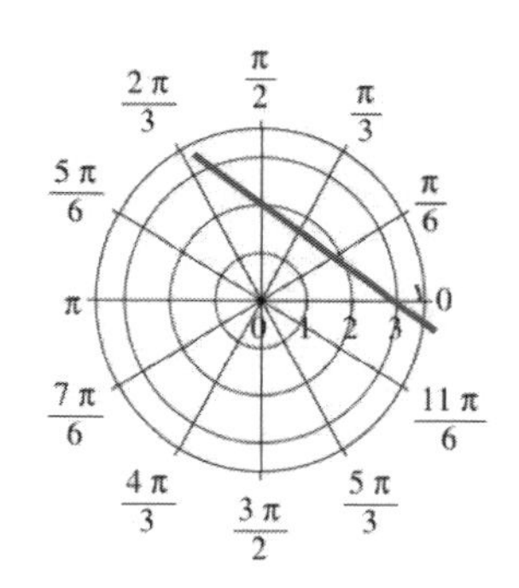

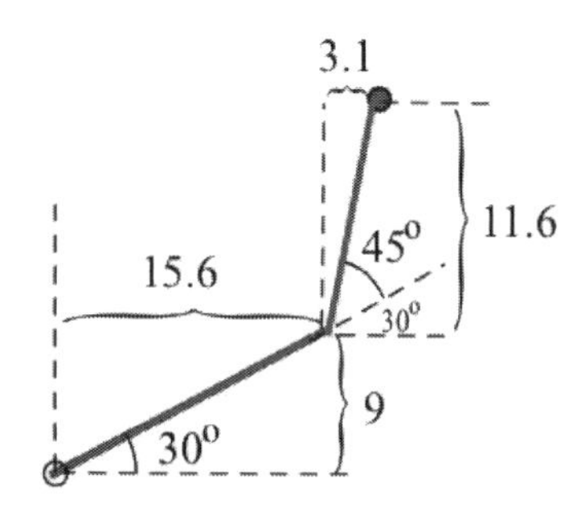

5. See margin figure. The points (almost) lie on a straight line.

6. $r^2 = x^2 + y^2$ and $r \cdot \sin(\theta) = y$, so:

$$r^2 = 4r \cdot \sin(\theta) \Rightarrow x^2 + y^2 = 4y$$

Putting this last equation into the standard form for a circle (by completing the square) yields $x^2 + (y-2)^2 = 4$, an equation for a circle with center at $(0, 2)$ and radius 2.

7. See margin figure. For point A, the "elbow," relative to O, the "shoulder": $x = 18\cos(30°) \approx 15.6$ inches and $y = 18\sin(30°) = 9$ inches. For point B, the "hand," relative to A: $x = 12\cos(75°) \approx 3.1$ inches and $y = 12\sin(75°) \approx 11.6$ inches. Then the rectangular coordinate location of B relative to O is $x \approx 15.6 + 3.1 = 18.7$ inches and $y \approx 9 + 11.6 = 20.6$ inches. The polar coordinate location of B relative to O is $r = \sqrt{x^2 + y^2} \approx 27.8$ inches and $\theta \approx 47.7°$ (or 0.83 radians).

11.2 *Calculus in Polar Coordinates*

The previous section introduced the polar coordinate system and discussed how to plot points, how to create graphs of functions (from data, a rectangular graph or a formula) and how to convert back and forth between the polar and rectangular systems. This section examines calculus in polar coordinates: rates of change, slopes of tangent lines, areas and lengths of curves.

The results we obtain may appear different than the corresponding results from earlier chapters, but they all follow from the approaches used in the rectangular coordinate system.

Polar Coordinates and Derivatives

In the rectangular coordinate system, the derivative $\frac{dy}{dx}$ measured both the rate of change of y with respect to x for a function $y = f(x)$ and the slope of the tangent line to the graph of $y = f(x)$. In the polar coordinate system other derivatives also commonly appear, and it is important that you learn to distinguish among them. If $r = g(\theta)$ then:

- $\frac{dr}{d\theta} = g'(\theta)$ measures the rate of change of r with respect to θ
- $\frac{dy}{dx}$ gives the slope $\frac{\Delta y}{\Delta x}$ of the tangent line to the graph of $r = g(\theta)$

The sign of $\frac{dr}{d\theta}$ tells us whether r is increasing or decreasing as θ increases.

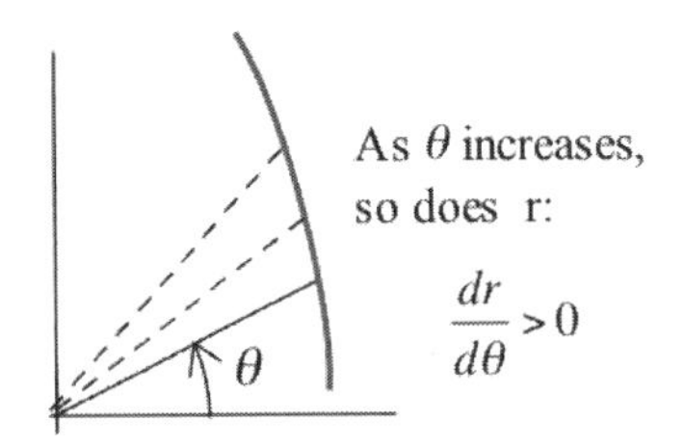

The derivative of a polar equation $r = g(\theta)$, $\frac{dr}{d\theta} = g'(\theta)$, tells us how r is changing with respect to (increasing) θ. For example, if $\frac{dr}{d\theta} > 0$ then the directed distance r is increasing as θ increases (see margin). However, $\frac{dr}{d\theta} = g'(\theta)$ is *not* the slope of the line tangent to the polar graph of $r = g(\theta)$. For the simple spiral $r = \theta$ (see second margin figure), $\frac{dr}{d\theta} = 1 > 0$ for all values of θ, but the slope of the tangent line, $\frac{dy}{dx}$, is sometimes positive and sometimes negative.

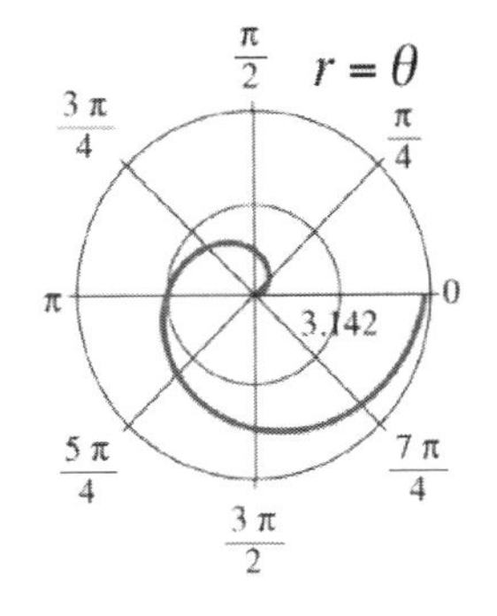

Similarly, $\frac{dx}{d\theta}$ tells us the rate of change of the x-coordinate of the graph with respect to (increasing) θ and $\frac{dy}{d\theta}$ tells us the rate of change of the y-coordinate of the graph with respect to (increasing) θ.

Example 1. State whether the values of $\frac{dx}{d\theta}, \frac{dy}{d\theta}, \frac{dr}{d\theta}$ and $\frac{dy}{dx}$ are positive (+), negative (−), zero (0) or undefined (U) at the points A and B on the graph in the bottom margin figure.

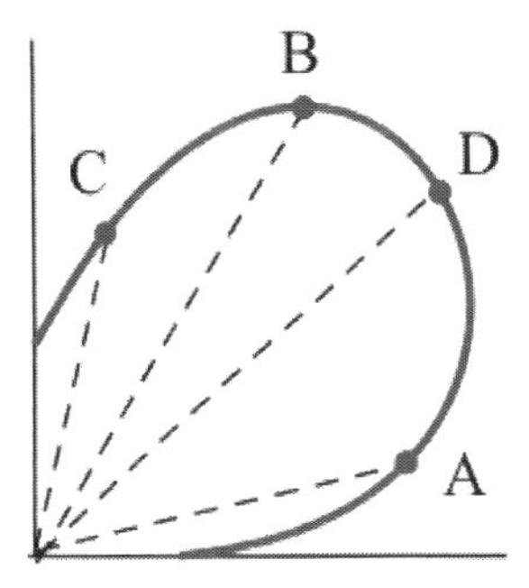

Solution. As θ increases near A, the x- and y-coordinates of the point on the graph are both increasing, the radius r (the distance from the point to the origin) is increasing, and the slope of the line tangent to the graph is positive, so all four derivatives are positive.

As θ increases near B: the x-coordinate is decreasing, so $\frac{dx}{d\theta} < 0$; the y-coordinate reaches a maximum, so $\frac{dy}{d\theta} = 0$; the radius r is getting

	$\frac{dx}{d\theta}$	$\frac{dy}{d\theta}$	$\frac{dr}{d\theta}$	$\frac{dy}{dx}$
A	+	+	+	+
B	−	0	−	0
C				
D				

smaller, so $\frac{dr}{d\theta} < 0$; and the tangent line is horizontal, so $\frac{dy}{dx} = 0$. We can collect these results in the margin table. ◀

Practice 1. Fill in the margin table for the points labeled C and D.

Slopes of Tangent Lines

If you know that $r = f(\theta)$ for some differentiable function f, you can calculate $\frac{dy}{dx}$, the slope of the tangent line to the graph of $r = f(\theta)$, by using the polar–rectangular conversion formulas and the Chain Rule:

$$\frac{dy}{d\theta} = \frac{dy}{dx} \cdot \frac{dx}{d\theta} \quad \Rightarrow \quad \frac{dy}{dx} = \frac{\frac{dy}{d\theta}}{\frac{dx}{d\theta}}$$

so to find the slope of the tangent line we need to compute $\frac{dx}{d\theta}$ and $\frac{dy}{d\theta}$. From the polar–rectangular conversion formulas, we know that:

$$x = r\cos(\theta) = f(\theta) \cdot \cos(\theta) \Rightarrow \frac{dx}{d\theta} = -f(\theta) \cdot \sin(\theta) + f'(\theta) \cdot \cos(\theta)$$

$$y = r\sin(\theta) = f(\theta) \cdot \sin(\theta) \Rightarrow \frac{dy}{d\theta} = f(\theta) \cdot \cos(\theta) + f'(\theta) \cdot \sin(\theta)$$

and hence:

$$\frac{dy}{dx} = \frac{f(\theta) \cdot \cos(\theta) + f'(\theta) \cdot \sin(\theta)}{-f(\theta) \cdot \sin(\theta) + f'(\theta) \cdot \cos(\theta)}$$

This result may be difficult to memorize, but you should be able to remember how to obtain the result using the conversion formulas, the Product Rule and the Chain Rule.

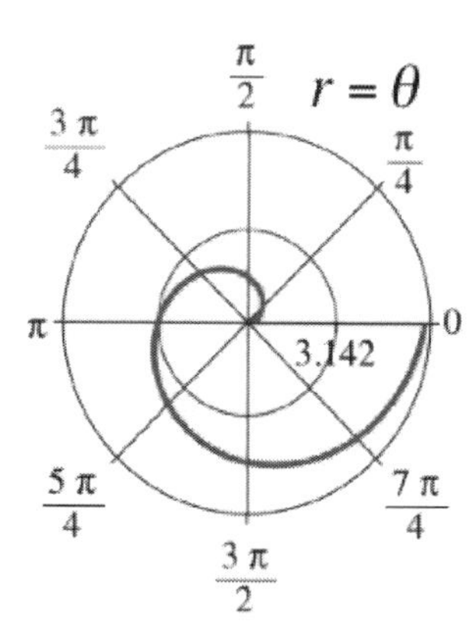

Example 2. Find the slopes of the lines tangent to the spiral $r = \theta$ at the points $P\left(\frac{\pi}{2}, \frac{\pi}{2}\right)$ and $Q(\pi, \pi)$.

Solution. Proceeding as above:

$$x = \theta \cdot \cos(\theta) \quad \Rightarrow \quad \frac{dx}{d\theta} = -\theta \cdot \sin(\theta) + \cos(\theta)$$

$$y = \theta \cdot \sin(\theta) \quad \Rightarrow \quad \frac{dy}{d\theta} = \theta \cdot \cos(\theta) + \sin(\theta)$$

$$\Rightarrow \quad \frac{dy}{dx} = \frac{\theta \cdot \cos(\theta) + \sin(\theta)}{-\theta \cdot \sin(\theta) + \cos(\theta)}$$

At the point P, $\theta = \frac{\pi}{2}$, so:

$$\frac{dy}{dx} = \frac{\frac{\pi}{2} \cdot 0 + 1}{-\frac{\pi}{2} \cdot 1 + 0} = -\frac{2}{\pi}$$

At the point Q, $\theta = \pi$, so:

$$\frac{dy}{dx} = \frac{\pi \cdot (-1) + 0}{-\pi \cdot 0 - 1} = \pi$$

The function $r = \theta$ is steadily increasing, but the slope of the line tangent to the polar graph can be negative or positive or 0 or even undefined (where?). ◀

Practice 2. Find equations for the lines tangent to the graph of $r = \theta$ at points P and Q in the preceding Example.

Practice 3. Compute the slopes of the lines tangent to the cardioid $r = 1 - \sin(\theta)$ (see margin for graph) when $\theta = 0, \frac{\pi}{4}$ and $\frac{\pi}{2}$.

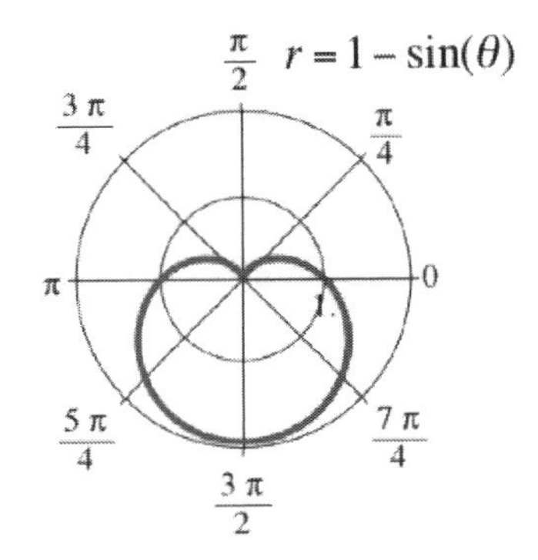

Areas in Polar Coordinates

The formulas for computing areas in rectangular and polar coordinates may appear quite different, but we obtain them the same way: partition a region into pieces, compute (approximate) areas of those pieces, add the small areas together to get a Riemann sum, and take the limit of that sum to get a definite integral. The chief difference here is the shape of the pieces: we use thin, almost-rectangular pieces in the Cartesian system and thin almost-sectors (pieces of pie) in the polar system.

We can obtain the formula for the area of a sector of a circle using proportions (see margin):

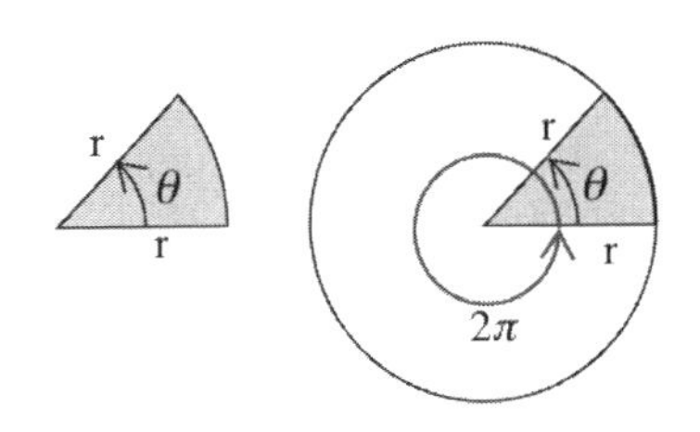

$$\frac{\text{area of sector}}{\text{area of whole circle}} = \frac{\text{sector angle}}{\text{angle of whole circle}} = \frac{\theta}{2\pi}$$

$$\Rightarrow \text{area of sector} = \frac{\theta}{2\pi}(\text{area of whole circle}) = \frac{\theta}{2\pi}\left(\pi r^2\right) = \frac{1}{2}r^2\theta$$

Given a region bounded by the polar curve $r = f(\theta)$ and the rays $\theta = \alpha$ and $\theta = \beta$, partition the θ-domain into n small pieces of angular "width" $\Delta\theta$. For the k-th polar "slice", choose an angle θ_k in that slice and approximate the area of the k-th slice with a sector of radius $f(\theta_k)$ and angle $\Delta\theta$. The area of this sector is $\frac{1}{2}[f(\theta_k)]^2\,\Delta\theta$, so the approximate area of the region is given by the Riemann sum:

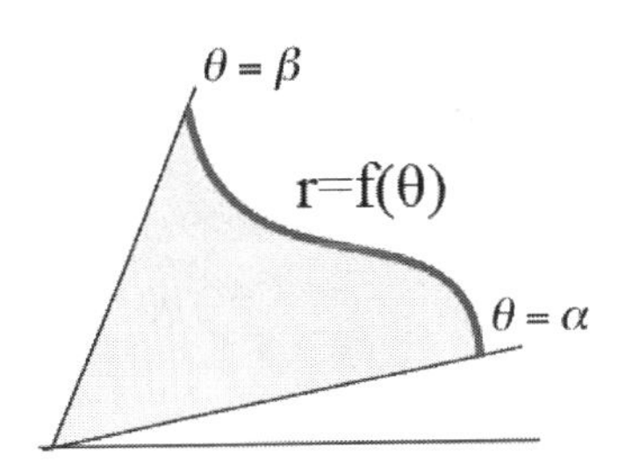

$$\sum_{k=1}^{n} \frac{1}{2}[f(\theta_k)]^2\,\Delta\theta \longrightarrow \int_{\alpha}^{\beta} \frac{1}{2}[f(\theta)]^2\,d\theta$$

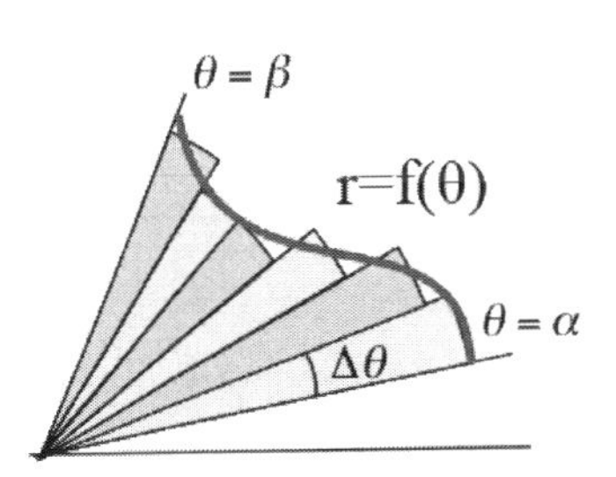

We can guarantee the convergence of the Riemann sum to the integral by requiring that $f(\theta)$ is continuous for $\alpha \le \theta \le \beta$.

> **Area in Polar Coordinates**
> If $f(\theta)$ is continuous on $[\alpha, \beta]$, the area of the region bounded by $r = f(\theta)$ and radial lines at angles $\theta = \alpha$ and $\theta = \beta$ is given by:
>
> $$\int_{\alpha}^{\beta} \frac{1}{2}[f(\theta)]^2\,d\theta = \int_{\alpha}^{\beta} \frac{1}{2}r^2\,d\theta$$

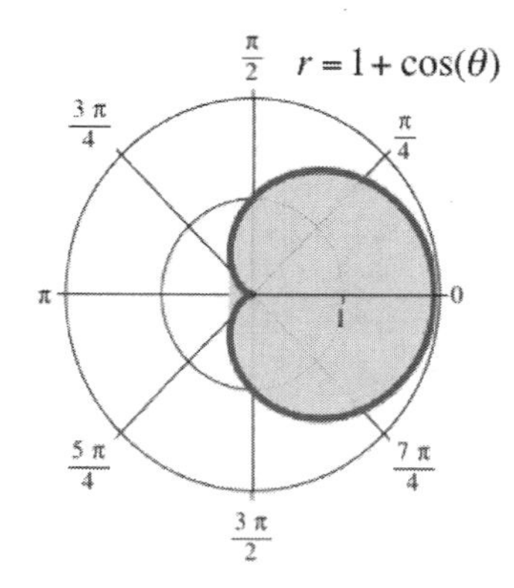

Example 3. Find the area of the region inside the cardioid $r = 1 + \cos(\theta)$ (see margin for graph).

Solution. A straightforward application of the area formula yields:

$$\int_0^{2\pi} \frac{1}{2}\left[1+\cos(\theta)\right]^2 \, d\theta = \int_0^{2\pi} \left[\frac{1}{2} + \cos(\theta) + \frac{1}{2}\cos^2(\theta)\right] d\theta$$
$$= \int_0^{2\pi} \left[\frac{1}{2} + \cos(\theta) + \frac{1}{2}\left(\frac{1}{2} + \frac{1}{2}\cos(2\theta)\right)\right] d\theta$$
$$= \int_0^{2\pi} \left[\frac{3}{4} + \cos(\theta) + \frac{1}{4}\cos(2\theta)\right] d\theta$$
$$= \left[\frac{3}{4}\theta + \sin(\theta) + \frac{1}{8}\sin(2\theta)\right]_0^{2\pi} = \frac{3\pi}{2}$$

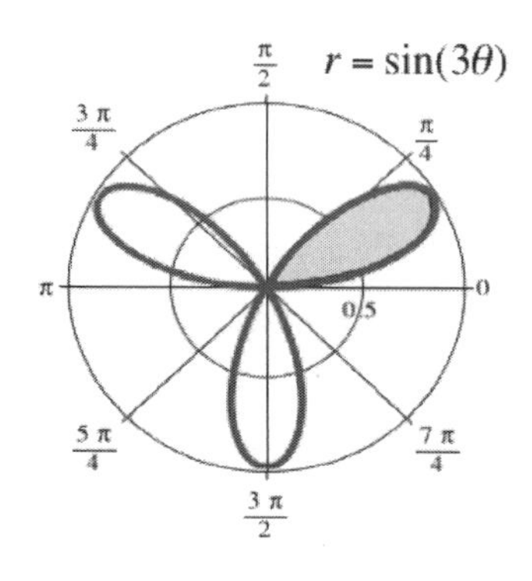

We could also have exploited the symmetry of the region, integrating instead from 0 to π (to get the area of the "top half" of the region) and then doubling the result. ◄

Practice 4. Find the area of the region inside one "petal" of the rose $r = \sin(3\theta)$ (see margin for graph).

We can also calculate the area *between* curves in polar coordinates. The area of the region (see margin) between the continuous curves $r = f(\theta)$ and $r = g(\theta)$ for $\alpha \le \theta \le \beta$, if $f(\theta) > g(\theta)$ on this interval, is:

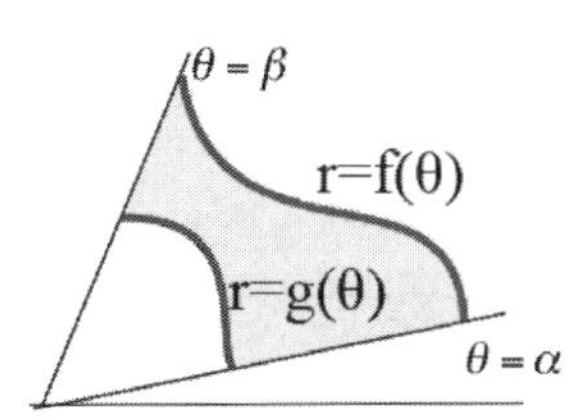

$$\int_\alpha^\beta \frac{1}{2}\left[(f(\theta))^2 - (g(\theta))^2\right] d\theta = \int_\alpha^\beta \frac{1}{2}\left[r_{\text{outer}}^2 - r_{\text{inner}}^2\right] d\theta$$

It is a good idea to sketch the graphs of the curves to help determine the endpoints of integration.

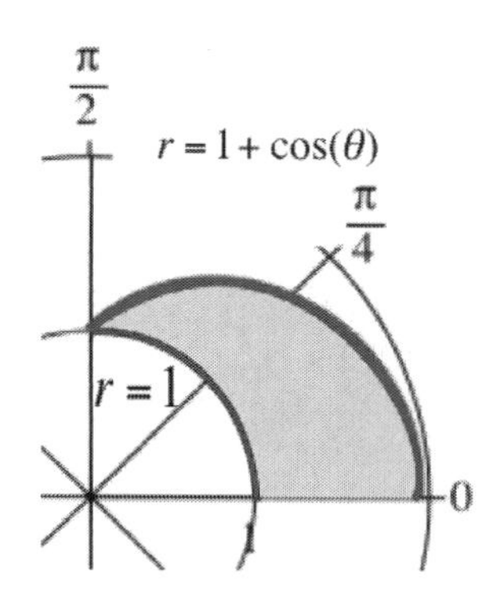

Example 4. Find the area of the shaded region in the margin figure.

Solution. The shaded region lies in the first quadrant, so $\alpha = 0$ and $\beta = \frac{\pi}{2}$. On that interval, $\cos(\theta) \ge 0 \Rightarrow 1 + \cos(\theta) \ge 1$, so $f(\theta) = 1 + \cos(\theta)$ generates the outer curve and $g(\theta) = 1$ generates the inner curve. The area of the region between these curves is therefore:

$$\int_0^{\frac{\pi}{2}} \frac{1}{2}\left[(1+\cos(\theta))^2 - (1)^2\right] d\theta = \int_0^{\frac{\pi}{2}} \left[\cos(\theta) + \frac{1}{2}\cos^2(\theta)\right] d\theta$$
$$= \int_0^{\frac{\pi}{2}} \left[\cos(\theta) + \frac{1}{4} + \frac{1}{4}\cos(2\theta)\right] d\theta = \left[\sin(\theta) + \frac{1}{4}\theta + \frac{1}{8}\sin(2\theta)\right]_0^{\frac{\pi}{2}}$$

which evaluates to $1 + \dfrac{\pi}{8} \approx 1.393$. ◄

Practice 5. Find the area of the region outside the cardioid $1 + \cos(\theta)$ and inside the circle $r = 2$.

Arclength in Polar Coordinates

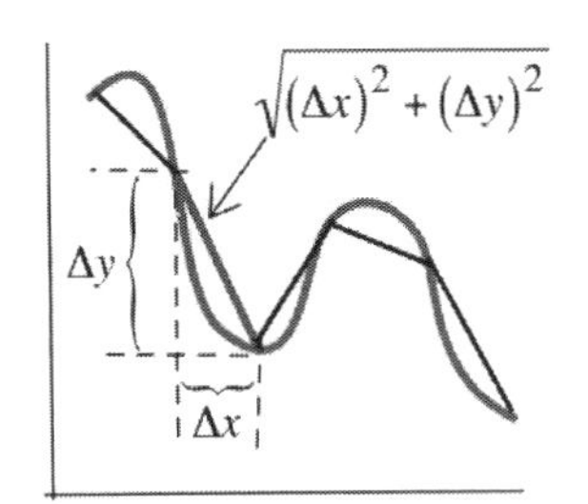

The formulas for calculating the lengths of curves in rectangular and polar coordinates look a bit different, but we can obtain both from the Pythagorean Theorem, following the method we used in Section 5.3 (see margin figure):

$$\text{length} \approx \sum_{k=1}^{n} \sqrt{(\Delta x)^2 + (\Delta y)^2} = \sum_{k=1}^{n} \sqrt{\left(\frac{\Delta x}{\Delta\theta}\right)^2 + \left(\frac{\Delta y}{\Delta\theta}\right)^2} \cdot \Delta\theta$$

If $r = f(\theta)$, a differentiable function of θ, then $x = f(\theta)\cdot\cos(\theta)$ and $y = f(\theta)\cdot\sin(\theta)$ will both be differentiable functions of θ so that $\frac{\Delta x}{\Delta\theta} \to \frac{dx}{d\theta}$ and $\frac{\Delta y}{\Delta\theta} \to \frac{dy}{d\theta}$ as $\Delta\theta \to 0$. Hence:

$$\sum_{k=1}^{n} \sqrt{\left(\frac{\Delta x}{\Delta\theta}\right)^2 + \left(\frac{\Delta y}{\Delta\theta}\right)^2} \cdot \Delta\theta \longrightarrow \int_{\theta=\alpha}^{\theta=\beta} \sqrt{\left(\frac{dx}{d\theta}\right)^2 + \left(\frac{dy}{d\theta}\right)^2}\, d\theta$$

Building on results from our earlier derivative computations:

$$\frac{dx}{d\theta} = -f(\theta)\sin(\theta) + f'(\theta)\cos(\theta) \Rightarrow \left(\frac{dx}{d\theta}\right)^2 = [f(\theta)]^2\sin^2(\theta) - 2f(\theta)f'(\theta)\sin(\theta)\cos(\theta) + \left[f'(\theta)\right]^2\cos^2(\theta)$$

$$\frac{dy}{d\theta} = f(\theta)\cos(\theta) + f'(\theta)\sin(\theta) \Rightarrow \left(\frac{dy}{d\theta}\right)^2 = [f(\theta)]^2\cos^2(\theta) + 2f(\theta)f'(\theta)\sin(\theta)\cos(\theta) + \left[f'(\theta)\right]^2\sin^2(\theta)$$

Adding these quantities and applying a square root yields:

$$\sqrt{\left(\frac{dx}{d\theta}\right)^2 + \left(\frac{dy}{d\theta}\right)^2} = \sqrt{[f(\theta)]^2 + [f'(\theta)]^2}$$

providing a compact formula for arclength in polar coordinates.

> **Arclength in Polar Coordinates**
> If $r = f(\theta)$, a differentiable function for $\alpha \le \theta \le \beta$, then the length of the graph of $r = f(\theta)$ between $\theta = \alpha$ and $\theta = \beta$ is:
>
> $$\int_{\theta=\alpha}^{\theta=\beta} \sqrt{[f(\theta)]^2 + [f'(\theta)]^2}\, d\theta = \int_{\theta=\alpha}^{\theta=\beta} \sqrt{r^2 + \left[\frac{dr}{d\theta}\right]^2}\, d\theta$$

Example 5. Find the length of the polar curve $r = \sqrt{\theta}$ for $\pi \le \theta \le 2\pi$.

Solution. $r = \sqrt{\theta} \Rightarrow \frac{dr}{d\theta} = \frac{1}{2\sqrt{\theta}}$ so the length is given by:

$$\int_{\pi}^{2\pi} \sqrt{\left(\sqrt{\theta}\right)^2 + \left(\frac{1}{2\sqrt{\theta}}\right)^2}\, d\theta = \int_{\pi}^{2\pi} \sqrt{\theta + \frac{1}{4\theta}}\, d\theta \approx 6.8287$$

Like most integrals arising from arclength computations, we are unable find an antiderivative of the integrand and compute and exact value for the length, so we resort to technology to provide an approximate numerical answer. ◀

11.2 Problems

In problems 1–4, fill in the table below to indicate whether the values of the indicated derivatives are are positive (+), negative (−), zero (0) or undefined (U) at each point.

	$\frac{dx}{d\theta}$	$\frac{dy}{d\theta}$	$\frac{dr}{d\theta}$	$\frac{dy}{dx}$
A				
B				
C				
D				
E				

1. See figure below left. 2. See figure below.

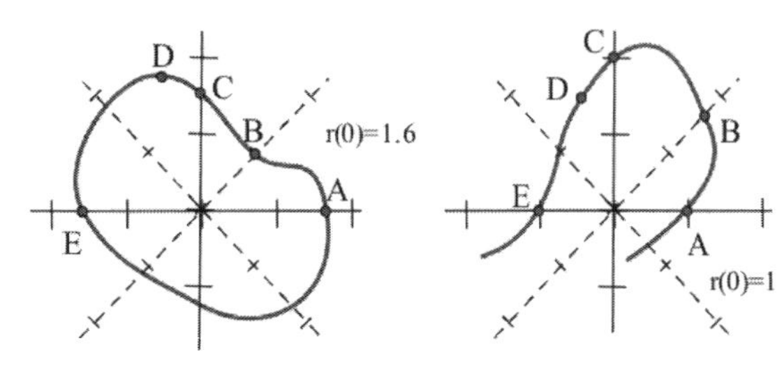

3. See figure below left. 4. See figure below.

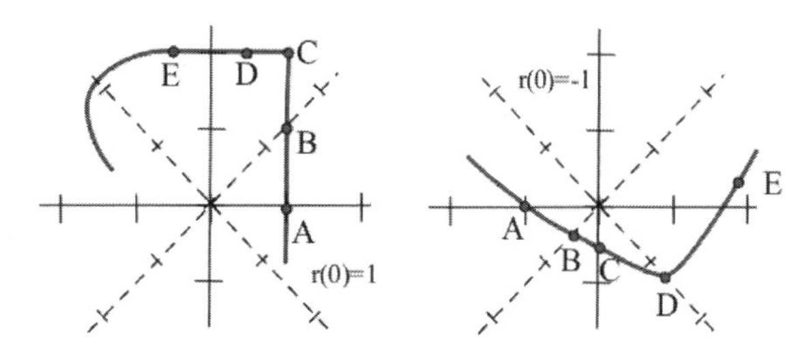

In Problems 5–8, sketch the graph of the given polar equation for $0 \le \theta \le 2\pi$; label the points A, B and C; and calculate the values of $\frac{dr}{d\theta}$ and $\frac{dy}{dx}$ at each of those points.

5. $r = 5$; $A\left(5, \frac{\pi}{4}\right)$, $B\left(5, \frac{\pi}{2}\right)$ and $C\left(5, \pi\right)$
6. $r = 2 + \cos(\theta)$; $A\left(2 + \sqrt{2}, \frac{\pi}{4}\right)$, $B\left(2, \frac{\pi}{2}\right)$ and $C\left(1, \pi\right)$
7. $r = 1 + \cos^2(\theta)$; $A\left(2, 0\right)$, $B\left(\frac{3}{2}, \frac{\pi}{4}\right)$ and $C\left(1, \frac{\pi}{2}\right)$
8. $r = \dfrac{6}{2 + \cos(\theta)}$; $A\left(2, 0\right)$, $B\left(3, \frac{\pi}{2}\right)$ and $C\left(\frac{24-6\sqrt{2}}{7}, \frac{\pi}{4}\right)$

9. Graph $r = 1 + 2\cos(\theta)$ for $0 \le \theta \le 2\pi$.
   (a) Show that the graph goes through the origin when $\theta = \frac{2\pi}{3}$ and $\theta = \frac{4\pi}{3}$.
   (b) Calculate $\dfrac{dy}{dx}$ when $\theta = \frac{2\pi}{3}$ and $\theta = \frac{4\pi}{3}$.
   (c) How can a curve have two different tangent lines (and slopes) at the origin?
10. Graph the cardiod $r = 1 + \sin(\theta)$ for $0 \le \theta \le 2\pi$.
   (a) At what points on the cardioid does $\frac{dx}{d\theta} = 0$?
   (b) At what points does $\frac{dy}{d\theta} = 0$?
   (c) At what points does $\frac{dr}{d\theta} = 0$?
   (d) At what points does $\frac{dy}{dx} = 0$?
11. Show that if a polar graph goes through the origin when the angle is θ_0 (and if $\frac{dr}{d\theta}$ exists there, but is not equal to 0) then the slope of the tangent line at the origin is $\tan\left(\theta_0\right)$.

In 12–20, represent the area of the given region as a definite integral. Then evaluate the integral exactly (if possible) or approximate using technology.

12. The shaded region in the figure below left.

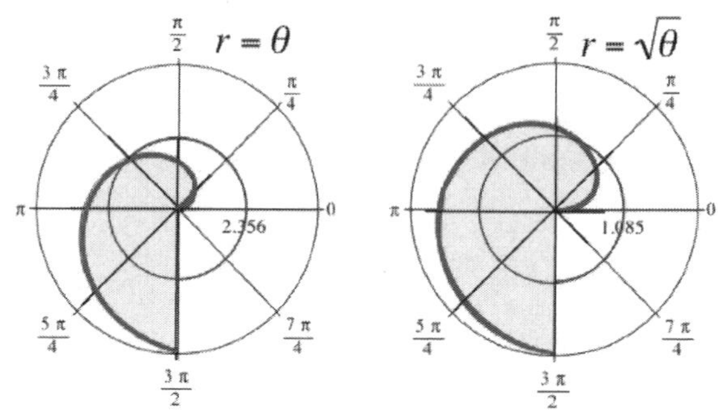

13. The shaded region in the figure above right.
14. The shaded region in the figure below left.

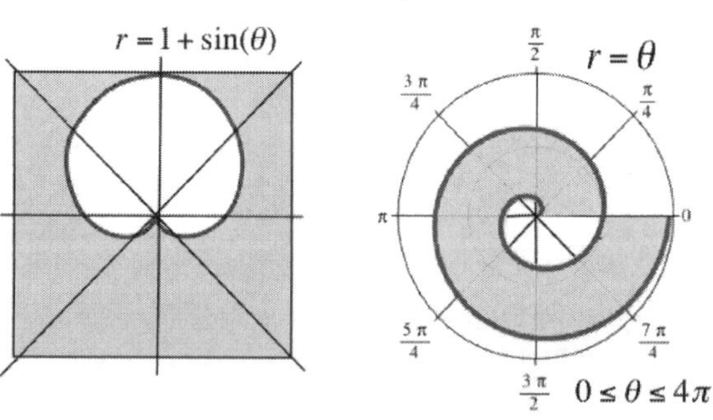

15. The shaded region in the figure above right.

16. The region inside the circle $r = 4\sin(\theta)$.
17. The region in the first quadrant outside the circle $r = 1$ and inside the cardiod $r = 1 + \cos(\theta)$.
18. The region in the second quadrant bounded by $r = \theta$ and $r = \theta^2$.
19. One "petal" of the graph of $r = \sin(3\theta)$.
20. One "petal" of the graph of $r = \sin(5\theta)$.
21. The "peanut" $r = 1.5 + \cos(2\theta)$.
22. The "peanut" $r = a + \cos(2\theta)$ (for $a > 1$).

In 23–30, represent the length of the curve as a definite integral. Then evaluate the integral exactly (if possible) or approximate using technology.

23. The spiral $r = \theta$ from $\theta = 0$ to $\theta = 2\pi$.
24. The spiral $r = \theta$ from $\theta = 2\pi$ to $\theta = 4\pi$.
25. The cardioid $r = 1 + \cos(\theta)$.
26. The circle $r = 4\sin(\theta)$ from $\theta = 0$ to $\theta = \pi$.
27. The circle $r = 5$ from $\theta = 0$ to $\theta = 2\pi$.
28. The "peanut" $r = 1.2 + \cos(2\theta)$.
29. One "petal" of $r = \sin(3\theta)$
30. One "petal" of $r = \sin(5\theta)$.
31. **Goat and Square Silo** (This problem does not require calculus.) One end of a 40-foot-long rope is attached to the middle of a wall of a 20-foot-square silo, and the other end is tied to a goat.

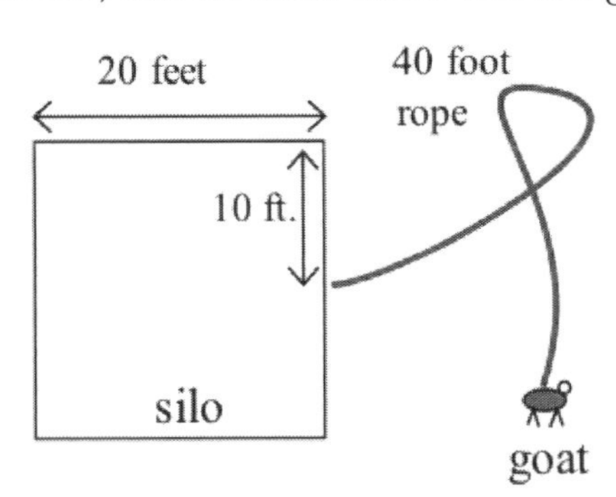

   (a) Sketch the region that the goat can reach.
   (b) Find the area of the region the goat can reach.
   (c) Can the goat reach a region with a bigger area if the rope is tied to the corner of the silo?

32. **Goat and Round Silo** (This problem does require calculus.) One end of a 10π-foot-long rope is attached to the wall of a round silo that has a radius of 10 feet, and the other end is tied to a goat.

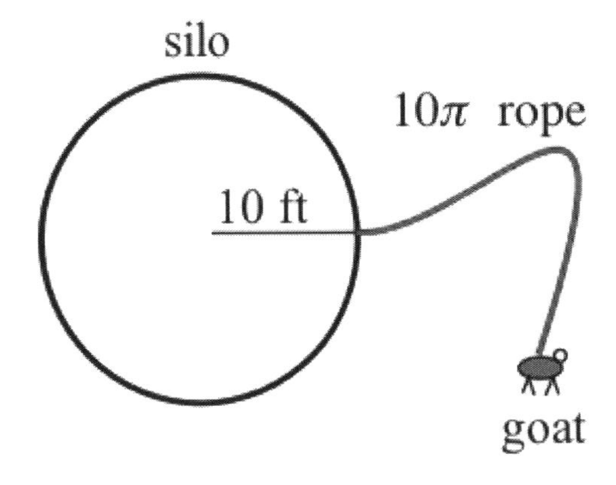

   (a) Sketch the region that the goat can reach.
   (b) Justify that the area of the shaded region shown below, as the goat goes around the silo from having θ feet of rope taut against the silo to having $\theta + \Delta\theta$ feet taut against the silo, is approximately:

$$\frac{1}{2}(10\pi - 10\theta)^2 \cdot \Delta\theta$$

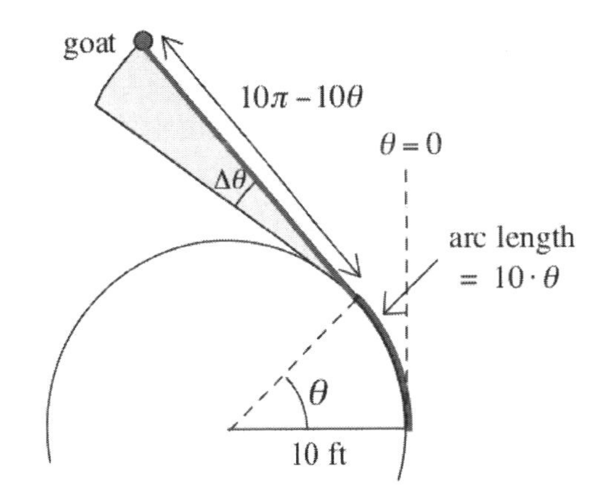

   (c) Use the preceding result to help calculate the area of the region that the goat can reach.

11.2 Practice Answers

	$\frac{dx}{d\theta}$	$\frac{dy}{d\theta}$	$\frac{dr}{d\theta}$	$\frac{dy}{dx}$
C	−	−	−	+
D	−	+	+	−

1. See margin table.

2. At P, $\theta = \frac{\pi}{2}$ and $r = \frac{\pi}{2}$, so $x = r\cos(\theta) = \frac{\pi}{2}\cos\left(\frac{\pi}{2}\right) = 0$ and $y = r\sin(\theta) = \frac{\pi}{2}\sin\left(\frac{\pi}{2}\right) = \frac{\pi}{2}$. From Example 2, we know that $\frac{dy}{dx} = -\frac{2}{\pi}$ at P, so an equation for the tangent line is $y = \frac{\pi}{2} - \frac{2}{\pi}(x - 0)$.

   At Q, $\theta = \pi$ and $r = \pi$, so $x = r\cos(\theta) = \pi\cos(\pi) = -\pi$ and $y = r\sin(\theta) = \pi\sin(\pi) = 0$. From Example 2, we know that $\frac{dy}{dx} = \pi$ at Q, so an equation for the tangent line is $y = 0 + \pi(x + \pi)$.

3. With $r = 1 - \sin(\theta)$, $x = r\cos(\theta) = (1 - \sin(\theta))\cos(\theta)$, so:

$$\begin{aligned}\frac{dx}{d\theta} &= (1 - \sin(\theta))(-\sin(\theta)) + \cos(\theta)(-\cos(\theta)) \\ &= -\sin(\theta) + \sin^2(\theta) - \cos^2(\theta) = -\sin(\theta) - \cos(2\theta)\end{aligned}$$

   Similarly, $y = r\sin(\theta) = (1 - \sin(\theta))\sin(\theta)$, so:

$$\begin{aligned}\frac{dy}{d\theta} &= (1 - \sin(\theta)) \cdot \cos(\theta) + \sin(\theta)(-\cos(\theta)) \\ &= \cos(\theta) - 2\sin(\theta)\cos(\theta) = \cos(\theta) - \sin(2\theta)\end{aligned}$$

   Therefore:

$$\frac{dy}{dx} = \frac{\frac{dy}{d\theta}}{\frac{dx}{d\theta}} = \frac{\cos(\theta) - \sin(2\theta)}{-\sin(\theta) - \cos(2\theta)}$$

   When $\theta = 0$, $\frac{dy}{dx} = -1$; when $\theta = \frac{\pi}{4}$, $\frac{dy}{dx} = \sqrt{2} - 1 \approx 0.414$; and when $\theta = \frac{\pi}{2}$, the derivative is undefined.

Looking at the graph of the cardioid, why should the slope be undefined there?

4. The "petals" of the rose $r = \sin(3\theta)$ intersect at the origin, where $r = 0 \Rightarrow \sin(3\theta) = 0 \Rightarrow 3\theta = k\pi \Rightarrow \theta = k \cdot \frac{\pi}{3}$ for any integer k. The shaded petal corresponds to $0 \le \theta \le \frac{\pi}{3}$, so its area is:

$$\int_0^{\frac{\pi}{3}} \frac{1}{2}\sin^2(3\theta)\,d\theta = \int_0^{\frac{\pi}{3}} \left[\frac{1}{4} - \frac{1}{4}\cos(6\theta)\right]d\theta = \left[\frac{1}{4}\theta - \frac{1}{24}\sin(6\theta)\right]_0^{\frac{\pi}{3}}$$

   which equals $\frac{\pi}{12} \approx 0.2618$.

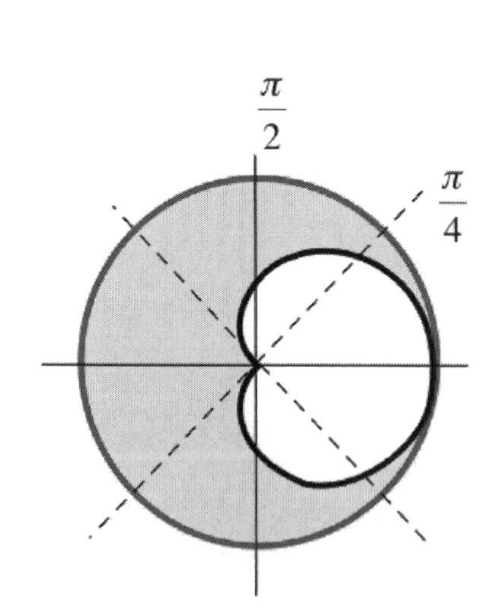

5. See margin for graph. The area of the region enclosed by the circle is $\pi \cdot 2^2 = 4\pi$, while the area of the region enclosed by the cardioid is $\frac{3\pi}{2}$ (using the result of Example 3). The shaded region therefore has area $4\pi - \frac{3\pi}{2} = \frac{5\pi}{2}$.

11.3 *Parametric Equations*

Some motions and paths are inconvenient, difficult or impossible for us to describe using a graph of the form $y = f(x)$.

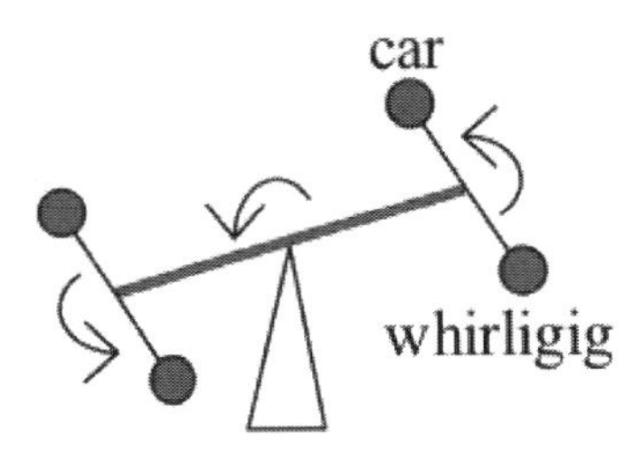

- A rider on a "whirligig" (see top margin figure) at a carnival travels in circles at the end of a rotating bar.
- A robot delivering supplies in a factory (second margin figure) must avoid obstacles.
- A fly buzzing around the room (third margin figure) or a molecule in a solution follow erratic paths.
- A stone caught in the tread of a rolling wheel has a smooth path with some sharp corners (see figure below).

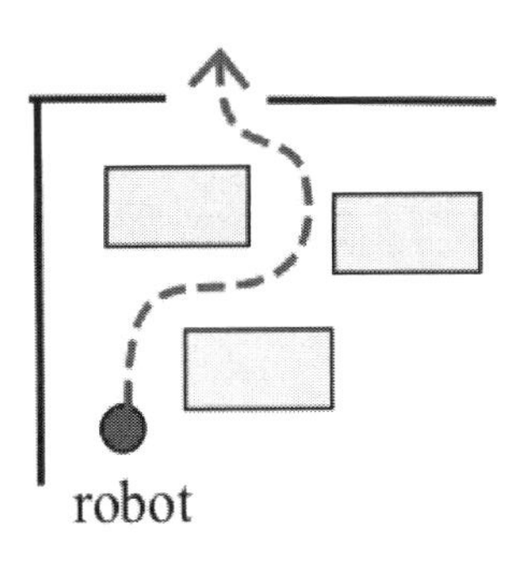

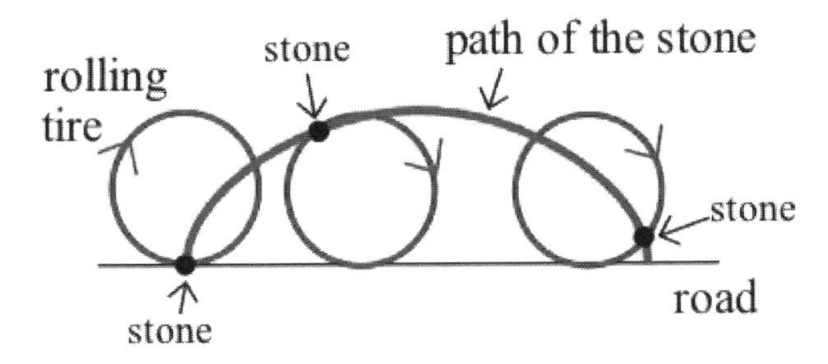

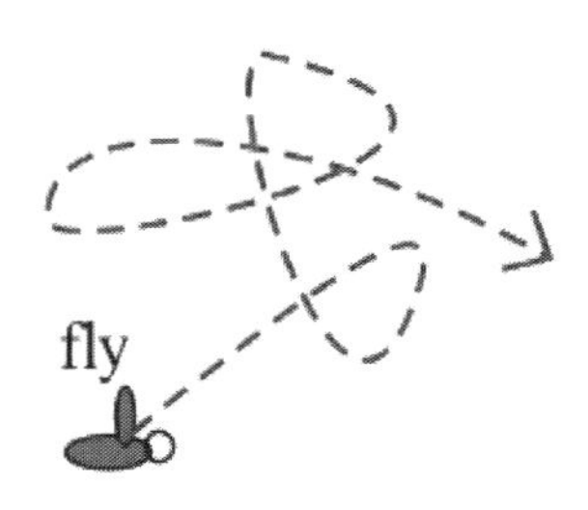

Parametric equations provide a way to describe all of these motions and paths. And parametric equations generalize easily to describe paths and motions in three (or more) dimensions.

We used parametric equations briefly in Sections 2.5 and 5.2. We consider them more carefully now, looking at functions given parametrically by data, graphs and formulas, and examining how to build formulas to describe certain motions parametrically (including the cycloid, one of the most famous curves in mathematics).

The next section uses calculus with parametric equations to find slopes of tangent lines, arclengths and areas.

In two dimensions, parametric equations describe the location of a point (x, y) on a graph or path as a function of a single independent variable t, a "parameter" often representing time. The coordinates x and y are functions of the variable t: $x = f(t)$ and $y = g(t)$ (see margin). (In three dimensions, we add a z-coordinate that is also a function of t: $z = h(t)$.) Among other applications, we can use parametric equations to analyze the forces acting on an object separately in each coordinate direction and then combine the results to determine the overall behavior of the object.

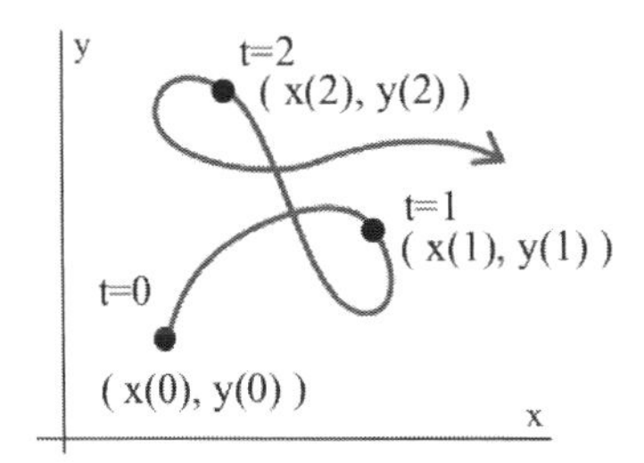

Graphing Parametric Curves

The data we need to create a graph can be given as a table of values, as graphs of $(t, f(t))$ and $(t, y(t))$, or as formulas for $f(t)$ and $g(t)$.

t	x	y
0	0	70
1	30	20
2	70	50
3	60	75
4	30	70
5	32	35
6	60	15
7	90	55
8	105	85
9	125	100
10	130	80
11	150	65
12	180	75
13	200	30

Example 1. The margin table records the location of a roller coaster car relative to its starting location. Use the data to sketch a graph of the car's path during the first seven seconds of motion.

Solution. The figure below plots the (x, y) locations of the car at one-second intervals from $t = 0$ to $t = 7$ seconds. We can connect these points using a smooth curve that shows one possible path of the car. ◀

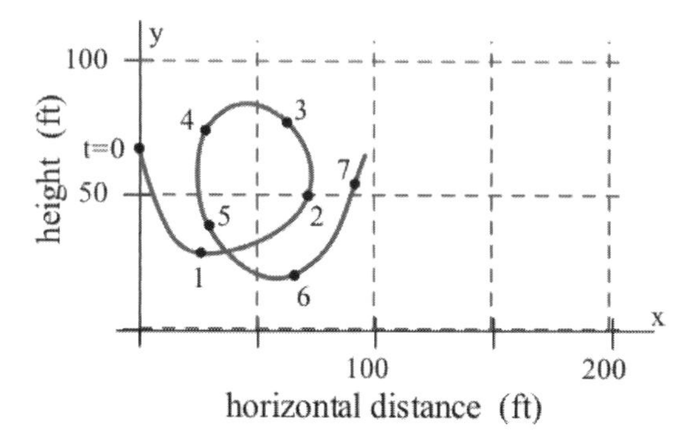

Practice 1. Use the remaining data in the margin table to sketch a possible path of the roller coaster car from $t = 7$ to $t = 13$ seconds.

Clearly the path of the roller coaster in the preceding Example is not the graph of a function $y = f(x)$. But every graph of the form $y = f(x)$ has an easy parametric representation: set $x(t) = t$ and $y(t) = f(t)$.

Sometimes a parametric graph can show patterns that are not clearly visible in individual graphs.

Example 2. The figures below are graphs of the populations of rabbits and foxes on an island. Use these graphs to sketch a parametric graph of rabbits (x-axis) versus foxes (y-axis) for $0 \leq t \leq 10$ years.

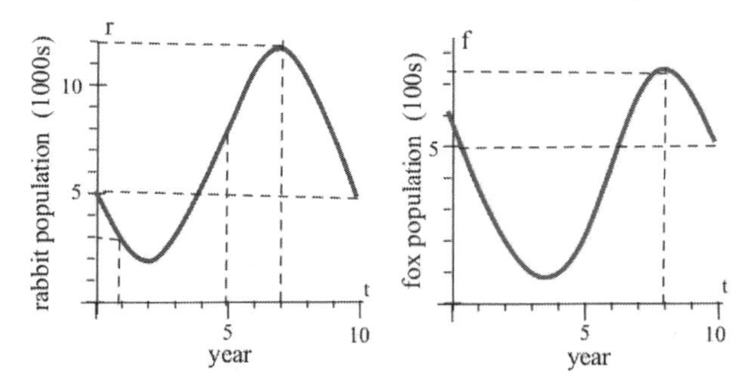

Solution. The separate rabbit and fox population graphs give us information about each population separately, but the parametric graph helps us see the effects of the interaction between the rabbits and the foxes more clearly. For each time t, you can read the rabbit and fox populations from the separate graphs (for example, when $t = 1$, there are roughly 3,000 rabbits and 400 foxes so $x \approx 3000$ and $y \approx 400$) and then combine this information to plot a single point.

If you repeat this process for a large number of values of t, you get a graph (see margin) of the "motion" of the rabbit and fox populations over a period of time. We can then ask questions about why the populations might exhibit this behavior. ◀

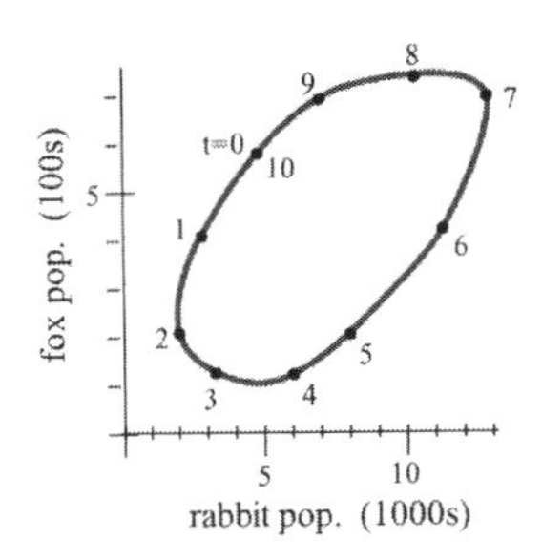

The type of graph created in the preceding Example is very common for "predator-prey" interactions. Some two-species populations approach a "steady state" or "fixed point" (see second margin figure), while others repeat a cyclical pattern over time (as in Example 2).

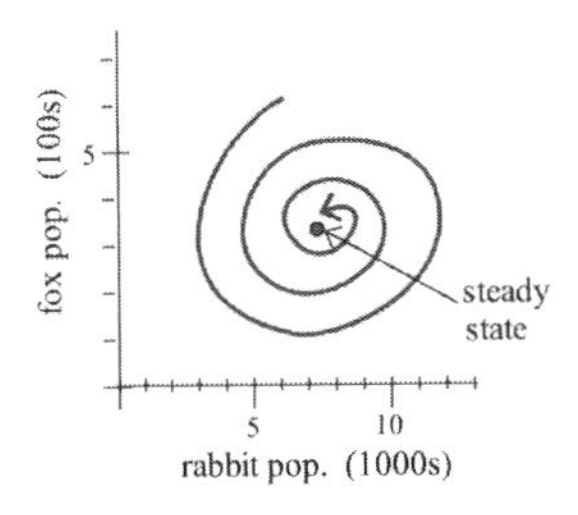

Practice 2. What would happen if the rabbit-fox graph touched the horizontal axis?

Example 3. Graph the parametric equations $x(t) = 2t - 2$ and $y(t) = 3t + 1$ in the xy-plane.

Solution. The margin table shows the values of x and y for several values of t and the graph shows these points plotted in the xy-plane. The graph appears to be a line. Often it is difficult or impossible to write y as a simple function of x, but in this situation we can do so:

t	x	y
−1	−4	−2
0	−2	1
1	0	4
2	2	7

$$x = 2t - 2 \Rightarrow t = \frac{1}{2}x + 1 \Rightarrow y = 3\left(\frac{1}{2}x + 1\right) + 1 = \frac{3}{2}x + 4$$

This agrees with what we see in the graph: a line with slope $\frac{3}{2}$ and y-intercept at $(0, 4)$. ◀

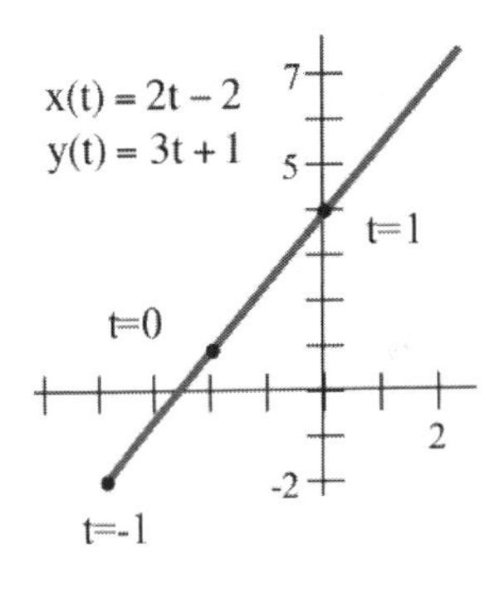

Practice 3. Graph $x(t) = 3 - t$ and $y(t) = t^2 + 1$ in the xy-plane. Then write y as a function of x alone and identify the shape of the graph.

Example 4. Graph $x(t) = 3\cos(t)$ and $y(t) = 2\sin(t)$ in the xy-plane for $0 \le t \le 2\pi$, then show that these parametric equations satisfy the relation $\frac{x^2}{9} + \frac{y^2}{4} = 1$ for all values of t.

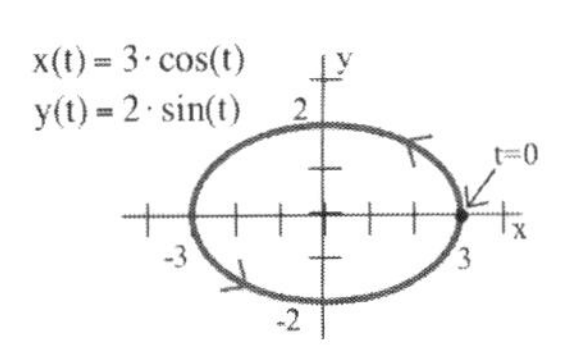

Solution. The graph, an ellipse, appears in the margin. Substituting $3\cos(t)$ for x and $2\sin(t)$ for y into the left-hand side of the given relation yields $\cos^2(t) + \sin^2(t) = 1$, as required. ◀

Practice 4. Graph the equations $x(t) = \sin(t)$ and $y(t) = 5\cos(t)$ in the xy-plane for $0 \le t \le 2\pi$, then show that these equations satisfy the relation $x^2 + \frac{y^2}{25} = 1$ for all values of t.

Example 5. Describe the motion of an object whose location at time t is given by $x(t) = -R \cdot \sin(t)$ and $y(t) = -R \cdot \cos(t)$.

Solution. At $t = 0$, the object starts at $x(0) = -R\sin(0) = 0$ and $y(0) = -R\cos(0) = -R$. By plotting $x(t)$ and $y(t)$ for several other

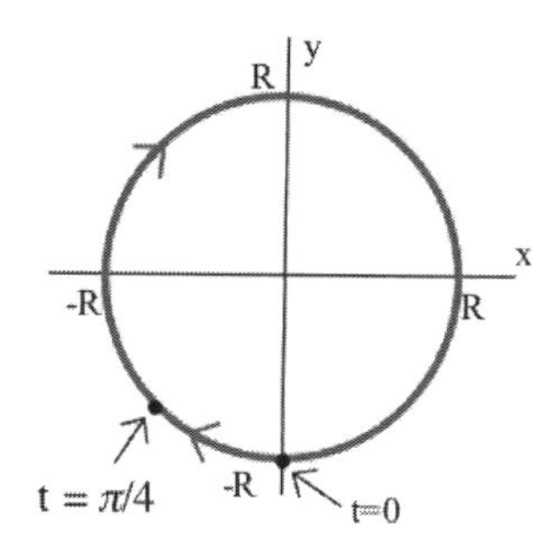

values of t (see margin figure), we can see that the object is rotating clockwise around the origin. Because:

$$x^2 + y^2 = [-R\sin(t)]^2 + [-R\cos(t)]^2 = R^2\left[\sin^2(t) + \cos^2(t)\right] = R^2$$

the object must traverse a circle of radius R centered at the origin. ◀

Practice 5. Each set of parametric equations below give the position of an object travelling around a circle of radius 1 centered at the origin.

(a) $x(t) = \cos(2t)$, $y(t) = \sin(2t)$

(b) $x(t) = -\cos(3t)$, $y(t) = \sin(3t)$

(c) $x(t) = \sin(4t)$, $y(t) = -\cos(4t)$

For each object, determine:

- the location of the object at time $t = 0$.
- whether the object is traveling clockwise or counterclockwise.
- the time it takes for the object to make one revolution.

Putting Motions Together

If we know how an object moves horizontally and how it moves vertically, we can combine these motions to see how the object moves through the xy-plane.

If you throw an object straight upward with an initial velocity of A feet per second, then its height after t seconds is $y(t) = A \cdot t - \frac{1}{2}g \cdot t^2$ feet where $g = 32$ feet/sec^2 is the (downward) acceleration of gravity (see below left). If you throw an object horizontally with an initial velocity of B feet per second, then its horizontal distance from the starting place after t seconds is $x(t) = B \cdot t$ feet (see below center).

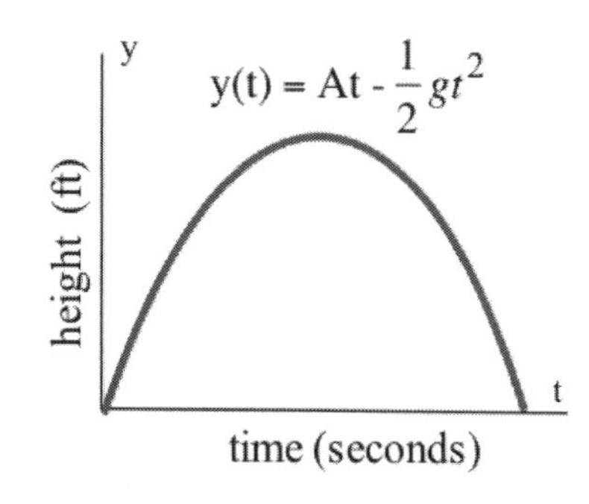

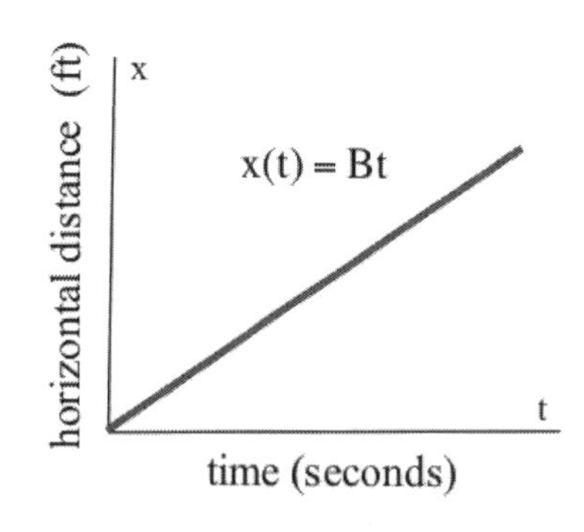

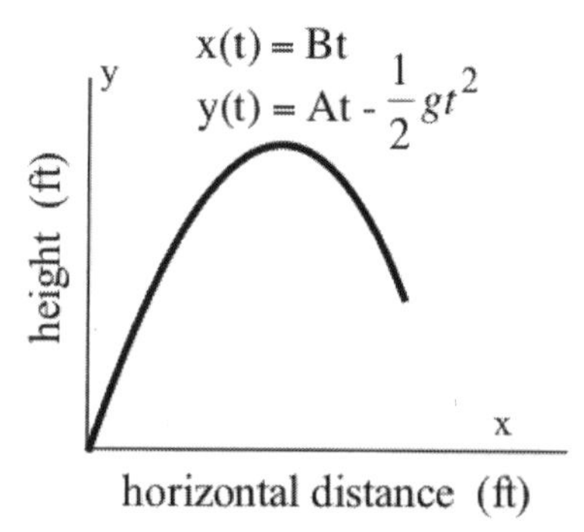

Example 6. Write parametric equations for the location at time t (above right) of an object thrown at an angle of 30° with the ground (horizontal) with an initial velocity 100 feet per second.

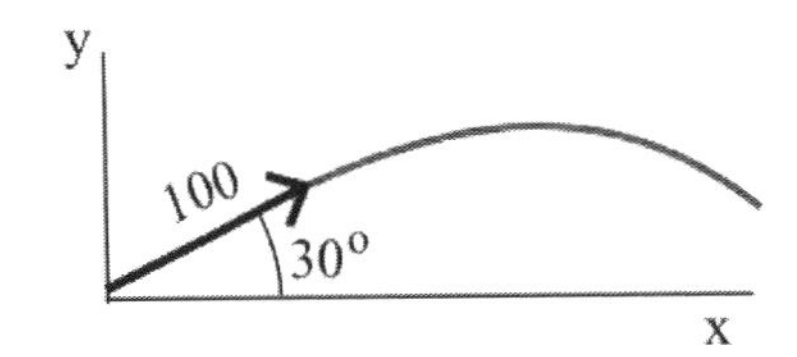

Solution. If the object travels 100 feet along a line at an angle of $30°$ to the horizontal ground (see margin), then it travels $100 \cdot \sin(30°) = 50$ feet upward and $100 \cdot \cos(30°) \approx 86.6$ feet sideways, so $A = 50$ and $B = 86.6$ (using the notation from the discussion above). The location of the object at time t is therefore given by $x(t) = 86.6t$ and $y(t) = 50t - \frac{1}{2}gt^2 = 50t - 16t^2$. ◀

Practice 6. You throw a ball upward at an angle of $45°$ with an initial velocity of 40 ft/sec.

(a) Write the parametric equations for the position of the ball as a function of time.

(b) Use the parametric equations to find when and then where the ball will hit the sloped ground (as shown in the margin figure).

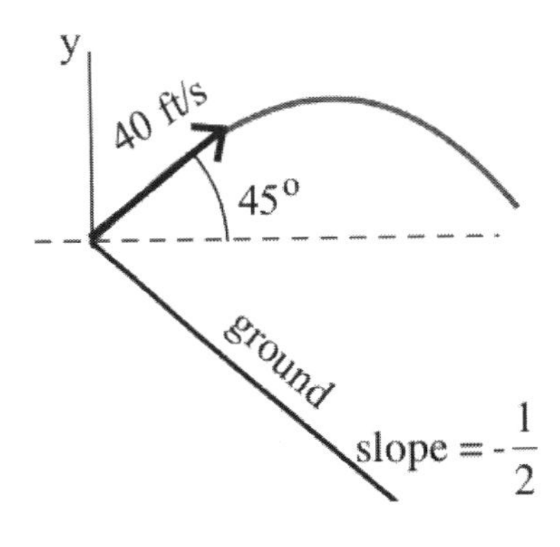

Sometimes we record the location or motion of an object using an instrument that is itself in motion (for example, tracking a pod of migrating whales from a moving ship) and we want to determine the path of the object independent of the location of the instrument. In that case, the "absolute" location of the object with respect to the origin is the sum of the relative location of the object (the pod of whales) with respect to the instrument (the ship) and the location of the instrument (the ship) with respect to the origin. The same approach allows us to describe the motion of linked objects, such as connected gears.

Example 7. A car on a carnival ride (see margin) makes one counterclockwise revolution (with radius $r = 8$ feet) about the pivot point A every two seconds. The pivot A is at the end of a longer arm (with radius $R = 20$ feet) that makes one counterclockwise revolution about its pivot point (the origin) every five seconds. If the ride begins with the two arms outstretched along the positive x-axis, sketch the path you think the car will follow. Then find a pair of parametric equations that describe the location of the car at time t.

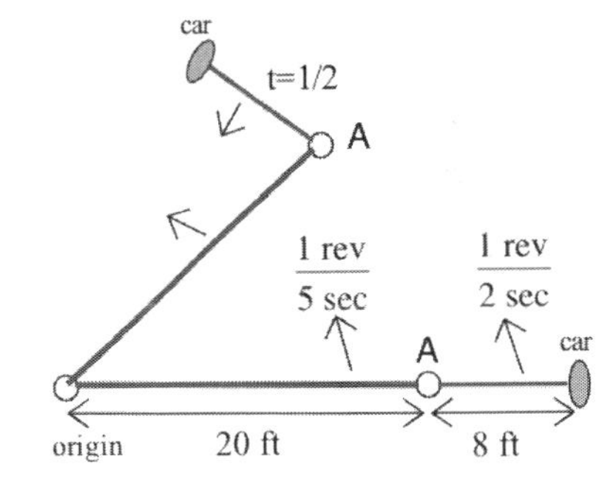

Solution. The location of the car relative to its pivot point A is given by $x_c(t) = 8\cos\left(\frac{2\pi}{2}t\right) = 8\cos(\pi t)$ and $y_c(t) = 8\sin\left(\frac{2\pi}{2}t\right) = 8\sin(\pi t)$.

The position of the pivot point A relative to the origin is given by $x_p(t) = 20\cos\left(\frac{2\pi}{5}t\right)$ and $y_p(t) = 20\sin\left(\frac{2\pi}{5}t\right)$, so the location of the car, relative to the origin, is given by:

$$x(t) = x_p(t) + x_c(t) = 8\cos(\pi t) + 20\cos\left(\frac{2\pi}{5}t\right)$$

$$y(t) = y_p(t) + y_c(t) = 8\sin(\pi t) + 20\sin\left(\frac{2\pi}{5}t\right)$$

Use technology to graph the path of the car for over the first five seconds and compare that path to your initial guess. ◀

The Cycloid

Our final example devlops parametric equations for a curve called a **cycloid**, one of the most famous and interesting curves in mathematics.

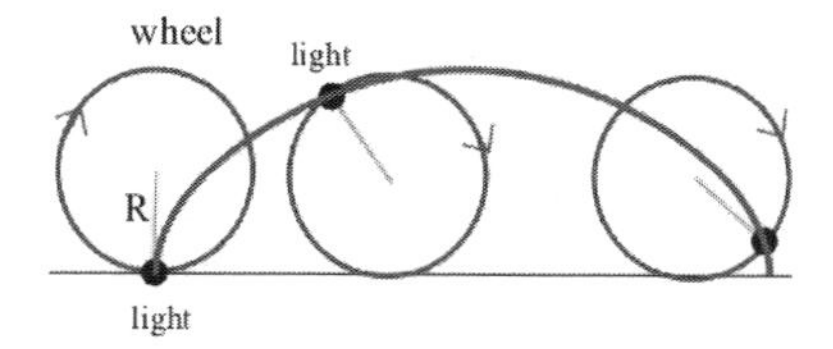

Example 8. A light is attached to the edge of a wheel of radius R, which rolls along a level road (see margin). Find parametric equations to describe the location of the light.

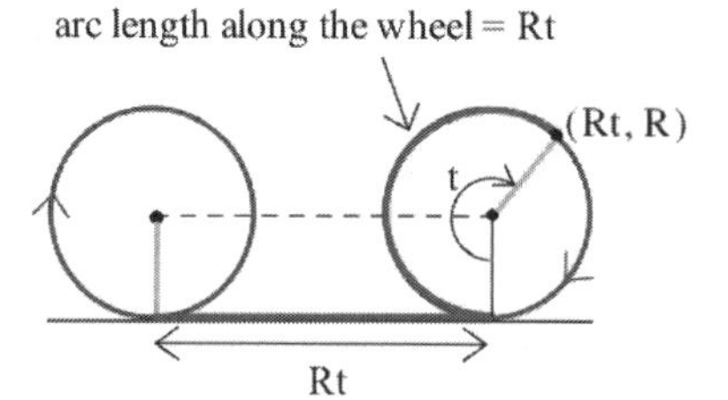

Solution. We can describe the location of the axle of the wheel, then the location of the light relative to the axle, and finally put the results together to get the location of the light.

The axle of the wheel is always R inches off the ground, so the y-coordinate of the axle is given by $y_a(t) = R$. When the wheel has rotated t radians about its axle, the wheel has rolled a distance of $R \cdot t$ along the road, so the x-coordinate of the axle is given by $x_a(t) = R \cdot t$.

The position of the light relative to the axle is given by $x_l(t) = -R\sin(t)$ and $y_l(t) = -R\cos(t)$ so the position of the light relative to the origin is given by:

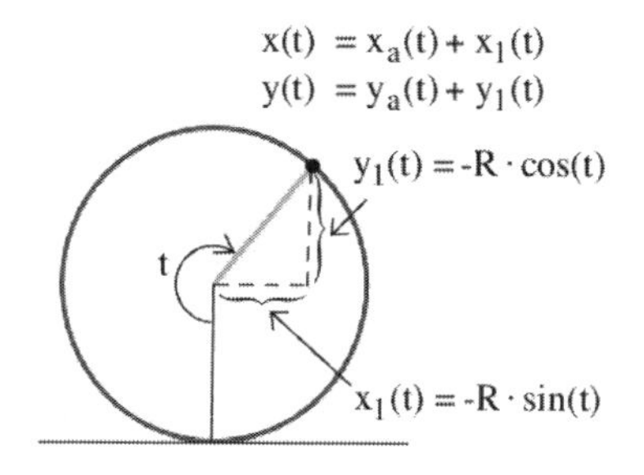

$$x(t) = x_a(t) + x_l(t) = Rt - R\sin(t) = R\left[t - \sin(t)\right]$$
$$y(t) = y_a(t) + y_l(t) = R - R\cos(t) = R\left[1 - \cos(t)\right]$$

Use technology (choose a value for R) to graph these equations. ◀

Many great mathematicians and physicists (Mersenne, Galileo, Newton, Bernoulli, Huygens and others) examined the cycloid, determined its properties and used it in physical applications. Marin Mersenne (1588–1648) thought the path might be part of an ellipse (it isn't). In 1634, Gilles Personne de Roberval (1602–1675) determined the parametric form of the cycloid and found the area under the cycloid, as did Descartes and Fermat, before Newton (1642–1727) was even born: they used various specialized geometric approaches to solve the area problem. Around the same time, Galileo determined the area experimentally by cutting a cycloidal region from a sheet of lead and balancing it against a number of disks (with the same radius as the circle that generated the cycloid) cut from the same material. How many disks do you think balance the cycloidal region's area?

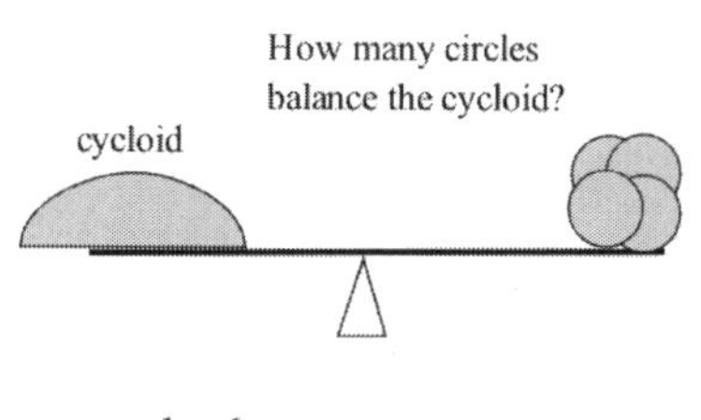

The cycloid's most amazing properties, however, involve motion along a cycloid-shaped path. Those discoveries had to wait for Newton and calculus.

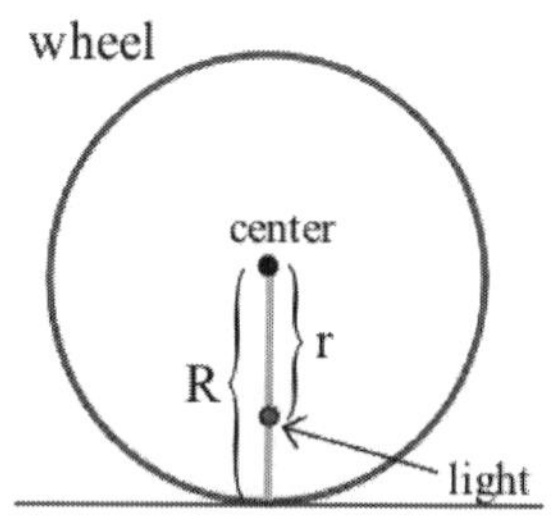

Practice 7. A light is attached r inches from the axle to a wheel of radius R inches ($r < R$) that rolls along a level road (see margin). Use the approach of Example 8 to find parametric equations to describe the location of the light. The resulting curve is called a **curate cycloid**.

11.3 Problems

For Problems 1–4, use the data in the table to create three graphs: (a) $(t, x(t))$ (b) $(t, y(t))$ and (c) the parametric graph $(x(t), y(t))$. (Connect the points with line segments to create the graph.)

1.

t	x	y
0	2	1
1	2	0
2	−1	0
3	1	−1

2.

t	x	y
0	0	1
1	1	1
2	1	−1
3	2	0

3.

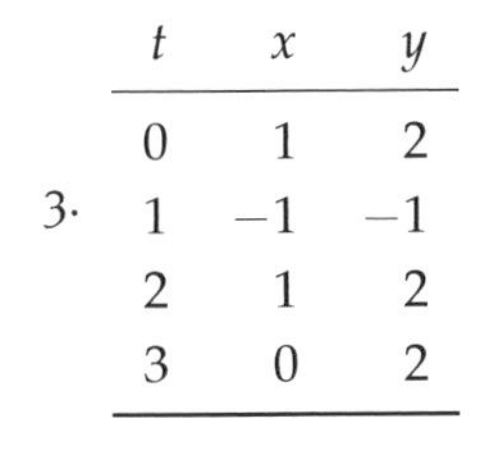

t	x	y
0	1	2
1	−1	−1
2	1	2
3	0	2

4.

t	x	y
0	0	1
1	−1	0
2	0	−2
3	3	1

For 5–8, use the given graphs of $(t, x(t))$ and $(t, y(t))$ to sketch the parametric graph $(x(t), y(t))$.

5.

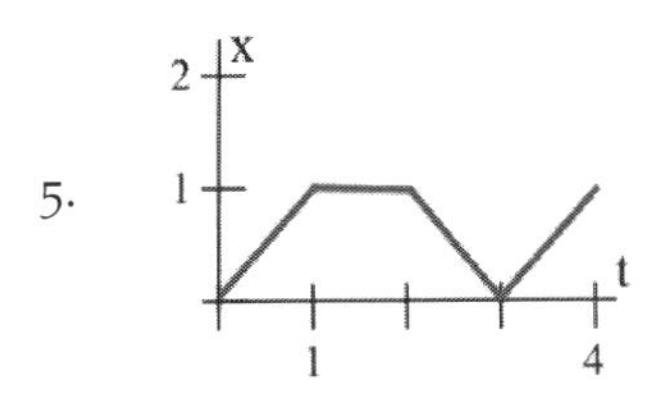

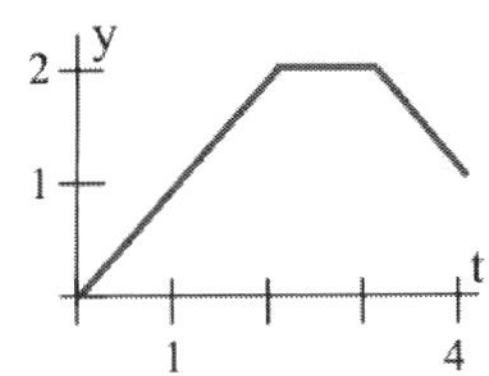

6.

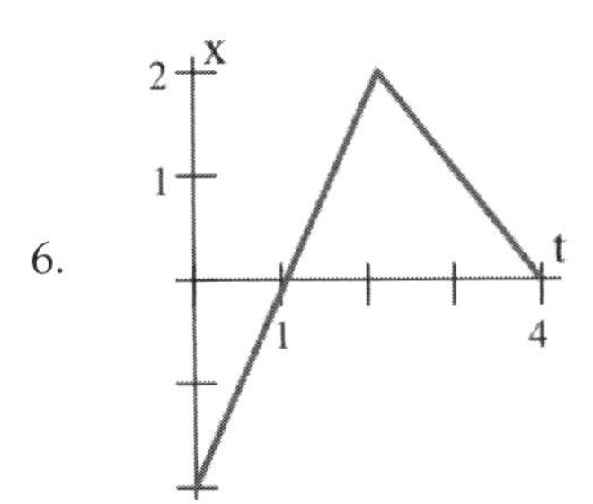

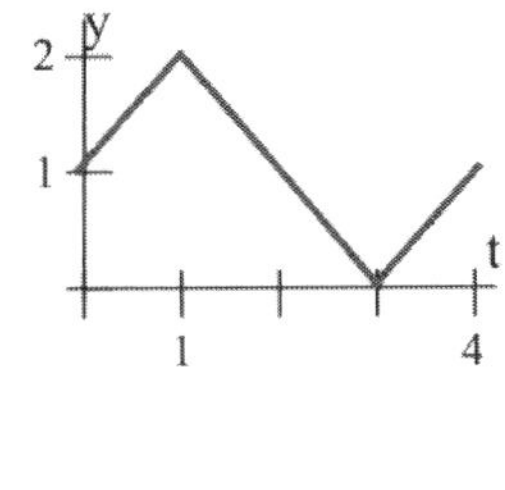

7.

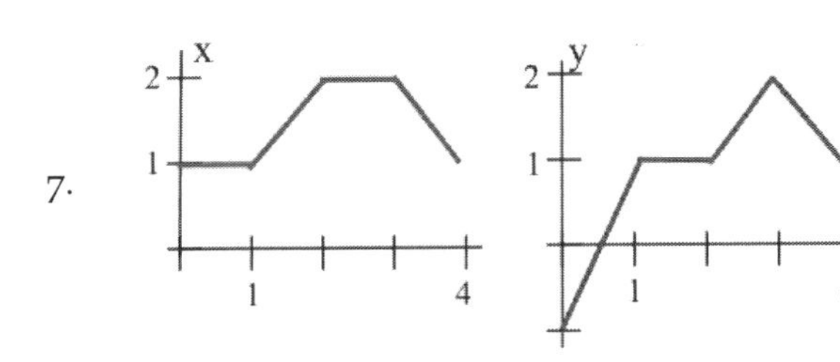

8.

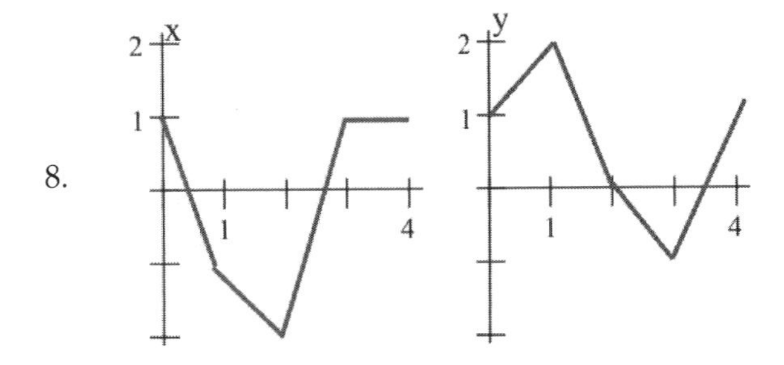

9. Graph the parametric equations $x(t) = 3t - 2$, $y(t) = 1 - 2t$. What shape is this graph?

10. Graph the parametric equations $x(t) = 2 - 3t$, $y(t) = 3 + 2t$. What shape is this graph?

11. Calculate the slope of the line through the points $P = (x(0), y(0))$ and $Q = (x(1), y(1))$ for the equations $x(t) = at + b$ and $y(t) = ct + d$.

12. Graph $x(t) = 3 + 2\cos(t)$, $y(t) = -1 + 3\sin(t)$ for $0 \le t \le 2\pi$. Describe the shape of the graph.

13. Graph $x(t) = -2 + 3\cos(t)$, $y(t) = 1 - 4\sin(t)$ for $0 \le t \le 2\pi$. Describe the shape of the graph.

14. Graph each set of parametric equations, then describe the similarities and the differences among these graphs.

    (a) $x(t) = t^2$, $y(t) = t$

    (b) $x(t) = \sin^2(t)$, $y(t) = \sin(t)$

    (c) $x(t) = t$, $y(t) = \sqrt{t}$.

15. Graph each set of parametric equations, then describe the similarities and the differences among these graphs.

    (a) $x(t) = t$, $y(t) = t$

    (b) $x(t) = \sin(t)$, $y(t) = \sin(t)$

    (c) $x(t) = t^2$, $y(t) = t^2$.

16. Graph the parametric equations:

$$x(t) = \left(4 - \frac{1}{t}\right)\cos(t), \quad y(t) = \left(4 - \frac{1}{t}\right)\sin(t)$$

for $t \ge 1$, then describe the behavior of the graph.

17. Graph $x(t) = \dfrac{\cos(t)}{t}$, $y(t) = \dfrac{\sin(t)}{t}$ for $t \geq \dfrac{\pi}{4}$, then describe the behavior of the graph.

18. Graph $x(t) = t + \sin(t)$, $y(t) = t^2 + \cos(t)$ for $0 \leq t \leq 2\pi$, then describe the shape of the graph.

Problems 19–22 refer to the rabbit–fox population graph below, which shows several different population cycles depending on the various numbers of rabbits and foxes. Wildlife biologists sometimes try to control animal populations by "harvesting" some of the animals, but this needs to be done with care. The thick dot on the graph is the fixed point for this two-species population.

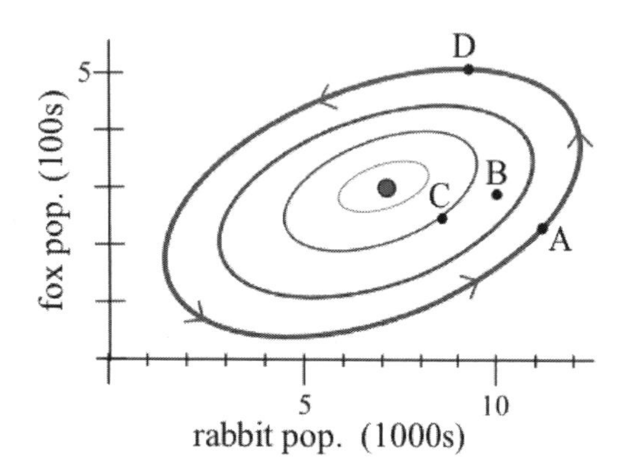

19. If there are currently 11,000 rabbits and 200 foxes (point A), and wildlife officials "harvest" 1,000 rabbits (removing them from the population), does the harvest shift the populations onto a cycle closer to or farther from the fixed point?

20. If there are currently 10,000 rabbits and 300 foxes (point B), and officials "harvest" 100 foxes, does the harvest shift the populations onto a cycle closer to or farther from the fixed point?

21. If there are currently 8,000 rabbits and 250 foxes (point C), and 1,000 rabbits die during a hard winter, does the wildlife biologist need to take action to main the population balance? Justify your response.

22. If there are currently 9,000 rabbits and 500 foxes (point D), and 2,000 rabbits die during a hard winter, does the wildlife biologist need to take action to main the population balance? Justify your response.

23. If $x(t) = at + b$ and $y(t) = ct + d$ with $a \neq 0$ and $c \neq 0$, write y as a function of x alone and show that the parametric graph $(x(t), y(t))$ is a line. What is the slope of that line?

24. Each set of parametric equations given below satisfy $x^2 + y^2 = 1$ and, for $0 \leq t \leq 2\pi$, describe the position of an object moving around a circle with radius 1 with center at the origin. Explain how the motions of the objects differ.
   (a) $x(t) = \cos(t)$, $y(t) = \sin(t)$
   (b) $x(t) = \cos(-t)$, $y(t) = \sin(-t)$
   (c) $x(t) = \cos(2t)$, $y(t) = \sin(2t)$
   (d) $x(t) = \sin(t)$, $y(t) = \cos(t)$
   (e) $x(t) = \cos\left(t + \frac{\pi}{2}\right)$, $y(t) = \sin\left(t + \frac{\pi}{2}\right)$

25. From a tall building, you observe a person walking along a straight path while twirling a light (parallel to the ground) at the end of a string.
   (a) If the person is walking slowly, sketch the path of the light.
   (b) How would the path of the light change if the person were running?
   (c) Sketch the path of the light for a person walking along a parabolic path.
   (d) Sketch the path of the light for a person running along a parabolic path.

26. William Tell aims his arrow directly at an apple and releases the arrow at exactly the same instant that the apple stem breaks. In a world without gravity (or air resistance), the apple remains in place after the stem breaks and the arrow flies straight to hit the apple (see below).

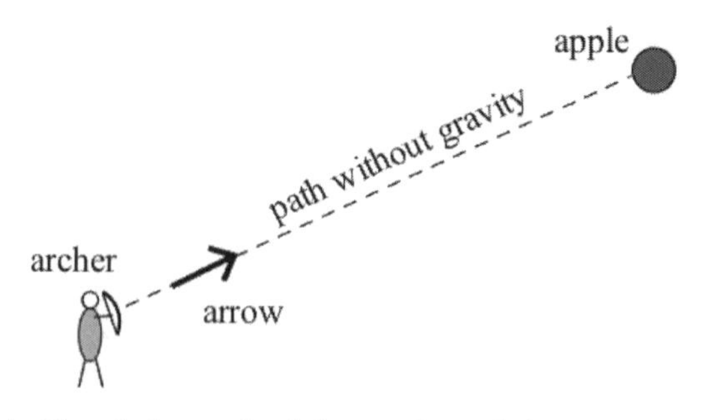

   (a) Sketch the path of the apple and the arrow in a world with gravity (but still no air resistance).
   (b) Does the arrow still hit the apple? Explain.

27. Find the radius R of a circle that generates a cycloid that starts at the point $(0,0)$ and:

    (a) passes through the point $(10\pi, 0)$ on its first complete revolution $(0 \le t \le 2\pi)$.

    (b) passes through the point $(5,2)$ on its first complete revolution. (Technology is helpful here.)

    (c) passes through the point $(2,3)$ on its first complete revolution. (Technology is helpful here.)

    (d) passes through the point $(4\pi, 8)$ on its first complete revolution.

28. Your friends are riding on the Ferris wheel illustrated below, and a t seconds after the ride begins, their location is given parametrically as:

$$\left(-20\sin\left(\frac{2\pi}{15}t\right), 30 - 20\cos\left(\frac{2\pi}{15}t\right)\right)$$

    (a) Is the Ferris wheel turning clockwise of counterclockwise?

    (b) How many seconds does it take the Ferris wheel to make one complete revolution?

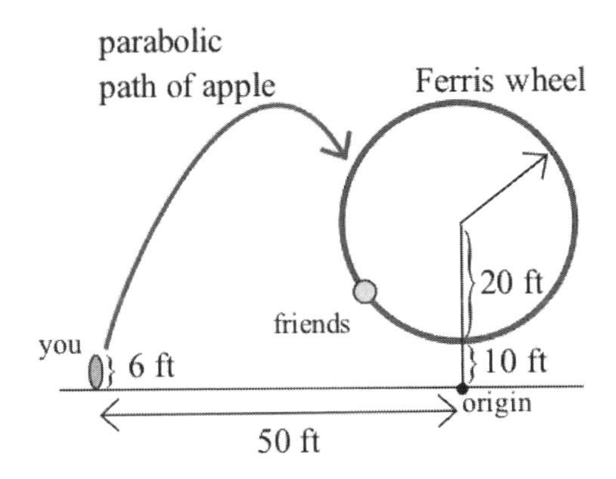

29. You are standing 50 feet to the left of the Ferris wheel from Problem 28. You toss an apple from a height of six feet above the ground at an angle of $45°$. Write parametric equations for the location of the apple (relative to the origin indicated in the figure above) at time t if:

    (a) you give the apple an initial velocity of 30 feet per second.

    (b) you give the apple an initial velocity of v feet per second.

30. Help—the Ferris wheel won't stop! To keep your friends on the Ferris wheel in Problems 28–29 from getting hungry, you toss an apple to them (at time $t = 0$). Find a formula for the distance between the apple and your friends at time t. Somehow (technology may be useful), find a value for the initial velocity v of the apple that will ensure that it comes close enough for your friends to catch it (within two feet should do the trick).

31. A wheel of radius R sits on a ledge, with a rod of length $1.5R$ attached to the center of the wheel and hanging down over the ledge. Find parametric equations for the path (called a **prolate cycloid**) of a light at the end of the rod.

32. A wheel of radius R rolls along the inside of a circle of radius $3R$. Find parametric equations for the path (called a **hypocycloid**) of a light on the edge of the wheel.

33. A wheel of radius R rolls along the outside of a circle of radius $3R$. Find parametric equations for the path (called an **epicycloid**) traced out by a light on the edge of the wheel.

11.3 Practice Answers

1. A possible path for the car appears in the margin.

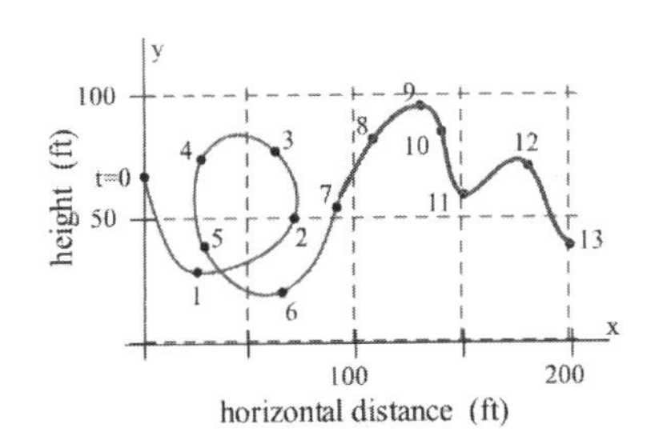

2. At the time the (rabbit, fox) parametric graph touches the horizontal axis there will be 0 foxes, so the fox population becomes extinct.

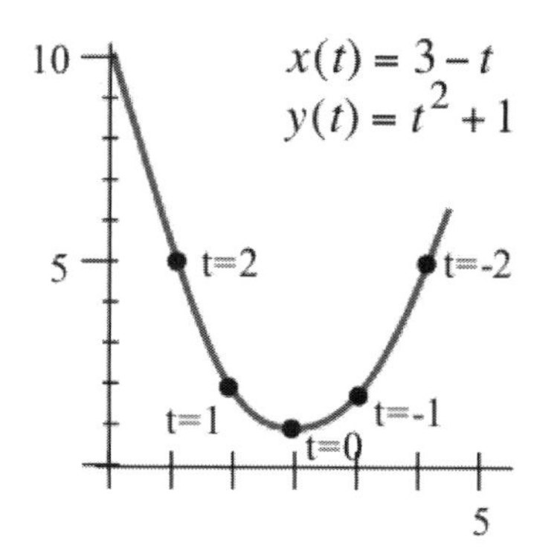

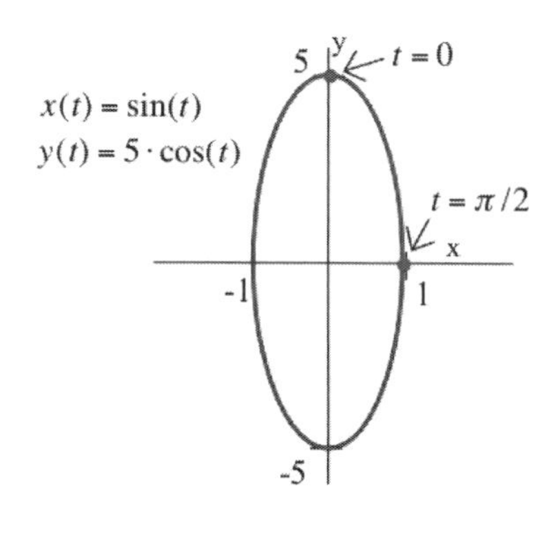

3. If $x = 3 - t$ and $y = t^2 + 1$ then $t = 3 - x$ and $y = (3 - x)^2 + 1 = x2 - 6x + 10$. The graph (see margin) is parabola, opening upward, with vertex at $(3, 1)$.

4. For all t:

$$\frac{x^2}{1} + \frac{y^2}{25} = \frac{\sin^2(t)}{1} + \frac{25\cos^2(t)}{25} = \sin^2(t) + \cos^2(t) = 1$$

A parametric graph of $x(t) = \sin(t)$ and $y(t) = 5\cos(t)$ appears in the second margin figure.

5. A starts at $(1, 0)$, travels counterclockwise, and takes $\frac{2\pi}{2} = \pi$ seconds to make one revolution. B starts at $(-1, 0)$, travels clockwise, and takes $\frac{2\pi}{3}$ seconds to make one revolution. C starts at $(0, -1)$, travels counterclockwise, and takes $\frac{2\pi}{4} = \frac{\pi}{2}$ seconds to make one revolution.

6. (a) $x(t) = 40\cos(45°)\,t = 20\sqrt{2}\,t$ and $y(t) = 40\sin(45°)\,t - 16t^2 = 20\sqrt{2}\,t - 16t^2$

   (b) Along the ground $y = -\frac{1}{2}x$, so the ball hits the ground when:

$$y(t) = -\frac{1}{2}x(t) \Rightarrow 20\sqrt{2}\,t - 16t^2 = -10\sqrt{2}\,t \Rightarrow t = \frac{15\sqrt{2}}{8}$$

   (assuming $t \neq 0$). The location of the ball is therefore given by:

$$x\left(\frac{15\sqrt{2}}{8}\right) = 20\sqrt{2}\left(\frac{15\sqrt{2}}{8}\right) = 75$$

$$y\left(\frac{15\sqrt{2}}{8}\right) = 20\sqrt{2}\left(\frac{15\sqrt{2}}{8}\right) - 16\left(\frac{15\sqrt{2}}{8}\right)^2 = -37.5$$

   so the ball hits the ground at location $(75, -37.5)$ after (approximately) 2.652 seconds.

7. The axle is located at $x_a(t) = Rt$ and $y_a(t) = R$ while the location of the light relative to the axle is given by $x_l(t) = -r\sin(t)$ and $y_l(t) = -r\cos(t)$, hence the position of the light relative to the origin is given by:

$$x(t) = x_a(t) + x_l(t) = Rt - r\sin(t)$$
$$y(t) = y_a(t) + y_l(t) = R - r\cos(t)$$

11.4 *Calculus with Parametric Equations*

The previous section discussed parametric equations, their graphs and some of their uses. This section examines some of the ideas and techniques of calculus as they apply to parametric equations: slope of a tangent line, speed, arclength and area. Treatments of slope, speed, and arclength for parametric equations previously appeared in Sections 2.5 and 5.3, so the presentation here is brief. The material on area (new to this section) is a variation on the Riemann-sum development of the integral. This section ends with an investigation of some of the properties of the cycloid.

Slope

Also see Section 2.5.

If $x(t)$ and $y(t)$ are differentiable functions of t, then the derivatives $\frac{dx}{dt}$ and $\frac{dy}{dt}$ measure the rates of change of x and y, respectively, with respect to t. The derivative $\frac{dy}{dx}$ measures the slope of the line tangent to the parametric graph $(x(t), y(t))$. To calculate $\frac{dy}{dx}$ we need to use the Chain Rule:

$$\frac{dy}{dt} = \frac{dy}{dx} \cdot \frac{dx}{dt} \quad \Rightarrow \quad \frac{dy}{dx} = \frac{\frac{dy}{dy}}{\frac{dx}{dt}}$$

as long as $\frac{dx}{dt} \neq 0$.

Example 1. The location of an object in a plane, relative to the orgin, is given by the parametric equations $x(t) = t^3 + 1$ feet and $y(t) = t^2 + t$ feet at time t seconds.

(a) Evaluate $x(t)$ and $y(t)$ at $t = -2, -1, 0, 1$ and 2, then graph the path of the object for $-2 \leq t \leq 2$.

(b) Evaluate $\frac{dy}{dx}$ for $t = -2, -1, 0, 1$ and 2. Do your calculated values for $\frac{dy}{dx}$ agree with the shape of your graph from part (a)?

Solution. (a) When $t = -2$, $x(-2) = (-2)^3 + 1 = -7$ and $y(-2) = (-2)^2 + (-2) = 2$. The other values for $x(t)$ and $y(t)$ appear in the margin table; a graph of $(x(t), y(t))$ appears below.

t	x	y	$\frac{dy}{dx}$
-2	-7	2	$-\frac{1}{4}$
-1	0	0	$-\frac{1}{3}$
0	1	0	UND
1	2	2	1
2	9	6	$\frac{5}{12}$

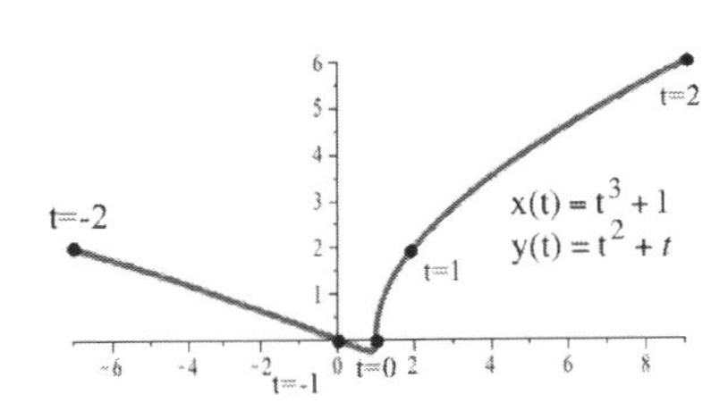

(b) $\frac{dy}{dt} = 2t + 1$ and $\frac{dx}{dt} = 3t^2$, so:

$$\frac{dy}{dx} = \frac{2t+1}{3t^2} \quad \Rightarrow \quad \left.\frac{dy}{dx}\right|_{t=2} = \frac{-3}{12} = -\frac{1}{4}$$

The other values for $\frac{dy}{dx}$ appear in the margin table (the value for $t = 0$ is undefined). ◀

Practice 1. Find an equation for the line tangent to the graph of the parametric equations from Example 1 at the point where $t = 3$.

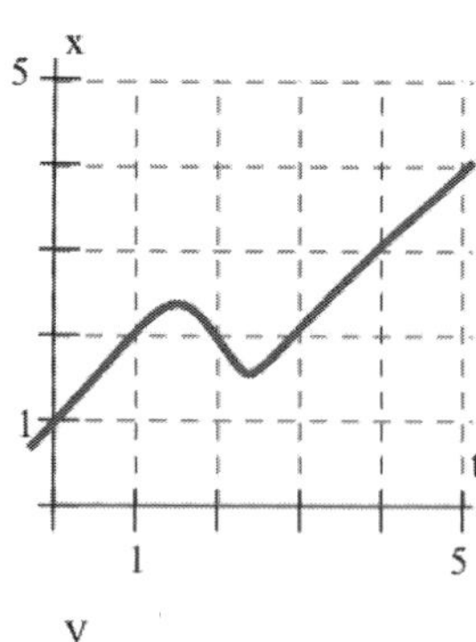

An object can "visit" the same location more than once, and a parametric graph pass through the same point more than once.

Example 2. The first two margin figures show the x- and y-coordinates of an object at time t.

(a) Sketch the parametric graph of $(x(t), y(t))$, the position of the object at time t.

(b) Give the coordinates of the object when $t = 1$ and $t = 3$.

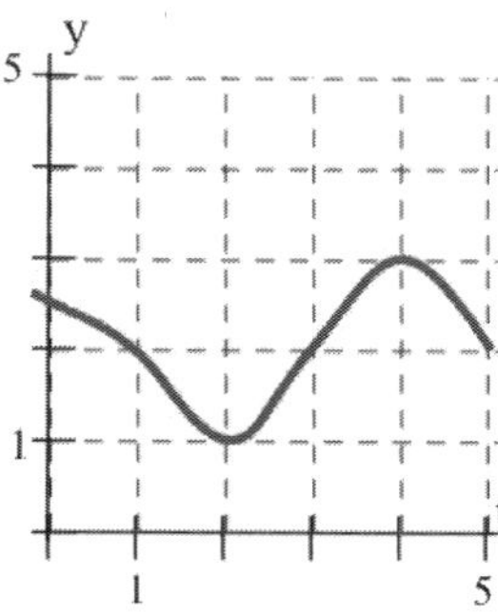

(c) Find the slopes of the tangent lines to the parametric graph when $t = 1$ and $t = 3$.

Solution. (a) By reading the x- and y-values on the graphs in margin figures, we can plot points on the parametric graph. The parametric graph appears in the bottom margin figure.

(b) When $t = 1$, $x = 2$ and $y = 2$ so the parametric graph goes through the point $(2, 2)$. When $t = 3$, the parametric graph goes through the same point $(2, 2)$, as observed in the parametric graph.

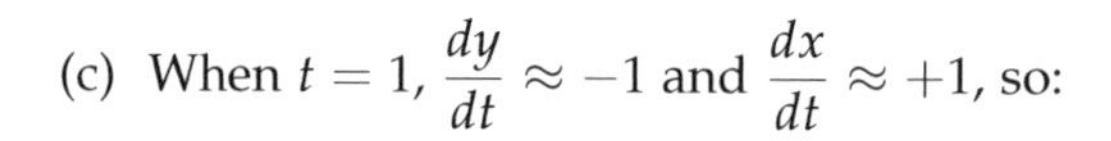

(c) When $t = 1$, $\frac{dy}{dt} \approx -1$ and $\frac{dx}{dt} \approx +1$, so:

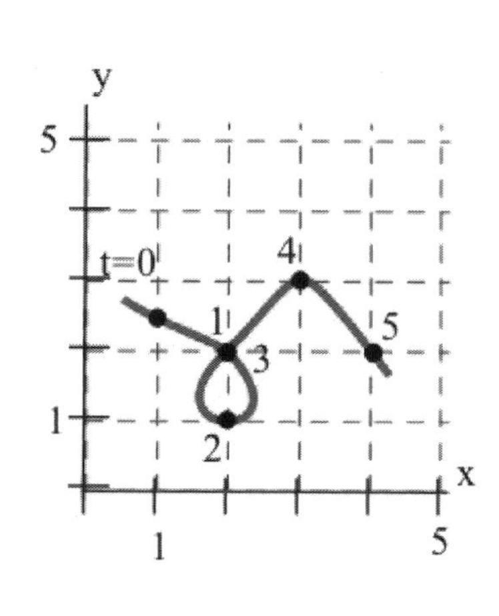

$$\frac{dy}{dx} = \frac{\frac{dy}{dt}}{\frac{dx}{dt}} \approx \frac{-1}{1} = -1$$

When $t = 3$, $\frac{dy}{dt} \approx +1$ and $\frac{dx}{dt} \approx +1$, so $\frac{dy}{dx} \approx 1$. These values agree with the appearance of the parametric graph. The object passes through the point $(2, 2)$ twice (when $t = 1$ and $t = 3$), but is traveling in a different direction each time. ◀

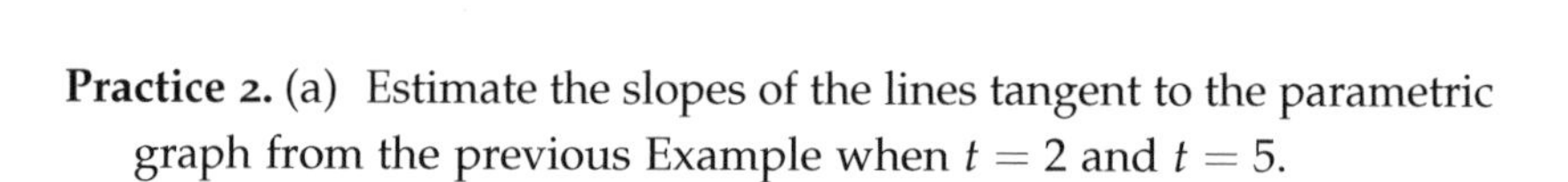

Practice 2. (a) Estimate the slopes of the lines tangent to the parametric graph from the previous Example when $t = 2$ and $t = 5$.

(b) At what time(s) does $\frac{dy}{dt} = 0$?

(c) When does the parametric graph have a maximum? A minimum?

(d) How are the maximum and minimum points on a parametric graph related to the derivatives of $x(t)$ and $y(t)$?

Speed

If you know how fast an object is moving in the x-direction $\left(\frac{dx}{dt}\right)$ and how fast it is moving in the y-direction $\left(\frac{dy}{dt}\right)$, it is straightforward to determine the **speed** of the object (how fast it is moving in the xy-plane).

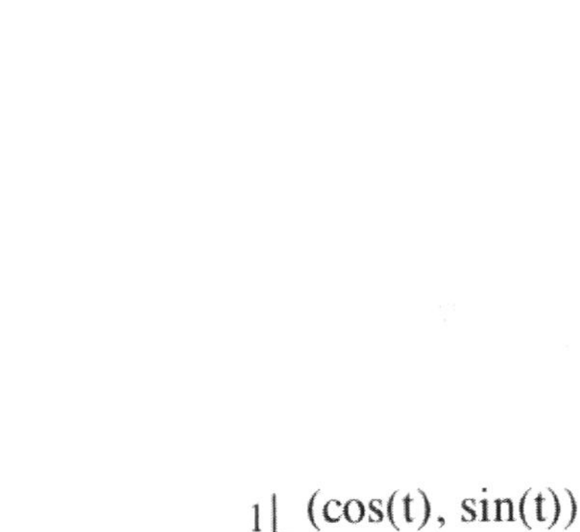

If, during a short interval of time Δt, the object's position changes by Δx in the x-direction and by Δy in the y-direction (see margin), then the object has moved a distance of $\sqrt{(\Delta x)^2 + (\Delta y)^2}$ in time Δt, so the average speed during this brief time interval is:

$$\frac{\text{distance moved}}{\text{time change}} = \frac{\sqrt{(\Delta x)^2 + (\Delta y)^2}}{\Delta t} = \sqrt{(\frac{\Delta x}{\Delta t})^2 + (\frac{\Delta y}{\Delta t})^2}$$

If $x(t)$ and $y(t)$ are differentiable functions of t, we can take the limit of the average speed (as Δt approaches 0) to get the instantaneous speed at time t:

$$\lim_{\Delta t \to 0} \sqrt{\left(\frac{\Delta x}{\Delta t}\right)^2 + \left(\frac{\Delta y}{\Delta t}\right)^2} = \sqrt{\left(\frac{dx}{dt}\right)^2 + \left(\frac{dy}{dt}\right)^2}$$

Example 3. At time t seconds the location (measured in feet) of an object in the xy-plane, relative to the origin is $(\cos(t), \sin(t))$. Sketch the path of the object and show that it is traveling at a constant speed.

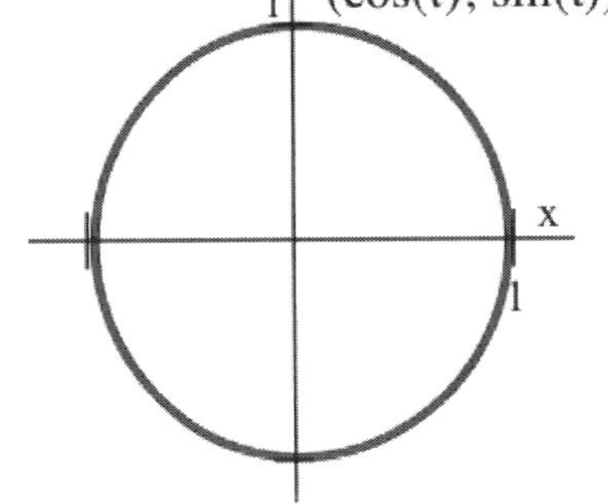

Solution. The object is moving in a circular path (see margin). The speed of the object is:

$$\sqrt{\left(\frac{dx}{dt}\right)^2 + \left(\frac{dy}{dt}\right)^2} = \sqrt{(-\sin(t))^2 + (\cos(t))^2} = 1$$

so its speed (1 foot per second) is indeed constant. ◀

Practice 3. Is the object in Example 2 traveling faster when $t = 1$ or when $t = 3$? When $t = 1$ or when $t = 2$?

Arclength

Also see Section 5.3.

In section 5.3 we approximated the total length L of a curve C by partitioning C into small pieces (see margin), approximating the length of each piece using the distance formula, and then adding the lengths of the pieces together to get:

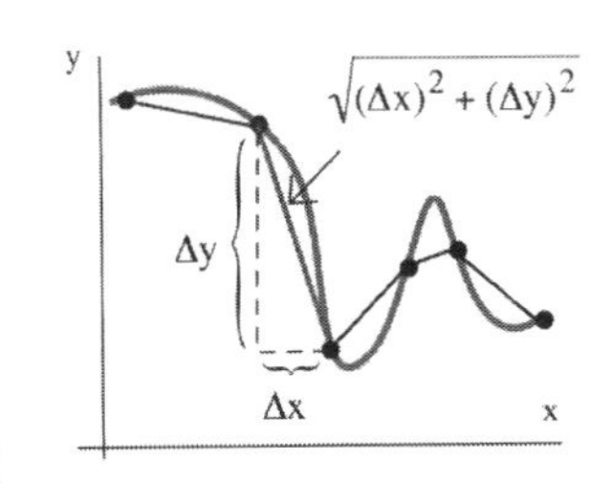

$$L \approx \sum \sqrt{(\Delta x)^2 + (\Delta y)^2} = \sum \sqrt{\left(\frac{\Delta x}{\Delta x}\right)^2 + \left(\frac{\Delta y}{\Delta x}\right)^2} \Delta x \longrightarrow \int_{x=a}^{x=b} \sqrt{1 + \left(\frac{dy}{dx}\right)^2}\, dx$$

where $x = a$ and $x = b$ correspond to the endpoints of C. We then used a similar approach for parametric equations:

$$L \approx \sum \sqrt{(\Delta x)^2 + (\Delta y)^2} = \sum \sqrt{\left(\frac{\Delta x}{\Delta t}\right)^2 + \left(\frac{\Delta y}{\Delta t}\right)^2}\, \Delta t$$

where $t = \alpha$ and $t = \beta$ correspond to the endpoints of C.

Arclength Formula (Parametric Version)

If	C is a curve given by $x = x(t)$ and $y = y(t)$ for $\alpha \le t \le \beta$
and	$x'(t)$ and $y'(t)$ exist and are continuous on $[\alpha, \beta]$
then	the length L of C is given by:

$$L = \int_\alpha^\beta \sqrt{\left(\frac{dx}{dt}\right)^2 + \left(\frac{dy}{dt}\right)^2}\, dt$$

Example 4. Find the length of the cycloid parametrized by $x(t) = R\,(t - \sin(t))$ and $y(t) = R\,(1 - \cos(t))$ for $0 \le t \le 2\pi$ (see below).

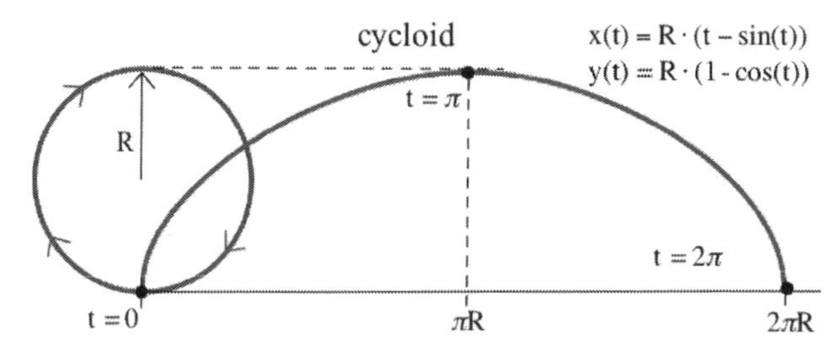

Solution. Computing $\frac{dx}{dt} = R\,(1 - \cos(t))$ and $\frac{dy}{dt} = R\sin(t)$ and using the parametric arclength formula yields:

$$\begin{aligned} L &= \int_0^{2\pi} \sqrt{\left(\frac{dx}{dt}\right)^2 + \left(\frac{dy}{dt}\right)^2}\, dt \\ &= \int_0^{2\pi} \sqrt{R\,(1 - \cos(t))^2 + (R\sin(t))^2}\, dt \\ &= R\int_0^{2\pi} \sqrt{1 - 2\cos(t) + \cos^2(t) + \sin^2(t)}\, dt = R\int_0^{2\pi} \sqrt{2 - 2\cos(t)}\, dt \end{aligned}$$

The resulting integral:

$$\int_0^{2\pi} \sqrt{2 - 2\cos(t)}\, dt$$

appears challenging, but in this instance clever use of a trigonometric identity allows us to find an exact value. In most instances, however, the integrals resulting from arclength computation will require numerical approximation (as we observed in Section 5.3).

Replacing θ with $\frac{t}{2}$ in the formula $\sin^2(\theta) = \frac{1}{2} - \frac{1}{2}\cos(2\theta)$ yields:

$$2 - 2\cos(t) = 4\sin^2\left(\frac{t}{2}\right)$$

so the integral becomes:

$$L = R\int_0^{2\pi} 2\sin\left(\frac{t}{2}\right)\, dt = \left[-4R\cos\left(\frac{t}{2}\right)\right]_0^{2\pi} = 8R$$

The length of a cycloid arch is 8 times the radius of the rolling circle that generated the cycloid. ◀

Practice 4. Represent the length of the ellipse parametrized by $x(t) = 3\cos(t)$ and $y = 2\sin(t)$ for $0 \le t \le 2\pi$ (see margin) as a definite integral, then use technology to approximate the value of the integral.

Area

When we first developed the definite integral, we approximated the area between the graph of a positive function $y = f(x)$ by partitioning the domain $a \le x \le b$ into n pieces of lengths Δx_k , finding the areas of the thin rectangles, and approximating the total area by adding the rectangle areas: $A \approx \sum y_k \Delta x_k$ (a Riemann sum). As $\Delta x \to 0$, the Riemann sum approached the definite integral $\int_{x=a}^{x=b} y\,dx$.

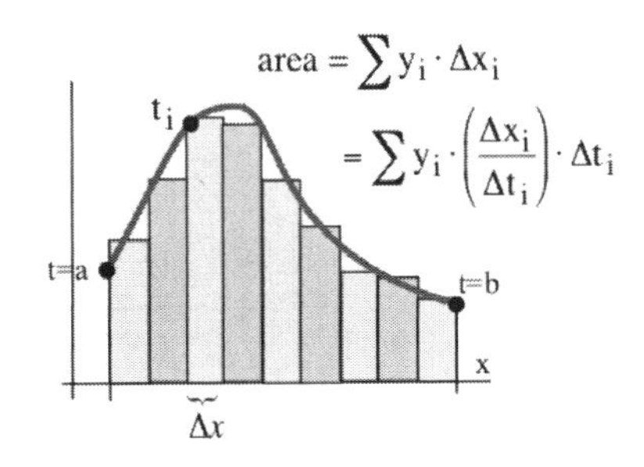

To compute the area of a similar region when the curve is defined by parametric equations, the process is similar, but the independent variable is now t and the domain is an interval $\alpha \le t \le \beta$. If $x(t)$ is an increasing function of t, any partition of the t-interval $[\alpha, \beta]$ into n pieces of lengths Δt_k induces a partition of the x-axis (see margin). We can use this induced partition of the x-axis to approximate the total area by:

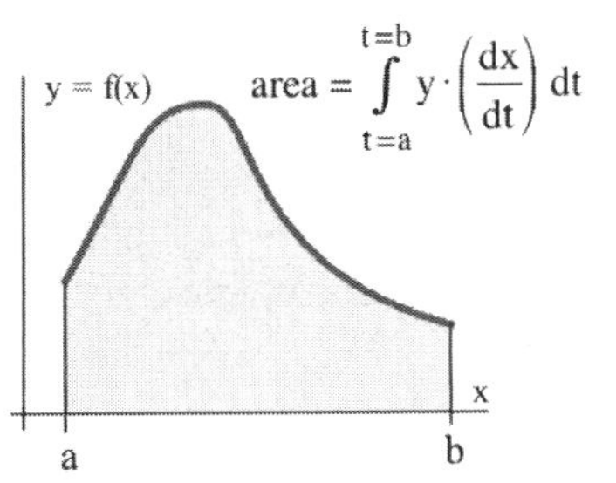

$$A \approx \sum y_k \Delta x_k = \sum y_k \cdot \frac{\Delta x_k}{\Delta t_k} \Delta t_k \longrightarrow \int_{t=\alpha}^{t=\beta} y \cdot \frac{dx}{dt}\,dt$$

as $\Delta t \to 0$.

Area with Parametric Equations

If	C is a curve given by $x = x(t)$ and $y = y(t)$ for $\alpha \le t \le \beta$
and	$x'(t)$ and $y'(t)$ exist and are continuous on $[\alpha, \beta]$
and	$y(t)$ and $x'(t)$ do not change sign on $[\alpha, \beta]$
then	the area between C and the x-axis is given by:

$$A = \left| \int_{\alpha}^{\beta} y(t) \cdot \frac{dx}{dt}\,dt \right|$$

The requirement that y not change sign prevents the parametric graph from being above the x-axis for some values of t and below the x-axis for other t-values. The requirement that $\frac{dx}{dt}$ not change sign prevents the graph from "turning around" (see below).

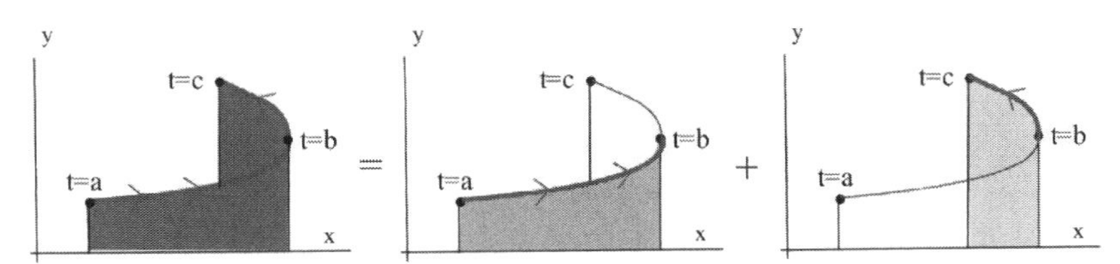

If either requirement is not satisfied, some of the area will be added (above center) and some will be subtracted (above right).

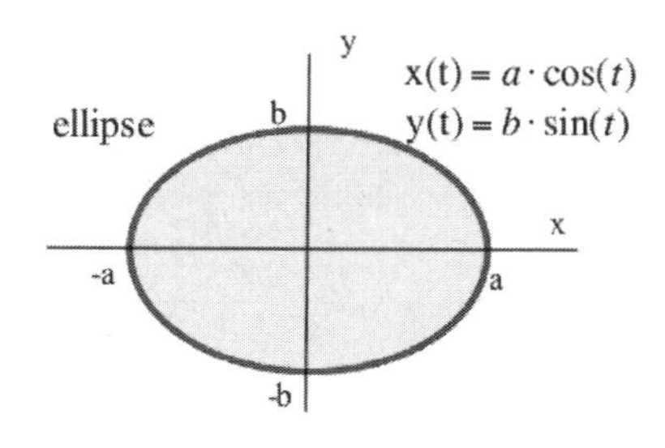

Example 5. Find the area of the region $\mathcal{R}$ in the first quadrant enclosed by the ellipse described by the parametric equations $x(t) = a\cos(t)$ and $y = b\sin(t)$ (for $a > 0$ and $b > 0$) (see margin).

Solution. In the first quadrant (where $0 \le t \le \frac{\pi}{2}$), $y = b\sin(t) > 0$ and $\frac{dx}{dt} = -a\sin(t) < 0$, so the area of $\mathcal{R}$ is given by:

$$A = \left|\int_{t=0}^{t=\frac{\pi}{2}} y(t)\cdot\frac{dx}{dt}\,dt\right| = \left|\int_{t=0}^{t=\frac{\pi}{2}} b\sin(t)\cdot[-a\sin(t)]\,dt\right|$$

Evaluating this integral yields:

$$A = ab\left|\int_{t=0}^{t=\frac{\pi}{2}} \sin^2(t)\,dt\right| = ab\left|\int_{t=0}^{t=\frac{\pi}{2}} \left[\frac{1}{2}-\frac{1}{2}\cos(2t)\right]dt\right|$$
$$= ab\left|\left[\frac{1}{2}t-\frac{1}{4}\sin(2t)\right]_{t=0}^{t=\frac{\pi}{2}}\right| = \frac{\pi ab}{4}$$

The area enclosed by the entire ellipse is πab; if $a = b$, the ellipse is a circle with radius $r = a = b$ with area πr^2 (as expected). ◀

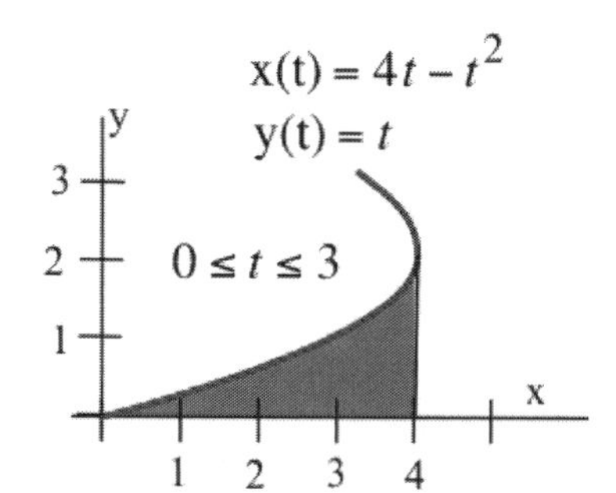

Practice 5. Let $x(t) = 4t - t^2$ and $y(t) = t$ (graphed in the margin).

(a) Represent the area of the shaded region as an integral.

(b) Evaluate the integral from part (a).

(c) Does $\int_0^3 t(4-2t)\,dt$ represent an area?

Properties of the Cycloid

In Section 11.3 we developed parametric equations for the cycloid: $x(t) = R(t - \sin(t))$ and $y(t) = R(1-\cos(t))$ For any $t \ge 0$, $y(t) \ge 0$ and $\frac{dx}{dt} = R(1-\cos(t)) \ge 0$ so the area between one arch of the cycloid and the x-axis is:

$$A = \left|\int_{t=0}^{t=2\pi} y(t)\cdot\frac{dx}{dt}\,dt\right| = \int_0^{2\pi}[R(1-\cos(t))]\cdot[R(1-\cos(t))]\,dt$$
$$= R^2\int_0^{2\pi}\left[1-2\cos(t)+\cos^2(t)dt\right]dt$$
$$= R^2\left[\frac{1}{2}t - 2\sin(t) + \frac{1}{2}t + \frac{1}{4}\sin(2t)\right]_0^{2\pi} = R^2[2\pi+\pi] = 3\pi R^2$$

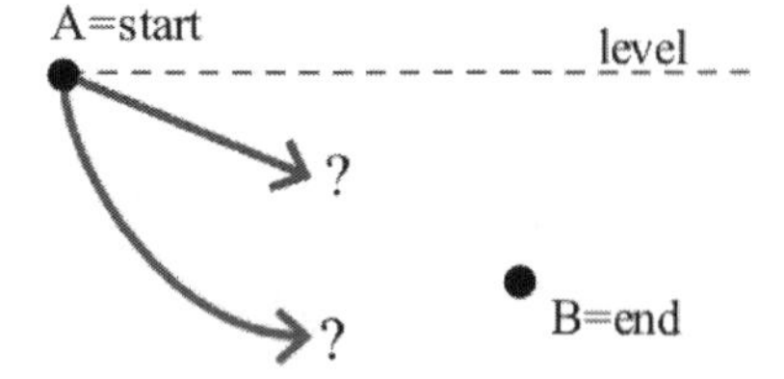

The area under one arch of a cycloid is 3 times the area of the circle that generates the cycloid. How does this compare with your guess from the end of Section 11.3?

You and a friend decide to hold a contest to see who can build a slide that gets a person from point A to point B (see margin) in the shortest time. What shape should you make your slide: a straight line, part of a circle, or something else? Assuming that the slide is frictionless and that the only acceleration is due to gravity, Johann Bernoulli (1667–1748) showed that the shortest-time ("**brachistochrone**" for "brachi," meaning "short," and "chrone," meaning time) path is part of a cycloid that starts at A and also goes through the point B. The margin figure shows the cycloidal paths for A and B as well as the cycloidal paths for two other "finish" points, C and D.

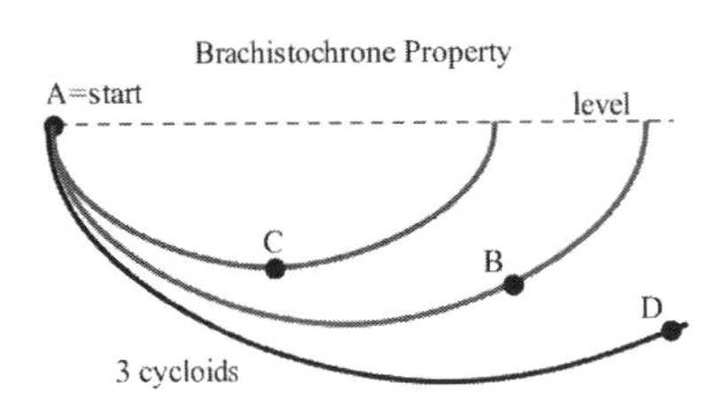

Even before Bernoulli solved the brachistochrone problem, the astronomer, physicist and mathematician Christiaan Huygens (1629–1695) attempted to design an accurate pendulum clock. On a standard pendulum clock (see margin), the path of the bob is part of a circle, and the period of the swing depends on the displacement angle of the bob; as friction slows the bob, the displacement angle gets smaller and the clock slows down. Huygens designed a clock (below left) whose bob swung in a curve so that the period of the swing did not depend on the displacement angle:

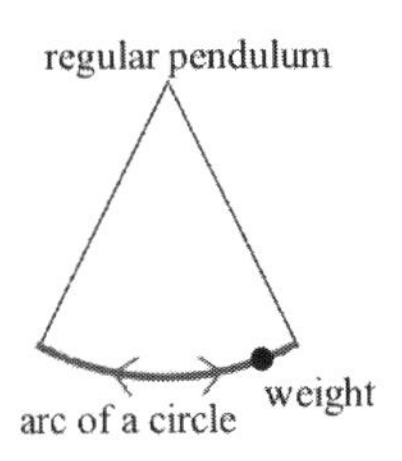

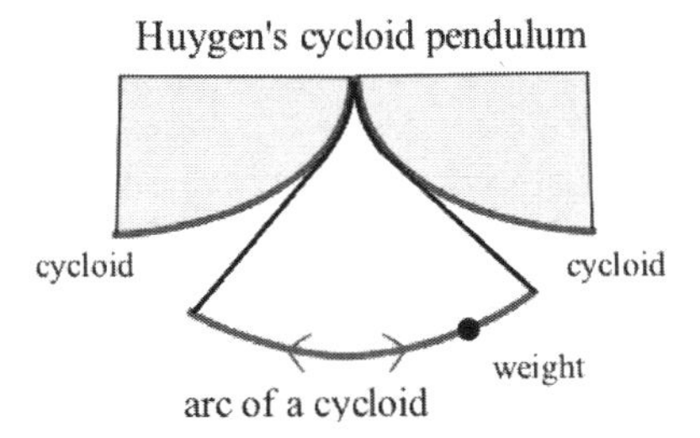

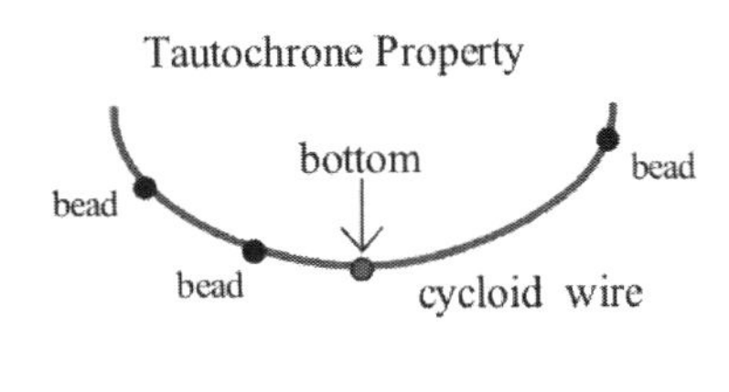

The curve Huygens found to solve the same-time ("**tautochrone**" for "tauto," meaning "same," and "chrone," meaning "time") problem was part of the cycloid. Beads strung on a wire in the shape of a cycloid (above right) reach the bottom in the same amount of time, no matter where along the wire (except the bottom point) you release them.

The brachistochane and tautochrone problems are examples from a field of mathematics called the **Calculus of Variations**. Typical optimization problems in calculus involve finding a point or number that maximizes or minimizes some quantity. Typical optimization problems in the Calculus of Variations involve finding a curve or function that maximizes or minimizes some quantity. For example, what curve or shape with a given length encloses the greatest area? (Answer: a circle.) Modern applications of Calculus of Variations include finding routes for airliners and ships to minimize travel time or fuel consumption depending on prevailing winds or currents.

11.4 Problems

For Problems 1–8, (a) sketch the parametric graph $(x(t), y(t))$, (b) find the slope of the line tangent to the graph at the given values of t, and (c) find the points (x, y) at which $\frac{dy}{dx}$ is either 0 or undefined.

1. $x(t) = t - t^2$, $y(t) = 2t + 1$; $t = 0, 1, 2$
2. $x(t) = t^3 + t$, $y(t) = t^2$; $t = 0, 1, 2$
3. $x(t) = 1 + \cos(t)$, $y(t) = 2 + \sin(t)$; $t = 0, \frac{\pi}{4}, \frac{\pi}{2}$
4. $x(t) = 1 + 3\cos(t)$, $y(t) = 2 + 2\sin(t)$; $t = 0, \frac{\pi}{4}, \frac{\pi}{2}$
5. $x(t) = \sin(t)$, $y(t) = \cos(t)$; $t = 0, \frac{\pi}{4}, \frac{\pi}{2}, 17.3$
6. $x(t) = 3 + \sin(t)$, $y(t) = 2 + \sin(t)$; $t = 0, \frac{\pi}{4}, \frac{\pi}{2}$
7. $x(t) = \ln(t)$, $y(t) = 1 - t^2$; $t = 1, 2, e$
8. $x(t) = \arctan(t)$, $y(t) = e^t$; $t = 0, 1, 2$

In Problems 9–12, use the given graphs of $x(t)$ and $y(t)$ to estimate (a) the slope of the line tangent to the parametric graph at $t = 0, 1, 2$ and 3, and (b) the points (x, y) at which $\frac{dy}{dx}$ is either 0 or undefined.

9.

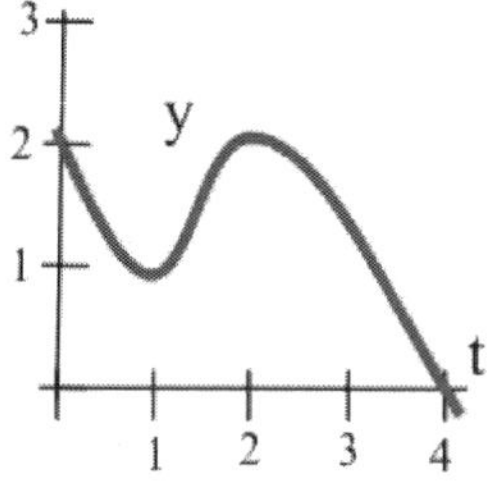

10.

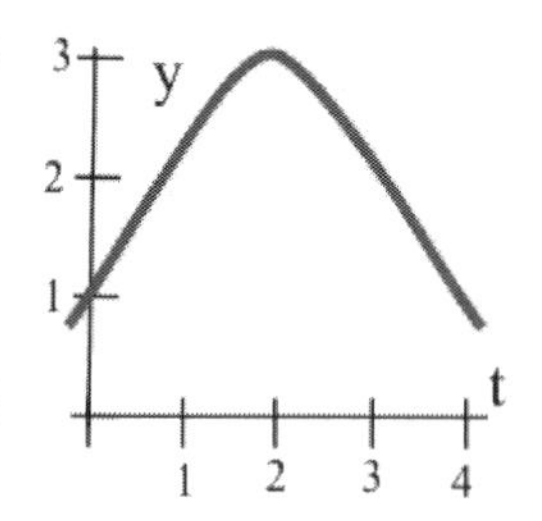

11.

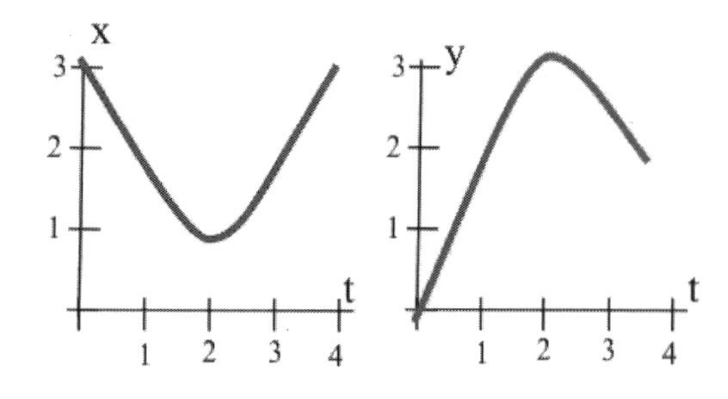

12.

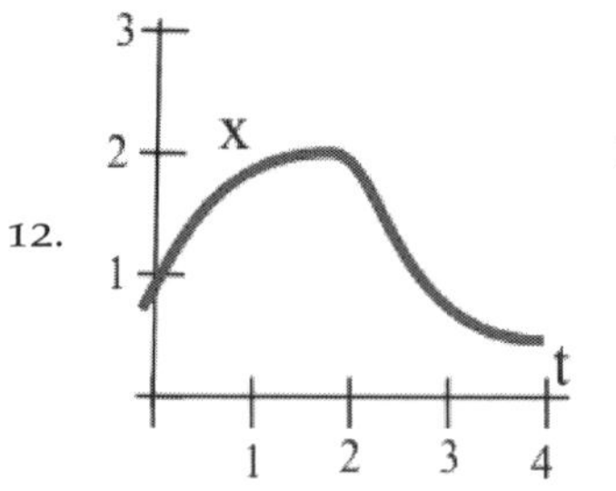

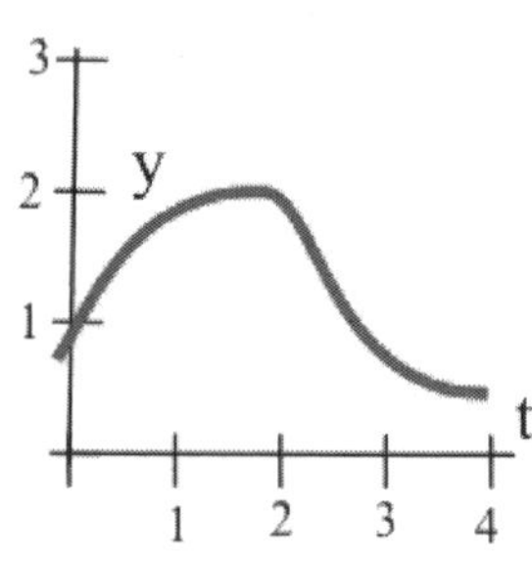

For Problems 13–20, use the given locations $x(t)$ and $y(t)$ of an object at time t seconds (measured in feet) to find the speed of the object at the given times.

13. $x(t) = t - t^2$, $y(t) = 2t + 1$; $t = 0, 1, 2$
14. $x(t) = t^3 + t$, $y(t) = t^2$; $t = 0, 1, 2$
15. $x(t) = 1 + \cos(t)$, $y(t) = 2 + \sin(t)$; $t = 0, \frac{\pi}{4}, \frac{\pi}{2}$
16. $x(t) = 1 + 3\cos(t)$, $y(t) = 2 + 2\sin(t)$; $t = 0, \frac{\pi}{4}, \frac{\pi}{2}, \pi$
17. $x(t)$ and $y(t)$ from Problem 9 at $t = 0, 1, 2, 3$ and 4
18. $x(t)$ and $y(t)$ from Problem 10 at $t = 0, 1, 2$ and 3
19. $x(t)$ and $y(t)$ from Problem 11 at $t = 0, 1, 2$ and 3
20. $x(t)$ and $y(t)$ from Problem 12 at $t = 0, 1, 2$ and 3
21. An object travels along a cycloidal path so that its location is given by $x(t) = R(t - \sin(t))$ and $y(t) = R(1 - \cos(t))$ after t seconds (with distances measured in feet).
    (a) Find the speed of the object at time t.
    (b) At what time is the object traveling fastest?
    (c) Where is the object on its cycloidal path when it is traveling fastest?
22. At time t seconds an object is located at $x(t) = 5\cos(t)$ and $y(t) = 2\sin(t)$ (measured in feet).
    (a) Find the speed of the object at time t.
    (b) At what time is the object traveling fastest?
    (c) Where is the object on its elliptical path when it is traveling fastest?

For 23–28 (a) represent the arclength of the parametric graph as a definite integral, and (b) evaluate the integral (using technology, if necessary).

23. $x(t) = t - t^2$, $y(t) = 2t + 1$ from $t = 0$ to 2
24. $x(t) = t^3 + t$, $y(t) = t^2$; $t = 0$ to 2

25. $x(t) = 1 + \cos(t)$, $y(t) = 2 + \sin(t)$; $t = 0$ to π

26. $x(t) = 1 + 3\cos(t)$, $y(t) = 2 + 2\sin(t)$; $t = 0$ to π

27. $x(t)$ and $y(t)$ given below; $t = 1$ to 3

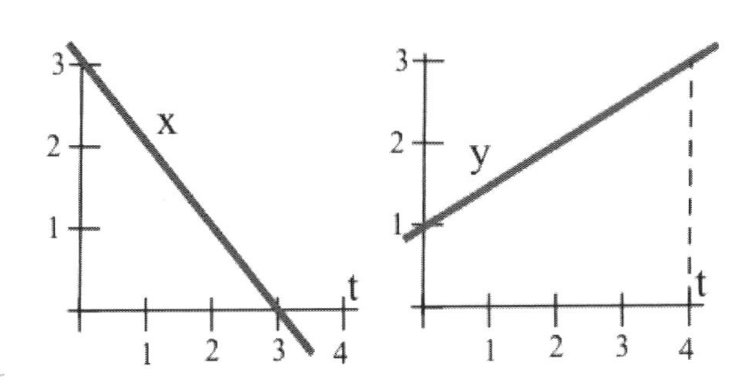

28. $x(t)$ and $y(t)$ from Problem 12; $t = 0$ to 2

For 29–32 (a) represent the area of the region between the parametric graph and the x-axis as a definite integral, and (b) evaluate the integral.

29. $x(t) = t^2$, $y(t) = 4t^2 - t^4$ for $0 \le t \le 2$

30. $x(t) = 1 + \sin(t)$, $y(t) = 2 + \sin(t)$ for $0 \le t \le \pi$

31. $x(t) = t^2$, $y(t) = 1 + \cos(t)$ for $0 \le t \le 2$

32. $x(t) = \cos(t)$, $y(t) = 2 - \sin(t)$ for $0 \le t \le \frac{\pi}{2}$

33. **"Cycloid" with a square wheel**: Find the area under one "arch" of the path of a point on the corner of a "rolling" square with side length R.

34. Find the area of the region between the x-axis and the curate cycloid $x(t) = R \cdot t - r \cdot \sin(t)$, $y(t) = R - r \cdot \cos(t)$ for $0 \le t \le 2\pi$.

11.4 Practice Answers

1. When $t = 3$, $x = 28$, $y = 12$ and (using results from Example 1):

$$\frac{dy}{dx} = \frac{2t+1}{3t^2} \quad \Rightarrow \quad \left.\frac{dy}{dx}\right|_{t=3} = \frac{7}{27}$$

so an equation for the tangent line is $y = 12 + \frac{7}{27}(x - 28)$.

2. (a) When $t = 2$, $\frac{dy}{dx} \approx 0$; when $t = 5$, $\frac{dy}{dx} \approx -1$.
   (b) When $t \approx 2$ and $t \approx 4$ (according to the $y(t)$ graph).
   (c) A minimum occurs when $t \approx 2$ and a maximum when $t \approx 4$.
   (d) If the parametric graph has a maximum or minimum at $t = t^*$, then $\frac{dy}{dt}$ equals 0 or is undefined when $t = t^*$.

3. When $t = 1$:

$$\text{speed} = \sqrt{\left(\frac{dx}{dt}\right)^2 + \left(\frac{dy}{dt}\right)^2} \approx \sqrt{1^2 + (-1)^2} = \sqrt{2} \approx 1.4\,\frac{\text{ft}}{\text{sec}}$$

When $t = 2$: speed $\approx \sqrt{(-1)^2 + 0^2} = 1\,\frac{\text{ft}}{\text{sec}}$

When $t = 3$: speed $\approx \sqrt{1^2 + 1^2} = \sqrt{2} \approx 1.4\,\frac{\text{ft}}{\text{sec}}$

4. Length $= \int_0^{2\pi} \sqrt{(-3\sin(t))^2 + (2\cos(t))^2}\,dt \approx 15.87$

5. (a) Area $= \int_0^2 t \cdot (4 - 2t)\,dt$ (b) $\int_0^2 \left[4t - 2t^2\right]\,dt = \left[2t^2 - \frac{2}{3}t^2\right]_0^2 = \frac{16}{3}$
   (c) No, it represents the area under the curve for $0 \le t \le 2$ minus the area under the curve for $2 \le t \le 3$.

A
Answers

Important Note about Precision of Answers: In many of the problems in this book you are required to read information from a graph and to calculate with that information. You should take reasonable care to read the graphs as accurately as you can (a small straightedge is helpful), but even skilled and careful people make slightly different readings of the same graph. That is simply one of the drawbacks of graphical information. When answers are given to graphical problems, the answers should be viewed as the best approximations we could make, and they usually include the word "approximately" or the symbol "$\approx$" meaning "approximately equal to." Your answers should be close to the given answers, but you should not be concerned if they differ a little. (Yes those are vague terms, but it is all we can say when dealing with graphical information.)

Section 9.1

1. (a) 32, 64 (b) $a_5 = 2^5$ (c) $a_n = 2^n$
3. (a) -1, 1 (b) $a_5 = (-1)^5$ (c) $a_n = (-1)^n$
5. (a) 120, 720 (b) $a_5 = 5!$ (c) $a_n = n!$
7. $1, \frac{3}{2}, \frac{11}{6}, \frac{25}{12}, \frac{137}{60}, \frac{49}{20}$
9. $1, \frac{1}{2}, \frac{3}{4}, \frac{5}{8}, \frac{11}{16}, \frac{21}{32}$
11. 1, 0, 1, 0, 1, 0
13. (a) $g(5) = -1$, $g(6) = 1$ (b) See below left.

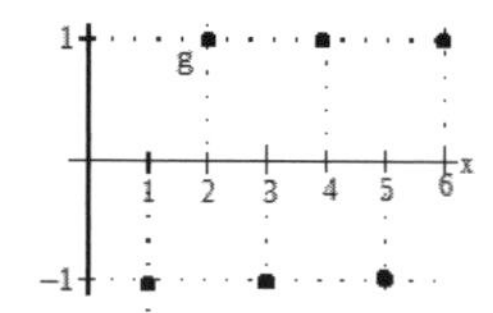

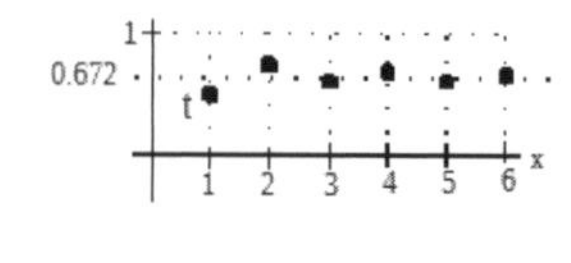

15. (a) $t(5) = \frac{21}{32}$, $t(6) = \frac{43}{64}$ (b) See above right.
17. $a_n = \frac{1}{n^2}$
19. $a_n = \frac{n-1}{n}$
21. $a_n = \frac{n}{2^n}$
23. $-1, 0, \frac{1}{3}, \frac{1}{2}, \frac{3}{5}, \frac{2}{3}$; see below left.

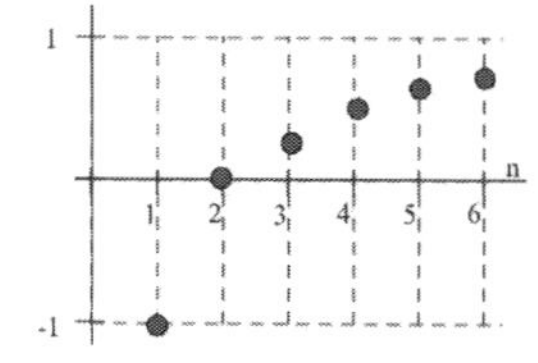

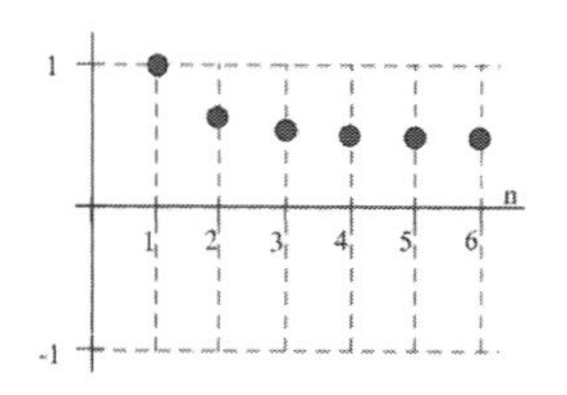

25. $1, \frac{2}{3}, \frac{3}{5}, \frac{4}{7}, \frac{5}{9}, \frac{6}{11}$; see above right.
27. $2, \frac{7}{2}, \frac{8}{3}, \frac{13}{4}, \frac{14}{5}, \frac{19}{6}$; see below left.

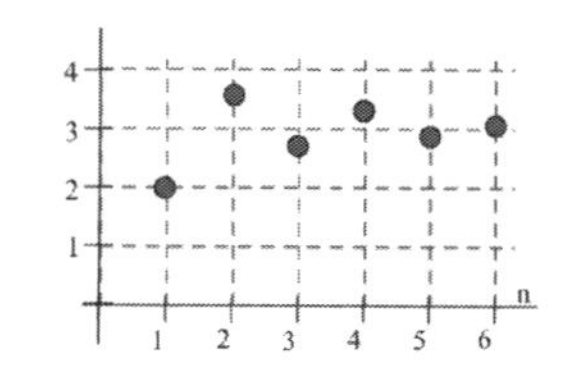

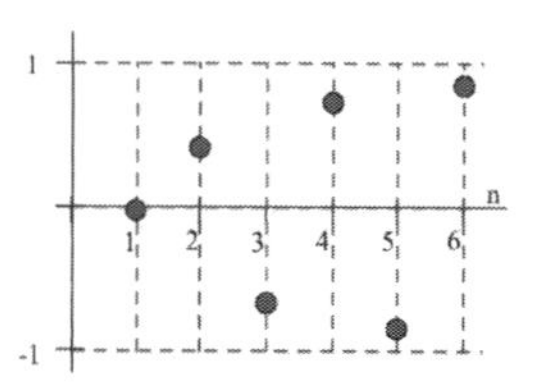

29. $0, \frac{1}{2}, -\frac{2}{3}, \frac{3}{4}, -\frac{4}{5}, \frac{5}{6}$; see above right.
31. $1, \frac{1}{2}, \frac{1}{6}, \frac{1}{24}, \frac{1}{120}, \frac{1}{720}$; see below left.

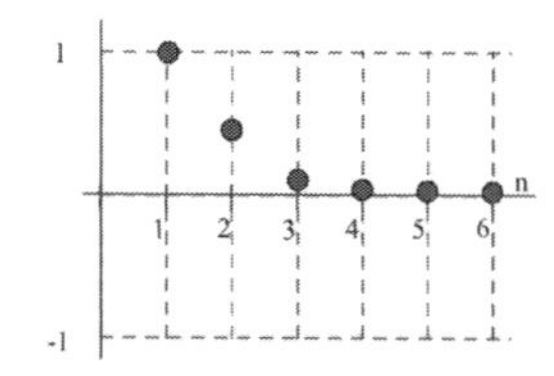

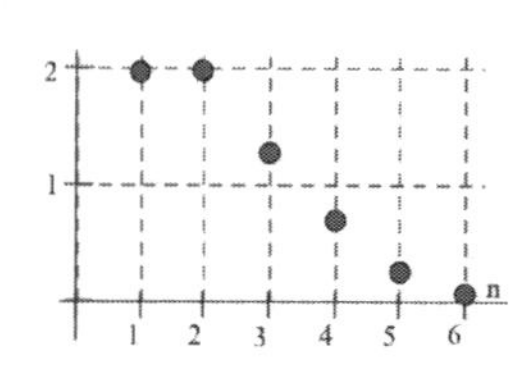

33. $2, 2, \frac{4}{3}, \frac{2}{3}, \frac{4}{15}, \frac{4}{45}$; see above right.
35. 2, -2, 2, -2, 2, -2, 2, -2, 2, -2
37. $\frac{\sqrt{3}}{2}, -\frac{\sqrt{3}}{2}, 0, \frac{\sqrt{3}}{2}, -\frac{\sqrt{3}}{2}, 0, \frac{\sqrt{3}}{2}, -\frac{\sqrt{3}}{2}, 0, \frac{\sqrt{3}}{2}$
39. 1, 3, 6, 10, 15, 21, 28, 36, 45, 55

Section 9.2

1. $\{a_n\}$ appears to converge; $\{b_n\}$ does not.
3. $\{f_n\}$ appears to converge; $\{e_n\}$ does not.
5. converges to 1
7. diverges (to ∞)
9. converges to $\frac{1}{2}$
11. converges to $\ln(3)$
13. diverges
15. converges to 0
17. converges to $e^{-1} = \frac{1}{e}$
19. converges to 1
21. Given $\epsilon > 0$, take $N \geq \frac{\sqrt{3}}{\sqrt{\epsilon}}$. Then:

$$n > N \Rightarrow n > \frac{\sqrt{3}}{\sqrt{\epsilon}} \Rightarrow n^2 > \frac{3}{\epsilon} \Rightarrow \frac{3}{n^2} < \epsilon$$

so that $|a_n - L| = \left|\frac{3}{n^2} - 0\right| = \frac{3}{n^2} < \epsilon$.

23. Given $\epsilon > 0$, take $N \geq \frac{1}{\epsilon}$. Then:

$$n > N \Rightarrow n > \frac{1}{\epsilon} \Rightarrow \frac{1}{n} < \epsilon$$

so that $|a_n - L| = \left|\frac{3n-1}{n} - 3\right| = \frac{1}{n} < \epsilon$.

25. The given sequence is a subsequence of $\left\{\frac{1}{n}\right\}$, which converges (to 0), so the given subsequence must also converge (to 0).
27. The sequence simplifies to $\{(-1)^n\}$: the subsequence of even-indexed terms is $\{1, 1, 1, \ldots\}$, which converges to 1, while the subsequence of odd-indexed terms is $\{-1, -1, -1, \ldots\}$, which converges to -1. Because the given sequence has two subsequences that converge to different limits, the original sequence diverges.
29. The given sequence is a subsequence of $\left\{\left(1 + \frac{5}{k}\right)^k\right\}$, which converges to e^5, so the given sequence also converges to e^5.
31. $a_{n+1} - a_n = \left[7 - \frac{2}{n+1}\right] - \left[7 - \frac{2}{n}\right] = \frac{2}{n} - \frac{2}{n+1} = \frac{2}{n(n+1)} > 0$ so $\{a_n\}$ is monotonically increasing.
33. $a_{n+1} - a_n = 2^{n+1} - 2^n == 2^n\,[2-1] = 2^n > 0$ so the sequence $\{a_n\}$ is monotonically increasing.
35. $a_{n+1} - a_n = \left[5 + \frac{7}{3^{n+1}}\right] - \left[5 + \frac{7}{3^n}\right] = \frac{7}{3^{n+1}} - \frac{7}{3^n} = \frac{-14}{3^{n+1}} < 0$ so $\{a_n\}$ is monotonically decreasing.

37. $a_n = \frac{n+1}{n!} \Rightarrow a_{n+1} = \frac{n+2}{(n+1)!} \Rightarrow \frac{a_{n+1}}{a_n} = \frac{n+2}{(n+1)!} \cdot \frac{n!}{n+1} = \frac{(n+2)\cdot n!}{(n+1)^2 \cdot n!} = \frac{n+2}{(n+1)^2} < 1$ when $n \geq 1$, so $\{a_n\}$ is monotonically decreasing.

39. $a_n = \left(\frac{5}{4}\right)^n \Rightarrow a_{n+1} = \left(\frac{5}{4}\right)^{n+1} \Rightarrow \frac{a_{n+1}}{a_n} = \frac{5}{4} > 1$, so $\{a_n\}$ is monotonically increasing.

41. $a_n = \frac{n}{e^n} \Rightarrow a_{n+1} = \frac{n+1}{e^{n+1}} \Rightarrow \frac{a_{n+1}}{a_n} = \frac{n+1}{e^{n+1}} \cdot \frac{e^n}{n} = \frac{(n+1)}{n \cdot e} < 1$ when $n \geq 1$, so $\{a_n\}$ is monotonically decreasing.

43. $f(x) = 5 - \frac{3}{x} \Rightarrow f'(x) = \frac{3}{x^2} > 0$, so $f(x)$ is always increasing, which means that $\left\{5 - \frac{3}{n}\right\}$ is monotonically increasing.

45. $f(x) = \cos\left(\frac{1}{x}\right) \Rightarrow f'(x) = \frac{1}{x^2}\sin\left(\frac{1}{x}\right)$, so $f'(x) > 0$ for $x \geq 1$, meaning that $f(x)$ is increasing and $\left\{\cos\left(\frac{1}{n}\right)\right\}$ is monotonically increasing.

47. $a_n = \frac{n+3}{n!} \Rightarrow a_{n+1} = \frac{n+4}{(n+1)!} \Rightarrow \frac{a_{n+1}}{a_n} = \frac{n+4}{(n+1)!} \cdot \frac{n!}{n+3} = \frac{(n+4)\cdot n!}{(n+1)(n+3)\cdot n!} = \frac{n+4}{(n+1)(n+3)} < 1$ when $n \geq 1$, so $\{a_n\}$ is monotonically decreasing.

49. $a_{n+1} - a_n = \left[1 - \frac{1}{2^{n+1}}\right] - \left[1 - \frac{1}{2^n}\right] = \frac{1}{2^n} - \frac{1}{2^{n+1}} = \frac{1}{2^{n+1}} > 0$ so the sequence $\left\{1 - \frac{1}{2^n}\right\}$ is monotonically increasing.

51. $a_n = \frac{n+1}{e^n} \Rightarrow a_{n+1} = \frac{n+2}{e^{n+1}} \Rightarrow \frac{a_{n+1}}{a_n} = \frac{n+2}{e^{n+1}} \cdot \frac{e^n}{n+1} = \frac{(n+2)}{(n+1)\cdot e} =< 1$ when $n \geq 1$, so $\left\{\frac{n+1}{e^n}\right\}$ is monotonically decreasing.

53. For $N = 4$: $a_1 = 4 \Rightarrow a_2 = \frac{1}{2}\left(4 + \frac{4}{4}\right) = 2.5 \Rightarrow a_3 = \frac{1}{2}\left(2.5 + \frac{4}{2.5}\right) = 2.05 \Rightarrow a_4 = \frac{1}{2}\left(2.05 + \frac{4}{2.05}\right) \approx 2.00061$. For $N = 9$: $a_1 = 9 \Rightarrow a_2 = \frac{1}{2}\left(9 + \frac{9}{9}\right) = 5 \Rightarrow a_3 = \frac{1}{2}\left(5 + \frac{9}{5}\right) = 3.2 \Rightarrow a_4 = \frac{1}{2}\left(3.2 + \frac{9}{3.2}\right) = 3.00625$. For $N = 5$: $a_1 = 5 \Rightarrow a_2 = \frac{1}{2}\left(5 + \frac{5}{5}\right) = 3 \Rightarrow a_3 = \frac{1}{2}\left(3 + \frac{5}{3}\right) \approx 2.333 \Rightarrow a_4 \approx \frac{1}{2}\left(2.333 + \frac{5}{2.333}\right) \approx 2.238$.

55. (a) Solving $0.01 = \frac{0.02}{0.02k+1}$ for k:

$$0.02k + 1 = \frac{0.02}{0.01} = 2 \Rightarrow 0.02k = 1 \Rightarrow k = \frac{1}{0.02}$$

or 50 generations.

(b) Solving $\frac{1}{2}p = \frac{p}{kp+1}$ for k in terms of p:

$$kp + 1 = \frac{p}{0.5p} = 2 \Rightarrow kp = 1 \Rightarrow k = \frac{1}{p}$$

57. (a) The first "few" grains can be anywhere on the x-axis. (b) After placing "a lot of grains," there will be a large pile of sand close to $x = 3$.

59. (a) $-1 \leq \sin(n) \leq 1$ for all integers n, so the first few grains will be scattered between -1 and $+1$ on the x-axis. (b) After placing "a lot of grains," the sand will be scattered "uniformly" along the interval from -1 to $+1$. (c) A formal proof of this fact is rather sophisticated, but the result is interesting: no two grains ever end up on the same point on the x-axis.

Section 9.3

1. $\sum_{k=1}^{\infty} \frac{1}{k}$ 3. $\sum_{k=1}^{\infty} \frac{2}{3k}$ 5. $\sum_{k=1}^{\infty} \left(-\frac{1}{2}\right)^k$

7. $s_1 = 1$, $s_2 = 1 + 4 = 5$, $s_3 = 5 + 9 = 14$, $s_4 = 14 + 16 = 30$; see below left.

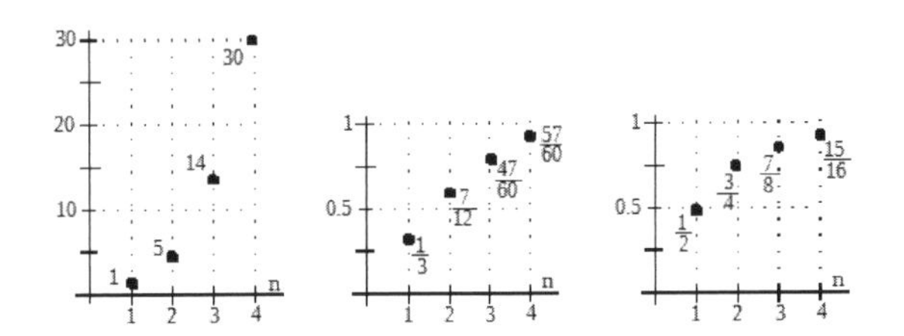

9. $s_1 = \frac{1}{3}$, $s_2 = \frac{7}{12}$, $s_3 = \frac{47}{60}$, $s_4 = \frac{19}{20}$; above center.

11. $s_1 = \frac{1}{2}$, $s_2 = \frac{3}{4}$, $s_3 = \frac{7}{8}$, $s_4 = \frac{15}{16}$; above right.

13. $a_1 = s_1 = 3$, $a_2 = s_2 - s_1 = 2 - 3 = -1$, $a_3 = s_3 - s_2 = 4 - 2 = 2$, $a_4 = s_4 - s_3 = 5 - 4 = 1$

15. $a_1 = 4$, $a_2 = 0.5$, $a_3 = -0.2$, $a_4 = 0.5$

17. $a_1 = 1$, $a_2 = 0.1$, $a_3 = 0.01$, $a_4 = 0.001$

19. $0.888\ldots = 0.8 + 0.08 + 0.008 + \cdots = \sum_{k=1}^{\infty} \frac{8}{10^k}$

21. $\sum_{k=1}^{\infty} \frac{5}{10^k}$ 23. $\sum_{k=1}^{\infty} \frac{a}{10^k}$ 25. $\sum_{k=1}^{\infty} \frac{17}{100^k}$

27. $\sum_{k=1}^{\infty} \frac{7}{100^k}$ 29. $\sum_{k=1}^{\infty} \frac{abc}{1000^k}$ 31. $\sum_{k=0}^{\infty} 30\,(0.8)^k$

33. 80%, 64%, 51.2%, $(0.8)^n \cdot 100\%$

35. $\lim_{k\to\infty} \left(\frac{1}{4}\right)^k = 0$, so $\sum_{k=1}^{\infty} \left(\frac{1}{4}\right)^k = 0$ may or may not converge. (Section 9.4 shows it converges.)

37. $\lim_{k\to\infty}\left(\frac{4}{3}\right)^k = \infty \neq 0$, so $\sum_{k=1}^{\infty}\left(\frac{4}{3}\right)^k$ diverges.

39. $\lim_{k\to\infty}\frac{\sin(k)}{k} = 0$, so $\sum_{k=1}^{\infty}\frac{\sin(k)}{k}$ may or may not converge. (Techniques you may learn in more advanced courses will show that it converges.)

41. $\lim_{k\to\infty}\cos(k)$ does not exist, so $\sum_{k=1}^{\infty}\cos(k)$ diverges.

43. $\lim_{k\to\infty}\frac{k^2-20}{k^5+4} = 0$, so $\sum_{k=1}^{\infty}\frac{k^2-20}{k^5+4}$ may or may not converge. (We'll see later that it converges.)

45. Let $s_n = \sum_{k=1}^{n} a_k$. Then:

$$cA = c\cdot\sum_{k=1}^{\infty} a_k = c\cdot\lim_{n\to\infty} s_n = \lim_{n\to\infty} c\cdot s_n$$
$$= \lim_{n\to\infty} c\cdot\sum_{k=1}^{n} a_k = \lim_{n\to\infty}\sum_{k=1}^{n} c\cdot a_k = \sum_{k=1}^{\infty} c\cdot a_k$$

47. If $s_n = \sum_{k=1}^{n} a_k$, $\sum_{k=1}^{\infty} a_k = A$ means that $\lim_{n\to\infty} s_n = A$. We also know that $\lim_{n\to\infty} s_{n-1} = A$ and that $a_n = s_n - s_{n-1}$, so:

$$\lim_{n\to\infty}, a_n = \lim_{n\to\infty}, [s_n - s_{n-1}]$$
$$= \lim_{n\to\infty}, s_n - \lim_{n\to\infty}, s_{n-1} = A - A = 0$$

Section 9.4

1. This is a geometric series with $|r| = \frac{2}{7} < 1$, so it converges to:

$$\frac{1}{1-r} = \frac{1}{1-\frac{2}{7}} = \frac{1}{\frac{5}{7}} = \frac{7}{5}$$

3. This is a geometric series with $|r| = \frac{4}{7} < 1$, so it converges to:

$$\frac{1}{1-r} = \frac{1}{1-\left(-\frac{4}{7}\right)} = \frac{1}{\frac{11}{7}} = \frac{7}{11}$$

5. This is a geometric series with $|r| = \frac{2}{7} < 1$, so it converges, but the index starts at $k = 1$, hence:

$$\sum_{k=1}^{\infty}\left(\frac{2}{7}\right)^k = \left[\sum_{k=0}^{\infty}\left(\frac{2}{7}\right)^k\right] - \left(\frac{2}{7}\right)^0$$
$$= \frac{1}{1-\frac{2}{7}} - 1 = \frac{7}{5} - 1 = \frac{2}{5}$$

7. This geometric series diverges: $|r| = \frac{7}{4} > 1$.

9. This is a geometric series with $|r| = \frac{2}{7} < 1$, so it converges, but the index starts at $k = 5$, hence:

$$\left[\sum_{k=0}^{\infty}\left(-\frac{2}{7}\right)^k\right] - \left[\sum_{k=0}^{4}\left(-\frac{2}{7}\right)^k\right]$$
$$= \frac{1}{1+\frac{2}{7}} - \left[1 - \frac{2}{7} + \frac{4}{49} - \frac{8}{343} + \frac{16}{2401}\right]$$
$$= -\frac{32}{21609} \approx -0.00148$$

11. This geometric series diverges: $|r| = \frac{\pi}{3} > 1$.

13. $\sum_{k=0}^{\infty}\left(\frac{1}{3}\right)^k = \frac{1}{1-\frac{1}{3}} = \frac{3}{2}$

15. $\sum_{k=3}^{\infty}\left(\frac{1}{2}\right)^k = \frac{1}{8}\cdot\sum_{k=3}^{\infty}\left(\frac{1}{2}\right)^k = \frac{1}{8}\cdot\frac{1}{1-\frac{1}{2}} = \frac{1}{4}$

17. $\sum_{k=1}^{\infty}\left(-\frac{2}{3}\right)^k = -\frac{2}{3}\cdot\sum_{k=0}^{\infty}\left(-\frac{2}{3}\right)^k = -\frac{2}{3}\cdot\frac{3}{5} = -\frac{2}{5}$

19. (a) $\frac{1}{1-\frac{1}{2}} - 1 = \frac{1}{\frac{1}{2}} - 1 = 2 - 1 = 1$

(b) $\frac{1}{1-\frac{1}{3}} - 1 = \frac{1}{\frac{2}{3}} - 1 = \frac{3}{2} - 1 = \frac{1}{2}$

(c) $\frac{1}{1-\frac{1}{a}} - 1 = \frac{1}{\frac{a-1}{a}} - 1 = \frac{a}{a-1} - \frac{a-1}{a-1} = \frac{1}{a-1}$

21. (a) $40(0.4)^n$ (b) $\sum_{n=0}^{\infty} 40(0.4)^n$ (c) $\frac{40}{1-0.4} \approx 66.67$ ft

23. (a) $\sum_{n=1}^{\infty}\left(\frac{1}{2}\right)^n$ (b) $\frac{1}{2}, \frac{1}{4}, \left(\frac{1}{2}\right)^n$ (c) All of it.

25. $\sum_{k=0}^{\infty}\left(\frac{1}{4}\right)^k = \frac{1}{1-\frac{1}{4}} = \frac{4}{3}$

27. (a) We can express the total area as:

$$1 + 3\cdot\frac{1}{9} + 3\cdot 4\cdot\frac{1}{9^2} + 3\cdot 4^2\cdot\frac{1}{9^3} + \cdots$$
$$= 1 + \frac{3}{9} + \frac{3}{9}\cdot\frac{4}{9} + + \frac{3}{9}\cdot\frac{4^2}{9^2} + \cdots$$
$$= 1 + \frac{1}{3}\left[1 + \frac{4}{9} + \left(\frac{4}{9}\right)^2 + \cdots\right]$$
$$= 1 + \frac{1}{3}\cdot\frac{1}{1-\frac{4}{9}} = 1 + \frac{1}{3}\cdot\frac{9}{5} = 1 + \frac{3}{5} = 1.6$$

(b) If L is the length of one side of the original triangle, its perimeter is $3L$. The first step replaces each original side with four smaller sides each $\frac{1}{3}$ the length of the original, so the perimeter after the first step is $3\cdot 4\cdot\frac{1}{3}L = 4L$. The second step replaces each of the 12 existing sides with four smaller sides each $\frac{1}{3}$ the length of the previous ones, so the perimeter is now $3 \cdot 4^2 \cdot \frac{1}{3}\left(\frac{1}{3}L\right) = 3L\left(\frac{4}{3}\right)^2$. The perimeter after n steps is $3L\left(\frac{4}{3}\right)^n$, which has limit ∞ as $n \to \infty$.

29. (a) The total height can be expressed as:

$$2 \cdot 1 + 2 \cdot \frac{1}{2} + 2 \cdot \frac{1}{4} + 2 \cdot \frac{1}{8} + \cdots = \frac{2}{1 - \frac{1}{2}} = 4$$

(b) The total surface area is:

$$\begin{aligned} &4\pi \cdot 1^2 + 4\pi\left(\frac{1}{2}\right)^2 + 4\pi\left(\frac{1}{4}\right)^2 + 4\pi\left(\frac{1}{8}\right)^2 + \cdots \\ &= 4\pi\left[1 + \frac{1}{4} + \left(\frac{1}{4}\right)^2 + \left(\frac{1}{4}\right)^3 + \cdots\right] \\ &= 4\pi \cdot \frac{1}{1 - \frac{1}{4}} = 4\pi \cdot \frac{4}{3} = \frac{16\pi}{3} \approx 16.755 \end{aligned}$$

(c) The total volume is:

$$\begin{aligned} &\frac{4}{3}\pi \cdot 1^3 + \frac{4}{3}\pi \cdot \left(\frac{1}{2}\right)^3 + \frac{4}{3}\pi \cdot \left(\frac{1}{4}\right)^3 + \cdots \\ &= \frac{4}{3}\pi\left[1 + \frac{1}{8} + \left(\frac{1}{8}\right)^2 + \cdots\right] \\ &= \frac{4}{3}\pi \cdot \frac{1}{1 - \frac{1}{8}} = \frac{4}{3}\pi \cdot \frac{8}{7} = \frac{32\pi}{21} \approx 4.787 \end{aligned}$$

31. We can rewrite $0.\overline{8} = 0.888\ldots$ as:

$$\frac{8}{10} \cdot \sum_{k=0}^{\infty}\left(\frac{1}{10}\right)^k = \frac{8}{10} \cdot \frac{1}{\frac{9}{10}} = \frac{8}{9}$$

Similarly, $0.\overline{9} = 0.999\ldots = \frac{9}{10} \cdot \frac{1}{\frac{9}{10}} = 1$ and $0.\overline{285714} = \frac{285714}{1000000} \cdot \frac{1}{\frac{999999}{1000000}} = \frac{285714}{999999}$.

33. The series converges precisely when:

$$\begin{aligned} &|2x+1| < 1 \Rightarrow -1 < 2x+1 < 1 \\ &\quad \Rightarrow -2 < 2x < 0 \Rightarrow -1 < x < 0 \end{aligned}$$

35. The series converges if and only if:

$$\begin{aligned} &|1-2x| < 1 \Rightarrow -1 < 2x-1 < 1 \\ &\quad \Rightarrow 0 < 2x < 2 \Rightarrow 0 < x < 1 \end{aligned}$$

37. The series converges when x satisfies:

$$|7x| < 1 \Rightarrow -1 < 7x < 1 \Rightarrow -\frac{1}{7} < x < \frac{1}{7}$$

39. The ratio is $\frac{x}{2}$, so the series converges when:

$$\left|\frac{x}{2}\right| < 1 \Rightarrow -1 < \frac{x}{2} < 1 \Rightarrow -2 < x < 2$$

41. The ratio is $2x$, so the series converges when:

$$|2x| < 1 \Rightarrow -1 < 2x < 1 \Rightarrow -\frac{1}{2} < x < \frac{1}{2}$$

43. The ratio is $\sin(x)$, so the series converges when: $|\sin(x)| < 1$, which holds true for all values of x except odd multiples of $\frac{\pi}{2}$.

45. The first student stated the formula correctly, but it is valid only when $|x| < 1$, so the second student should not have put $x = 2$ into the formula.

47. This is a telescoping sum: $s_4 = \frac{1}{3} - \frac{1}{5} = \frac{2}{15}$, $s_5 = \frac{1}{3} - \frac{1}{6} = \frac{1}{6}$, $s_n = \frac{1}{3} - \frac{1}{n+1}$

49. This is a telescoping sum: $s_4 = 1^3 - 5^3 = -124$, $s_5 = 1^3 - 6^3 = -215$, $s_n = 1^3 - (n+1)^3$

51. This is a telescoping sum: $s_4 = f(1) - f(5)$, $s_5 = f(1) - f(6)$, $s_n = f(1) - f(n+1)$

53. $s_4 = \sin(1) - \sin\left(\frac{1}{5}\right) \approx 0.643$
$s_5 = \sin(1) - \sin\left(\frac{1}{6}\right) \approx 0.676$
$s_n = \sin(1) - \sin\left(\frac{1}{n+1}\right) \to \sin(1) \approx 0.841$

55. $s_4 = \frac{1}{4} - \frac{1}{25} = 0.21$; $s_5 = \frac{1}{4} - \frac{1}{36} = \frac{2}{9}$
$s_n = \frac{1}{4} - \frac{1}{(n+1)^2} \to \frac{1}{4}$

57. On your own.

59. On your own.

61. (a) $\frac{3}{4}$ (b) $\frac{c}{(c-2)^2}$

Section 9.5

1. Sum.

3. $\sum_{k=1}^{\infty} f(k)$

5. $\sum_{k=2}^{\infty} f(k)$

7. $f(1) + f(2)$

9. $f(2) + f(3)$

11. $f(1) + f(2) + f(3) < \int_1^4 f(x)\,dx < f(2) + f(3) + f(4) < \int_2^5 f(x)\,dx$

13. (a) You did well. (b) You may have done well or you may have done poorly. (c) You may have done well or poorly. (d) You did poorly.

15. (a) Unknown is good. (b) Unknown might be good or might be bad. (c) Unknown might be good or might be bad. (d) Unknown is bad.

17. $\sum_{k=2} \frac{1}{k^3 - 5}$ 19. $\sum_{k=1} \frac{1}{k^2 + 5k}$

21. (a) $k + 4$ (b) $\frac{k+4}{k+3}$ (c) $\frac{k+4}{k+3}$

23. (a) $\frac{3}{k+1}$ (b) $\frac{\frac{3}{k}}{\frac{3}{k+1}}$ (c) $\frac{k}{k+1}$

25. (a) 2^{k+1} (b) $\frac{2^{k+1}}{2^k}$ (c) 2 27. (a) x^{k+1} (b) x (c) x

29. converges; $\frac{1}{2}$ 31. diverges; 2

33. diverges; 1 35. diverges; $\frac{k}{k+1}$

37. $<$ 39. $>$ 41. $>$

43. $s_3 < s_5 < s_4 < s_6$ 45. $s_5 < s_6 < s_4 < s_3$

47. $s_1 = 2$, $s_2 = 1$, $s_3 = 1.9$, $s_4 = 1.1$, $s_5 = 1.8$, $s_6 = 1.2$, $s_7 = 1.7$, $s_8 = 1.3$; "funnel-shaped":

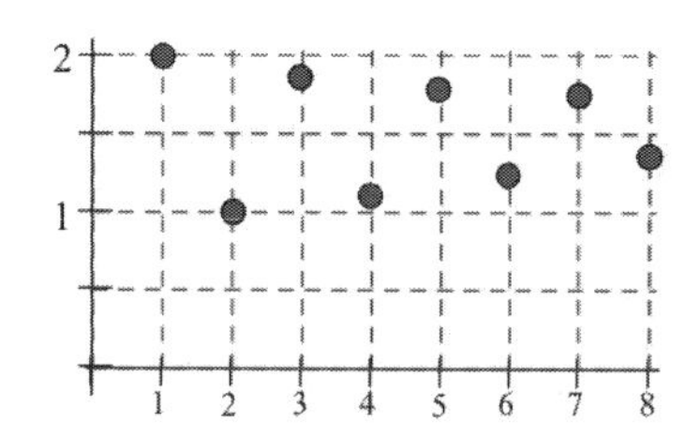

49. $s_1 = -2$, $s_2 = -0.5$, $s_3 = -1.3$, $s_4 = -0.7$, $s_5 = -1.1$, $s_6 = -0.9$, $s_7 = 1.1$, $s_8 = 1.0$; initially "funnel-shaped":

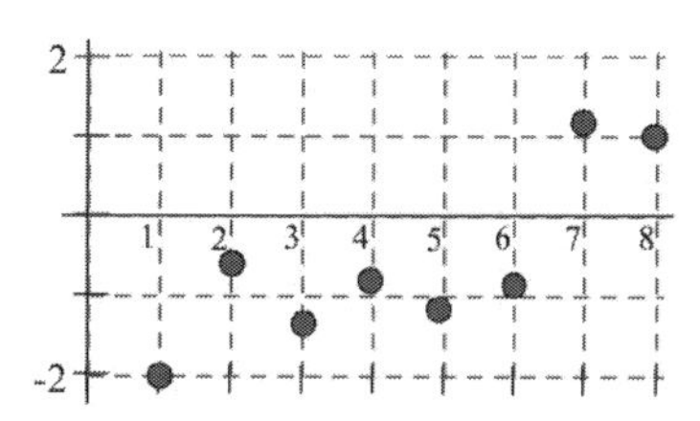

51. The terms a_k need to alternate in sign.

53. (a) D, E, F (b) A, D (c) D

55.
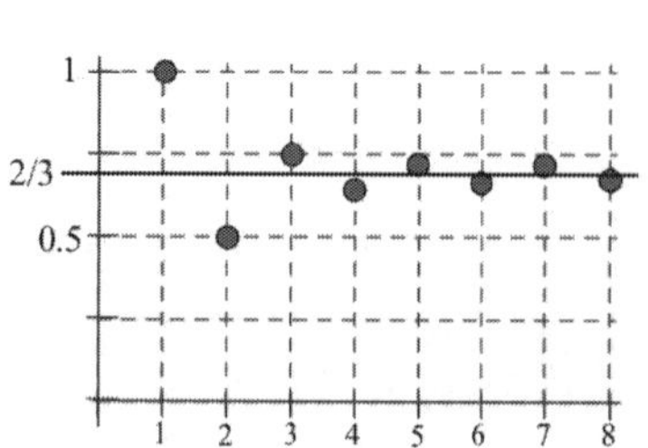

57.
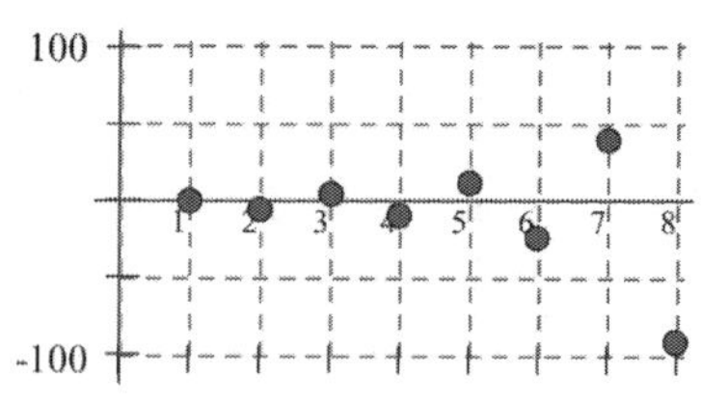

Section 9.6

1. $f(x) = (2x+5)^{-1}$ is positive, continuous and decreasing on $[1, \infty)$, and:

$$\lim_{M\to\infty} \int_1^M \frac{1}{2x+5}\,dx = \lim_{M\to\infty} \left[\frac{\ln(2x+5)}{2}\right]_1^M = \infty$$

so $\int_1^\infty \frac{1}{2x+5}\,dx$ and $\sum_{k=1}^{\infty} \frac{1}{2k+5}$ both diverge.

3. $f(x) = (2x+5)^{-\frac{3}{2}}$ is positive, continuous and decreasing on $[1, \infty)$, and:

$$\int_1^M (2x+5)^{-\frac{3}{2}}\,dx = \left[\frac{-1}{\sqrt{2x+5}}\right]_1^M \to \frac{1}{\sqrt{7}}$$

as $M \to \infty$ so $\int_1^\infty (2x+5)^{-\frac{3}{2}}\,dx$ converges, hence $\sum_{k=1}^{\infty} \frac{1}{(2k+5)^{\frac{3}{2}}}$ converges.

5. On $[2, \infty)$, $f(x) = \frac{1}{x \cdot [\ln(x)]^2}$ is positive, continuous and decreasing, and:

$$\int_1^M \frac{1}{1 \cdot [\ln(x)]^2}\,dx = \left[\frac{-1}{\ln(x)}\right]_1^M \to \frac{1}{\ln(2)}$$

as $M \to \infty$ so $\int_1^\infty \frac{1}{1 \cdot [\ln(x)]^2}\,dx$ converges, hence $\sum_{k=2}^{\infty} \frac{1}{k \cdot [\ln(k)]^2}$ converges.

7. $f(x) = \frac{1}{1+x^2}$ is positive, continuous and decreasing (everywhere), and:

$$\int_1^M \frac{1}{1+x^2}\,dx = [\arctan(x)]_1^M \to \frac{\pi}{2} - \frac{\pi}{4} = \frac{\pi}{4}$$

as $M \to \infty$ so $\int_1^M \frac{1}{1+x^2}\,dx$ converges, hence $\sum_{k=1}^{\infty} \frac{1}{k^2+1}$ converges.

9. $\displaystyle\sum_{k=1}^{\infty}\left[\frac{1}{k}-\frac{1}{k+3}\right]$ is a telescoping series:

$$\left[1-\frac{1}{4}\right]+\left[\frac{1}{2}-\frac{1}{5}\right]+\left[\frac{1}{3}-\frac{1}{6}\right]+\left[\frac{1}{4}-\frac{1}{7}\right]+\left[\frac{1}{5}-\frac{1}{8}\right]+\cdots=1+\frac{1}{2}+\frac{1}{3}$$

so the series converges to $\frac{11}{6}$. The Integral Test also works because $f(x)=\dfrac{1}{x}-\dfrac{1}{x+3}$ is positive, continuous and decreasing (you should verify this), and:

$$\lim_{M\to\infty}\int_1^M\left[\frac{1}{x}-\frac{1}{x+3}\right]dx=\lim_{M\to\infty}\Big[\ln(x)-\ln(x+3)\Big]_1^M=\lim_{M\to\infty}\left[\ln\left(\frac{M}{M+3}\right)-\ln(1)+\ln(4)\right]=\ln(4)$$

Because the improper integral converges, the series converges as well, but the Integral Test does not tell us the sum of the series. In this instance, the "telescoping series" method is both easier and more precise.

11. Applying the Integral Test to $\displaystyle\sum_{k=1}^{\infty}\frac{1}{k(k+5)}$ using $f(x)=\dfrac{1}{x(x+5)}$ (you should verify that this function is positive, continuous and decreasing) and Partial Fraction Decomposition:

$$\lim_{M\to\infty}\int_1^M\frac{1}{5}\left[\frac{1}{x}-\frac{1}{x+5}\right]dx=\lim_{M\to\infty}\frac{1}{5}\Big[\ln(x)-\ln(x+5)\Big]_1^M=\lim_{M\to\infty}\frac{1}{5}\left[\ln\left(\frac{M}{M+5}\right)+\ln(6)\right]=\frac{\ln(6)}{5}$$

Because the improper integral converges, the series converges as well, but we can use the same partial fraction decomposition to turn the series into a telescoping series and find its exact value:

$$\left[1-\frac{1}{6}\right]+\left[\frac{1}{2}-\frac{1}{7}\right]+\left[\frac{1}{3}-\frac{1}{8}\right]+\left[\frac{1}{4}-\frac{1}{9}\right]+\left[\frac{1}{5}-\frac{1}{10}\right]+\left[\frac{1}{6}-\frac{1}{11}\right]+\cdots=1+\frac{1}{2}+\frac{1}{3}+\frac{1}{4}+\frac{1}{5}=\frac{137}{60}$$

13. Applying the Integral Test to $\displaystyle\sum_{k=1}^{\infty}ke^{-k^2}$ using $f(x)=xe^{-x^2}$ (verify it is positive, continuous and decreasing):

$$\lim_{M\to\infty}\int_1^M xe^{-x^2}\,dx=\lim_{M\to\infty}\left[\frac{-1}{2e^{x^2}}\right]_1^M=\lim_{M\to\infty}\left[\frac{-1}{2e^{M^2}}+\frac{1}{2e}\right]=\frac{1}{2e}$$

Because the improper integral converges, the series converges as well.

15. Applying the Integral Test to $\displaystyle\sum_{k=1}^{\infty}\frac{1}{\sqrt{6k+10}}$ using $f(x)=\dfrac{1}{\sqrt{6x+10}}$ (verify it is positive, continuous and decreasing):

$$\lim_{M\to\infty}\int_1^M(6x+10)^{-\frac{1}{2}}\,dx=\lim_{M\to\infty}\left[\frac{1}{3}\sqrt{6x+10}\right]_1^M=\lim_{M\to\infty}\frac{1}{3}\left[\sqrt{6M+10}-4\right]=\infty$$

Because the improper integral diverges, the series diverges as well.

17. converges ($p=4>1$)

19. diverges ($p=\frac{1}{5}\le 1$)

21. diverges ($p=1\le 1$)

23. converges ($p=\frac{3}{2}>1$)

25. converges ($p=\frac{4}{3}>1$)

27. diverges ($p=\frac{2}{3}\le 1$)

29. $\displaystyle\int_1^{11}\frac{1}{x^3}\,dx\le s_{10}\le 1+\int_1^{10}\frac{1}{x^3}\,dx\ \Rightarrow\ 0.4958677<s_{10}<1.495$

$\displaystyle\int_1^{101}\frac{1}{x^3}\,dx\le s_{100}\le 1+\int_1^{100}\frac{1}{x^3}\,dx\ \Rightarrow\ 0.0.499951<s_{100}<1.49995$

$\displaystyle\int_1^{1000001}\frac{1}{x^3}\,dx\le s_{1000000}\le 1+\int_1^{1000000}\frac{1}{x^3}\,dx\ \Rightarrow\ 0.5000000<s_{1000000}<1.5000000$

31. $\ln(11) < s_{10} < 1 + \ln(10)$,
$4.6151 < s_{100} < 5.6052$,
$13.8155 < s_{1000000} < 14.8155$

33. $\arctan(11) - \frac{\pi}{4} < s_{10} < \frac{1}{2} + \arctan(11) - \frac{\pi}{4}$,
$0.7755 < s_{100} < 1.2754$,
$0.7854 < s_{1000000} < 1.2854$

35. $s_{10} = \sum_{k=1}^{10} \frac{1}{k^4} \approx 1.08203658 \Rightarrow s_{11} \approx 1.08210488$ and $\int_{11}^{\infty} \frac{1}{x^4}\,dx = \frac{1}{3993} \approx 0.00025044$, so:

$$1.08228702 < \sum_{k=1}^{\infty} \frac{1}{k^4} < 1.08235532$$

Using $n = 20$ yields:

$$1.08232058 < \sum_{k=1}^{\infty} \frac{1}{k^4} < 1.08232572$$

37. $\int_{11}^{\infty} \frac{1}{x^2+1}\,dx \approx 0.09065989$, $s_{10} \approx 0.98179282$ and $s_{11} \approx 0.98998954$, so:

$$1.07245271 < \sum_{k=1}^{\infty} \frac{1}{k^2+1} < 1.08064943$$

Using $n = 20$ yields:

$$1.07552492 < \sum_{k=1}^{\infty} \frac{1}{k^2+1} < 1.07778736$$

39. Using $n = 10$: $2.5984 < \sum_{k=1}^{\infty} \frac{1}{k\sqrt{k}} < 2.6258$
Using $n = 20$: $2.6071 < \sum_{k=1}^{\infty} \frac{1}{k\sqrt{k}} < 2.6175$

41. Use the substitution $u = \ln(x) \Rightarrow du = \frac{1}{x}\,dx$ and the Integral Test to see that the improper integral:

$$\int_2^{\infty} \frac{1}{x \cdot [\ln(x)]^q}\,dx$$

diverges for $q \leq 1$ and converges for $q > 1$, hence the series does as well.

43. converges ($q = 3 > 1$)

45. diverges ($\ln\left(k^3\right) = 3\ln(k)$ so $q = 1 \leq 1$)

Section 9.7

1. $0 \leq \cos^2(k) \leq 1 \Rightarrow 0 \leq \frac{\cos^2(k)}{k^2} \leq \frac{1}{k^2}$ so $\sum_{k=1}^{\infty} \frac{\cos^2(k)}{k^2}$ converges by BCT with $\sum_{k=1}^{\infty} \frac{1}{k^2}$.

3. $n - 1 < n \Rightarrow \frac{5}{n-1} \geq \frac{5}{n}$ so $\sum_{n=3}^{\infty} \frac{5}{n-1}$ diverges by comparison with $\sum_{n=3}^{\infty} \frac{5}{n}$, which diverges because it is a multiple of p-series with $p = 1$.

5. $3 + \cos(m) \geq 2$ so $\sum_{m=1}^{\infty} \frac{3+\cos(m)}{m}$ diverges by comparison with $\sum_{m=1}^{\infty} \frac{2}{m}$, which diverges because it is a multiple of the harmonic series.

7. For $k \geq 3$, $\ln(k) > 1$ and $\frac{\ln(k)}{k} > \frac{1}{k}$, so $\sum_{k=2}^{\infty} \frac{\ln(k)}{k}$ diverges by comparison with $\sum_{k=2}^{\infty} \frac{1}{k}$.

9. $k \geq 9 \Rightarrow 0 < \frac{k+9}{k \cdot 2^k} \leq \frac{2k}{k \cdot 2^k} = \frac{1}{2^{k-1}}$ so $\sum_{k=1}^{\infty} \frac{k+9}{k \cdot 2^k}$ converges by comparison with $\sum_{k=1}^{\infty} \frac{1}{2^{k-1}}$ (a geometric series with ratio $\frac{1}{2} < 1$).

11. $k \geq 2 \Rightarrow k! \geq k(k-1) \geq k\left(k - \frac{1}{2}k\right) = \frac{1}{2}k^2 \Rightarrow 0 < \frac{1}{k!} < \frac{2}{k^2}$ so $\sum_{k=1}^{\infty} \frac{1}{k!}$ converges by comparison with $\sum_{k=1}^{\infty} \frac{2}{k^2}$ (a p-series with $p = 2 > 1$).

13. Using the LCT with the harmonic series:

$$\lim_{k\to\infty} \frac{\frac{k+1}{k^2+4}}{\frac{1}{k}} = \lim_{k\to\infty} \frac{k^2+k}{k^2+4} = 1$$

so $\sum_{k=3}^{\infty} \frac{k+1}{k^2+4}$ because $\sum_{k=3}^{\infty} \frac{1}{k}$ diverges.

15. Diverges by LCT with the harmonic series.

17. Converges by LCT with the p-series $\sum_{k=1}^{\infty} \frac{1}{k^3}$:

$$\lim_{k\to\infty} \frac{\frac{k^3}{(1+k^2)^3}}{\frac{1}{k^3}} = \lim_{k\to\infty} \frac{k^6}{k^6+2k^4+3k^2+1} = 1$$

19. Diverges by LCT with harmonic series.

21. Converges by LCT with the p-series $\sum_{k=1}^{\infty} \frac{1}{k^3}$

23. Converges by LCT with the p-series $\sum_{n=3}^{\infty} \frac{1}{n^2}$
25. Diverges by LCT with the p-series $\sum_{k=1}^{\infty} \frac{1}{\sqrt{k}}$
27. Converges by LCT with the p-series $\sum_{k=2}^{\infty} \frac{1}{k^3}$
29. Converges by LCT with the p-series $\sum_{k=1}^{\infty} \frac{1}{k^2}$
31. Diverges by LCT with the p-series $\sum_{k=1}^{\infty} \frac{1}{\sqrt{k}}$
33. Diverges by LCT with harmonic series.
35. Diverges by LCT with harmonic series.
37. Diverges by LCT with harmonic series.
39. Diverges by LCT with the p-series $\sum_{k=1}^{\infty} \frac{1}{\sqrt{k}}$
41. Converges by LCT with $\sum_{k=1}^{\infty} \frac{1}{3^k}$
43. Diverges by Test for Divergence.
45. Converges by LCT with $\sum_{k=1}^{\infty} \frac{1}{e^k}$
47. Converges by LCT with the p-series $\sum_{n=1}^{\infty} \frac{1}{n^2}$
49. Diverges by LCT with harmonic series.
51. Converges by LCT with the p-series, p=3.
53. Diverges by Test for Divergence.
55. Diverges by Test for Divergence.
57. Converges: geometric series with $r = \frac{1}{3}$.
59. Diverges by Test for Divergence.
61. Converges: geometric series with $r = e^{-1}$.
63. Converges: geometric series with $r = \frac{\pi^2}{e^3} < \frac{1}{2}$.
65. Converges by BCT with the p-series, $p = 3$.
67. Converges by Integral Test.
69. Diverges by LCT with harmonic series.
71. For $x \geq 5$, $\ln(x) < \sqrt{x}$; to verify this, note that:

$$\mathbf{D}\left(\ln(x)\right) = \frac{1}{x} \leq \frac{1}{2\sqrt{x}} = \mathbf{D}\left(\sqrt{x}\right)$$

and $\ln(5) < \sqrt{5}$. Hence for $k \geq 5$:

$$\frac{\ln(k)}{k^2} < \frac{\sqrt{k}}{k^2} = \frac{1}{k^{\frac{3}{2}}}$$

so $\sum_{k=2}^{\infty} \frac{\ln(k)}{k^2}$ converges by BCT with $\sum_{k=2}^{\infty} \frac{1}{k^{\frac{3}{2}}}$.

73. Diverges by LCT with harmonic series.
75. Diverges by LCT with harmonic series.
77. Diverges by Test for Divergence.

Section 9.8

1. (a) See below left. (b) Alternating (so far).

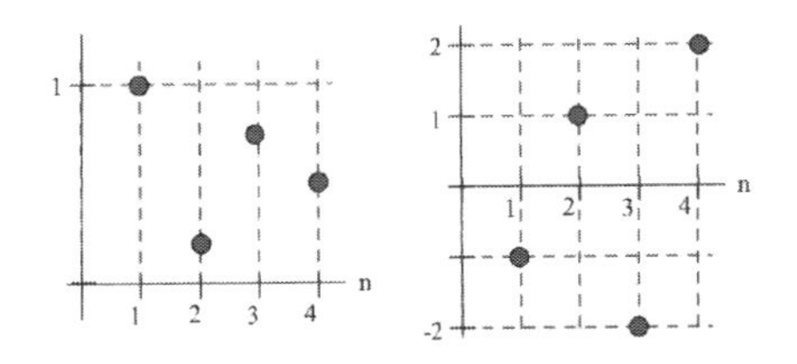

3. (a) See above right. (b) Alternating (so far).
5. (a) See below. (b) Not alternating.

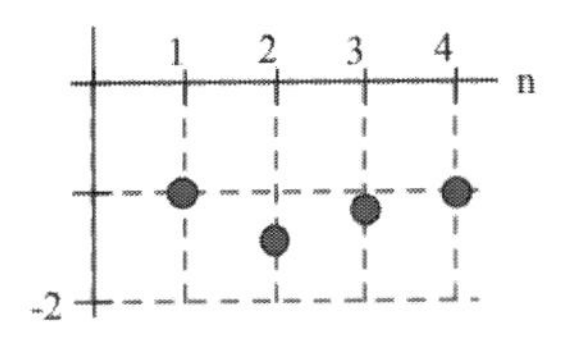

7. Alternating: $a_1 = 2$, $a_2 = -1$, $a_3 = 2$, $a_4 = -1$, $a_5 = 2$
9. Not alternating: $a_1 = 2$, $a_2 = 1$, $a_3 = -0.9$, $a_4 = 0.8$, $a_5 = -0.1$
11. Not alternating: $a_1 = -1$, $a_2 = 2$, $a_3 = -1.2$, $a_4 = 0.2$, $a_5 = 0.2$
13. A has decreasing partial sums, so terms are all negative; C has increasing partial sums, so terms are all positive.
15. B has increasing partial sums, so terms are all positive (they do not alternate).
17. Converges by AST. 19. Converges by AST.
21. Diverges by Test for Divergence.
23. Converges by AST. 25. Converges by AST.
27. Diverges by Test for Divergence.
29. The AST does *not* apply to this series because the terms are all negative, but it does converge; factor out -2 and use the LCT with the resulting series and a geometric series.
31. Converges because all terms are 0.

33. (a) $s_4 = 1 - \frac{1}{4} + \frac{1}{9} - \frac{1}{16} = \frac{115}{144} \approx 0.79861$
(b) $|a_5| = \frac{1}{25} = 0.04$ (c) $0.75861 < S < 0.83861$

35. (a) $s_4 = \frac{1}{\ln(2)} - \frac{1}{\ln(3)} + \frac{1}{\ln(4)} - \frac{1}{\ln(5)} \approx 0.6325$
(b) $|a_5| = \frac{1}{\ln(6)} \approx 0.5581$ (c) $0.0744 < S < 1.1906$

37. (a) $s_4 \approx 0.20992$ (b) $|a_5| = (0.8)^6 \approx 0.26214$
(c) $-0.05222 < S < 0.47206$

39. (a) $s_4 \approx 0.441836$ (b) $|a_5| = \sin\left(\frac{1}{5}\right) \approx 0.198669$
(c) $0.243167 < S < 0.640505$

41. (a) $s_4 = -1 + \frac{1}{8} - \frac{1}{27} + \frac{1}{64} \approx -0.896412$
(b) $|a_5| = 0.008$ (c) $-0.904412 < S < -0.888412$

43. $\frac{1}{(N+1)+6} \leq \frac{1}{100} \Rightarrow N+7 \geq 100 \Rightarrow N \geq 93$

45. We need $\frac{2}{\sqrt{(N+1)+21}} \leq \frac{1}{100} \Rightarrow \sqrt{N+22} \geq 200 \Rightarrow$
$N + 22 \geq 40000 \Rightarrow N \geq 39978$

47. $\left(\frac{1}{3}\right)^{N+1} \leq \frac{1}{500} \Rightarrow N+1 \geq \frac{\ln(500)}{\ln(3)} \approx 5.66$, so use $N = 5$.

49. We need $\frac{1}{(N+1)^4} \leq \frac{1}{1000} \Rightarrow (N+1)^4 \geq 1000 \Rightarrow$
$N + 1 > 5.62 \Rightarrow N > 4.62$, so use $N = 5$.

51. $\frac{1}{(N+1)+\ln(N+1)} \leq \frac{1}{25} \Rightarrow (N+1) + \ln(N+1) \geq 25$; this will certainly be true if $N+1 \geq 25 \Rightarrow N \geq 24$ (but some experimenting with a calculator shows that $N = 21$ works while $N = 20$ does not).

53. (a) $S(0.3) = 0.3 - \dfrac{(0.3)^3}{3!} + \dfrac{(0.3)^5}{5!} - \dfrac{(0.3)^7}{7!} + \cdots$
(b) $s_3 = 0.3 - \frac{(0.3)^3}{3!} + \frac{(0.3)^5}{5!} \approx 0.29552025$
(c) $|S - s_3| \leq \frac{(0.3)^7}{7!} \approx 0.000000043$

55. (a) $S(0.1) = 0.1 - \dfrac{(0.1)^3}{3!} + \dfrac{(0.1)^5}{5!} - \dfrac{(0.1)^7}{7!} + \cdots$
(b) $s_3 = 0.1 - \frac{(0.1)^3}{3!} + \frac{(0.1)^5}{5!} \approx 0.09983342$
(c) $|S - s_3| \leq \frac{(0.1)^7}{7!} \approx 2 \times 10^{-11}$

57. (a) $C(1) = 1 - \dfrac{1}{2!} + \dfrac{1}{4!} - \dfrac{1}{6!} + \cdots$
(b) $s_3 = 1 - \frac{1}{2!} + \frac{1}{4!} \approx 0.5416667$
(c) $|S - s_3| \leq \frac{1}{6!} \approx 0.0013889$

59. (a) $1 - \dfrac{(-0.2)^2}{2!} + \dfrac{(-0.2)^4}{4!} - \dfrac{(-0.2)^6}{6!} + \cdots$
(b) $s_3 = 1 - \frac{(-0.2)^2}{2!} + \frac{(-0.2)^4}{4!} \approx 0.980066667$
(c) $|S - s_3| \leq \frac{(-0.2)^6}{6!} \approx 9 \times 10^{-8}$

61. (a) $1 + (-1) + \dfrac{(-1)^2}{2!} + \dfrac{(-1)^3}{3!} + \dfrac{(-1)^4}{4!} + \cdots$
(b) $s_3 = 1 - 1 + \frac{1}{2} = 0.5$
(c) $|S - s_3| \leq \frac{1}{3!} = \frac{1}{6} \approx 0.16667$

63. (a) $1 + (-0.2) + \dfrac{(-0.2)^2}{2!} + \dfrac{(-0.2)^3}{3!} + \dfrac{(-0.2)^4}{4!} + \cdots$
(b) $s_3 = 1 - 0.2 + \frac{0.04}{2} = 0.82$
(c) $|S - s_3| \leq \frac{(0.2)^4}{6} \approx 0.0013333$

Section 9.9

1. $\displaystyle\sum_{k=1}^{\infty} \left| \frac{(-1)^{k+1}}{k+2} \right| = \sum_{k=1}^{\infty} \frac{1}{k+2}$, which diverges (by LCT with the harmonic series) so $\displaystyle\sum_{k=1}^{\infty} \frac{(-1)^{k+1}}{k+2}$ is not absolutely convergent, however the AST applies, so it converges conditionally.

3. $\displaystyle\sum_{n=1}^{\infty} \left| (-1)^{n+1} \cdot \frac{5}{n^3} \right| = 5 \sum_{n=1}^{\infty} \frac{1}{n^3}$, which converges, so $\displaystyle\sum_{n=1}^{\infty} (-1)^{n+1} \cdot \frac{5}{n^3}$ converges absolutely.

5. $\displaystyle\sum_{k=0}^{\infty} \left| (-0.5)^k \right| = \sum_{k=0}^{\infty} (0.5)^k$ is a convergent geometric series, so $\displaystyle\sum_{k=0}^{\infty} (-0.5)^k$ converges absolutely.

7. $\displaystyle\sum_{k=1}^{\infty} \left| \frac{(-1)^{k+1}}{k^2} \right| = \sum_{k=1}^{\infty} \frac{1}{k^2}$ converges (by the P-Test), so $\displaystyle\sum_{k=1}^{\infty} \frac{(-1)^{k+1}}{k^2}$ converges absolutely.

9. $\displaystyle\sum_{n=2}^{\infty} \left| (-1)^n \cdot \frac{\ln(n)}{n} \right| = \sum_{n=2}^{\infty} \frac{\ln(n)}{n}$ diverges (by BCT with the harmonic series), but the AST says $\displaystyle\sum_{n=2}^{\infty} (-1)^n \cdot \frac{\ln(n)}{n}$ converges conditionally.

11. $\displaystyle\sum_{k=1}^{\infty} \left| \frac{(-1)^k}{k + \ln(k)} \right| = \sum_{k=1}^{\infty} \frac{1}{k + \ln(k)}$ diverges by BCT with $\displaystyle\sum_{k=1}^{\infty} \frac{1}{2k}$, but the AST applies to $\displaystyle\sum_{k=1}^{\infty} \frac{(-1)^k}{k + \ln(k)}$, so it converges conditionally.

13. $\displaystyle\sum_{k=1}^{\infty} \left| (-1)^k \cdot \sin\left(\frac{1}{k}\right) \right| = \sum_{k=1}^{\infty} \sin\left(\frac{1}{k}\right)$ diverges (by LCT with the harmonic series), but the AST says $\displaystyle\sum_{k=1}^{\infty} (-1)^k \cdot \sin\left(\frac{1}{k}\right)$ converges conditionally.

15. $\sum_{k=1}^{\infty} \sqrt{k} \sin\left(\frac{1}{k^2}\right)$ converges by LCT with $\sum_{k=1}^{\infty} \frac{1}{k^{\frac{3}{2}}}$, so $\sum_{k=1}^{\infty} (-1)^k \sqrt{k} \sin\left(\frac{1}{k^2}\right)$ converges absolutely.

17. $\sum_{m=2}^{\infty} (-1)^m \cdot \frac{\ln(m)}{\ln(m^3)}$ diverges, because the terms do not approach 0.

19. Converges conditionally.

21. Diverges by Test for Divergence.

23. Converges absolutely (all terms are 0).

25. Converges conditionally.

27. Diverges by Test for Divergence.

29. Converges absolutely.

31. $\frac{n!}{(n+1)!} = \frac{n!}{(n+1) \cdot n!} = \frac{1}{n}$

33. $\frac{(n-1)!}{(n+1)!} = \frac{(n-1)!}{(n+1) \cdot n \cdot (n-1)!} = \frac{1}{n(n+1)}$

35. $\frac{n!}{(n+2)!} = \frac{n!}{(n+2)(n+1) \cdot n!} = \frac{1}{(n+1)(n+2)}$

37. $\frac{2 \cdot n!}{n! \cdot (n+1)(n+2) \cdots (2n)} = \frac{2}{(n+1)(n+2) \cdots (2n)}$

39. $\frac{n \cdot n \cdot n \cdots n \cdot n}{1 \cdot 2 \cdot 3 \cdots (n-1) \cdot n} = \frac{n}{1} \cdot \frac{n}{2} \cdot \frac{n}{3} \cdots \frac{n}{n-1} \cdot \frac{n}{n}$

41. $\frac{\frac{1}{k+1}}{\frac{1}{k}} = \frac{k}{k+1} \to 1$, so the Ratio Test is inconclusive; series diverges (harmonic series).

43. $\frac{\frac{1}{(k+1)^3}}{\frac{1}{k^3}} = \left(\frac{k}{k+1}\right)^3 \to 1$, so the Ratio Test is inconclusive; series converges (P-Test).

45. $\frac{\left(\frac{1}{2}\right)^{k+1}}{\left(\frac{1}{2}\right)^k} = \frac{1}{2} < 1$; absolutely convergent (AC).

47. $\frac{1^{n+1}}{1^n} = 1$, so the Ratio Test is inconclusive; diverges (by Test for Divergence).

49. $\frac{\frac{1}{(k+1)!}}{\frac{1}{k!}} = \frac{k!}{(k+1)!} = \frac{1}{k+1} \to 0 < 1$; AC.

51. $\frac{\frac{2^{k+1}}{(k+1)!}}{\frac{2^k}{k!}} = 2 \cdot \frac{k!}{(k+1)!} = \frac{2}{k+1} \to 0 < 1$; AC.

53. $\frac{\left(\frac{1}{2}\right)^{3k+3}}{\left(\frac{1}{2}\right)^{3k}} = \frac{1}{8} < 1$; converges absolutely.

55. $\frac{(0.9)^{2k+3}}{(0.9)^{2k+1}} = 0.81 < 1$; converges absolutely.

57. $\left|\frac{(-1.1)^{k+1}}{(-1.1)^k}\right| = 1.1 > 1$; diverges.

59. $\left|\frac{(x-5)^{k+1}}{(x-5)^k}\right| = |x-5| < 1 \Rightarrow 4 < x < 6$. At $x = 4$ and $x = 6$ the series diverges (by the Test for Divergence), so the series converges absolutely on $(4, 6)$ and diverges elsewhere.

61. $\left|\frac{\frac{(x-5)^{k+1}}{(k+1)^2}}{\frac{(x-5)^k}{k^2}}\right| = \left(\frac{k}{k+1}\right)^2 |x-5| \to |x-5| < 1 \Rightarrow$ $4 < x < 6$. At $x = 4$ and $x = 6$ the series converges absolutely (by the P-Test), so the series converges absolutely on $[4, 6]$ and diverges elsewhere.

63. $\left|\frac{\frac{(x-2)^{k+1}}{(k+1)!}}{\frac{(x-2)^k}{k!}}\right| = \frac{1}{k+1} \cdot |x-2| \to 0 < 1$ for all x, so the series converges absolutely on $(-\infty, \infty)$.

65. Converges absolutely on $\left[\frac{11}{2}, \frac{13}{2}\right]$.

67. Converges absolutely on $(-\infty, \infty)$.

69. $\left|\frac{\frac{(x+1)^{2k+2}}{k+1}}{\frac{(x+1)^{2k}}{k}}\right| = \frac{k}{k+1} \cdot (x+1)^2 \to (x+1)^2 < 1 \Rightarrow$ $|x+1| < 1 \Rightarrow -2 < x < 0$; at $x = -2$ the series diverges and at $x = 0$ the series diverges, so it converges absolutely on $(-2, 0)$.

71. Converges absolutely on $[4, 6]$.

73. Converges absolutely on $(-\infty, \infty)$.

75. Converges absolutely on $(-\infty, \infty)$.

77. Converges absolutely on $(-\infty, \infty)$.

79. $\sqrt[k]{\left(\frac{2}{7}\right)^k} = \frac{2}{7} < 1$; absolutely convergent.

81. $\sqrt[k]{\frac{1}{k^3}} \to 1$, so Root Test inconclusive; absolutely convergent by P-Test.

83. $\sqrt[k]{\frac{1}{k^k}} = \frac{1}{k} \to 0 < 1$; absolutely convergent.

85. $\sqrt[k]{\left(\frac{1}{2}-\frac{2}{k}\right)^k} = \frac{1}{2}-\frac{2}{k} \to \frac{1}{2} < 1$; AC.

87. $\sqrt[k]{\left(\frac{2+k}{k}\right)^k} = \frac{2+k}{k} \to 1$, so Root Test inconclusive; diverges by Test for Divergence.

89. $\sqrt[k]{|\cos(k\pi)|^k} = 1$, so Root Test is inconclusive; diverges by Test for Divergence.

91. $L = \frac{2}{3} < 1$; absolutely convergent.

93. $\sqrt[k]{\frac{(2k)^k}{k^{2k}}} = \frac{2k}{k^2} = \frac{2}{k} \to 0 < 1$; AC.

95. $\frac{1}{2} - 1 + \frac{1}{4} + \frac{1}{6} + \frac{1}{8} + \frac{1}{10} + \frac{1}{12} + \frac{1}{14} + \frac{1}{16} - \frac{1}{3} + \frac{1}{18} + \frac{1}{20} + \frac{1}{22} + \frac{1}{24} + \frac{1}{26}$

97. $\frac{1}{\sqrt{2}} + \frac{1}{2} - 1 + \frac{1}{\sqrt{6}} + \frac{1}{\sqrt{8}} + \frac{1}{\sqrt{10}} - \frac{1}{\sqrt{3}} + \frac{1}{\sqrt{12}} + \frac{1}{\sqrt{14}} - \frac{1}{\sqrt{5}} + \frac{1}{4} - \frac{1}{\sqrt{7}} + \frac{1}{\sqrt{18}} + \frac{1}{\sqrt{20}} - \frac{1}{3}$

99. $\frac{1}{\sqrt{2}} - 1 + \frac{1}{2} + \frac{1}{\sqrt{6}} - \frac{1}{\sqrt{3}} + \frac{1}{\sqrt{8}} + \frac{1}{\sqrt{10}} - \frac{1}{\sqrt{5}} + \frac{1}{\sqrt{12}} - \frac{1}{\sqrt{7}} + \frac{1}{\sqrt{14}} - \frac{1}{3} + \frac{1}{4} + \frac{1}{\sqrt{18}} - \frac{1}{\sqrt{11}}$

101. On your own.

103. On your own.

Section 10.1

1. This is a geometric series with ratio x, so it converges precisely when $|x| < 1$; the interval of convergence is $(-1, 1)$. (Graph it yourself.)

3. Applying the Ratio Test:
$$\left|\frac{3^{k+1} \cdot x^{k+1}}{3^k \cdot x^k}\right| = |3x|$$
for all values of x, so the series converges when $|3x| < 1 \Rightarrow |x| < \frac{1}{3}$ and diverges when $|x| > \frac{1}{3}$. At $x = \frac{1}{3}$ the series becomes $\sum_{k=1}^{\infty} 1$, which diverges by the Test for Divergence; at $x = -\frac{1}{3}$, the series becomes $\sum_{k=1}^{\infty} (-1)^k$, which also diverges by the Test for Divergence. The interval of convergence is therefore $\left(-\frac{1}{3}, \frac{1}{3}\right)$. (The graph is left to you.)

5. Applying the Ratio Test:
$$\left|\frac{\frac{x^{k+1}}{k+1}}{\frac{x^k}{k}}\right| = \frac{k}{k+1} \cdot |x| \longrightarrow |x|$$
so the series converges when $|x| < 1$ and diverges when $|x| > 1$. At $x = 1$ the series becomes the harmonic series, which diverges; at $x = -1$, the series becomes the alternating harmonic series, which converges conditionally (by the Alternating Series Test). The interval of convergence is therefore $[-1, 1)$. (The graph is left to you.)

7. Applying the Ratio Test:
$$\left|\frac{(k+1) \cdot x^{k+1}}{k \cdot x^k}\right| = \frac{k+1}{k} \cdot |x| \longrightarrow |x|$$
so the series converges when $|x| < 1$ and diverges when $|x| > 1$. At $x = 1$ the series becomes $\sum_{k=1}^{\infty} k$, which diverges by the Test for Divergence; at $x = -1$, the series becomes $\sum_{k=1}^{\infty} k \cdot (-1)^k$, which also diverges by the Test for Divergence. The interval of convergence is therefore $(-1, 1)$.

9. Applying the Ratio Test:
$$\left|\frac{(k+1) \cdot x^{2k+3}}{k \cdot x^{2k+1}}\right| = \frac{k+1}{k} \cdot x^2 \longrightarrow x^2$$
so the series converges when $x^2 < 1 \Rightarrow |x| < 1$ and diverges when $|x| > 1$. At $x = 1$ the series becomes $\sum_{k=1}^{\infty} k$, which diverges by the Test for Divergence; at $x = -1$, the series becomes $\sum_{k=1}^{\infty} -k$, which also diverges by the Test for Divergence. The interval of convergence is therefore $(-1, 1)$.

11. Applying the Ratio Test:
$$\left|\frac{\frac{x^{k+1}}{(k+1)!}}{\frac{x^k}{k!}}\right| = \frac{k! \cdot |x|}{(k+1)!} = \frac{k! \cdot |x|}{(k+1) \cdot k!} = \frac{|x|}{k+1} \longrightarrow 0$$
for any x, so the interval of convergence is therefore $(-\infty, \infty)$.

13. Applying the Ratio Test:
$$\left|\frac{(k+1) \cdot \frac{x^{2k+2}}{4^{2k+2}}}{k \cdot \frac{x^{2k}}{4^{2k}}}\right| = \frac{(k+1) \cdot x^2}{16k} \longrightarrow \frac{x^2}{16}$$
so the series converges when $\frac{x^2}{16} < 1 \Rightarrow x^2 < 16 \Rightarrow |x| < 4$ and diverges when $|x| > 4$. At $x = \pm 4$ the series becomes $\sum_{k=1}^{\infty} k$, which diverges, so the interval of convergence is $(-4, 4)$.

15. Applying the Ratio Test, $\left|\frac{x^{k+1}}{2^{k+1}} \cdot \frac{2^k}{x^k}\right| = \frac{|x|}{2}$ for all x, so the series converges when $\frac{|x|}{2} < 1 \Rightarrow |x| < 2$ and diverges when $|x| > 2$. At $x = 2$ the series becomes $\sum_{k=1}^{\infty} 1$, which diverges; at $x = -2$, it becomes $\sum_{k=1}^{\infty} (-1)^k$, which also diverges. The interval of convergence is $(-2, 2)$.

17. $R = \frac{1}{2}(1 - (-1)) = 1$

19. $R = 1$ 21. $R = 1$ 23. $R = 4$

25. $\sum_{k=0}^{\infty} \frac{x^k}{5^k}$ is one possibility.

27. $\sum_{k=0}^{\infty} \frac{x^k}{2^k \cdot k^2}$ is one possibility.

29. $R = \frac{1}{2}(5 - (-5)) = 5$ 31. $R = 2$

33. This is a geometric series with ratio x, so it converges precisely when $|x| < 1$, hence its interval of convergence is $(-1, 1)$. On that interval:

$$\sum_{k=0}^{\infty} x^k = \frac{1}{1-x}$$

35. This is a geometric series with ratio $2x$, so it converges precisely when $|2x| < 1$, hence its interval of convergence is $\left(-\frac{1}{2}, \frac{1}{2}\right)$. On that interval:

$$\sum_{k=0}^{\infty} (2x)^k = \frac{1}{1-2x}$$

37. This is a geometric series with ratio x, so its interval of convergence is $(-1, 1)$. On that interval:

$$\sum_{k=1}^{\infty} x^k = \left[\sum_{k=0}^{\infty} x^k\right] - 1 = \frac{1}{1-x} - 1 = \frac{x}{1-x}$$

39. Geometric with ratio x^3, converges if $|x^3| < 1 \Rightarrow |x| < 1$, so interval of convergence is $(-1, 1)$ and:

$$\sum_{k=0}^{\infty} (x^3)^k = \frac{1}{1-x^3}$$

41. This is a geometric series with ratio $4x$, so it converges precisely when $|4x| < 1$, hence its interval of convergence is $\left(-\frac{1}{4}, \frac{1}{4}\right)$. On that interval:

$$\sum_{k=0}^{\infty} (4x)^k = \frac{1}{1-4x}$$

Section 10.2

1. This is a geometric series with ratio $x + 2$, so it converges precisely when:

$$|x+2| < 1 \Rightarrow -1 < x+2 < 1 \Rightarrow -3 < x < -1$$

The interval of convergence is $(-3, -1)$, so $R = \frac{1}{2}(-1 - (-3)) = 1$. (The graph is left to you.)

3. This is a geometric series with ratio $x + 5$, so it converges when:

$$|x+5| < 1 \Rightarrow -1 < x+5 < 1 \Rightarrow -6 < x < -4$$

and diverges everywhere else. The interval of convergence is $(-6, -4)$, so $R = 1$.

5. Applying the Ratio Test:

$$\left|\frac{\frac{(x-2)^{k+1}}{k+1}}{\frac{(x-1)^k}{k}}\right| = \frac{k}{k+1} \cdot |x-2| \longrightarrow |x-2|$$

so the series converges when:

$$|x-2| < 1 \Rightarrow -1 < x-2 < 1 \Rightarrow 1 < x < 3$$

and diverges when $x < 1$ or $x > 3$. At $x = 3$ the series becomes the harmonic series, which diverges; at $x = 1$, the series becomes the alternating harmonic series, which converges conditionally. The interval of convergence is therefore $[1, 3)$, hence $R = 1$.

7. Applying the Ratio Test:

$$\left|\frac{\frac{(x-7)^{2k+3}}{(k+1)^2}}{\frac{(x-7)^{2k+1}}{k^2}}\right| = \left(\frac{k}{k+1}\right)^2 \cdot (x-7)^2 \longrightarrow (x-7)^2$$

so the series converges when:

$$(x-7)^2 < 1 \Rightarrow |x-7| < 1$$
$$\Rightarrow -1 < x-7 < 1 \Rightarrow 6 < x < 8$$

and diverges when $x < 6$ or $x > 8$. At $x = 8$, the series becomes $\sum_{k=1}^{\infty} \frac{1}{k^2}$, which converges (by the P-test, with $p = 2$); at $x = 6$, the series becomes

$\sum_{k=1}^{\infty} \frac{-1}{k^2}$, which likewise converges. The interval of convergence is $[6,8]$, hence $R = 1$.

9. This is a geometric series with ratio $2x-6$, so it converges precisely when:

$$|2x-6| < 1 \Rightarrow -1 < 2x-6 < 1 \Rightarrow 5 < 2x < 7 \Rightarrow 2.5 < x < 3.5$$

The interval of convergence is $(2.5, 3.5)$; $R = 0.5$.

11. Applying the Ratio Test:

$$\left| \frac{\frac{(x-5)^{k+1}}{(k+1)!}}{\frac{(x-5)^k}{k!}} \right| = \frac{k! \cdot |x-5|}{(k+1)!} = \frac{|x-5|}{k+1} \longrightarrow 0$$

for any x, so the interval of convergence is $(-\infty, \infty)$ and $R = \infty$.

13. Applying the Ratio Test:

$$\left| \frac{(k+1)! \cdot (x-7)^{k+1}}{k! \cdot x^k} \right| = (k+1) \cdot |x-7|$$

which has a limit of ∞ as $k \to \infty$ for all x except $x = 7$ (in which case the limit is 0, so the series converges). The interval of convergence is therefore the single point$\{7\}$ and $R = 0$.

15. The center of the interval is $x = 5$ but the power series is centered at $x = 4$.

17. The interval of convergence must be centered at $x = 7$ so the only candidates for it are: $(5,9)$, $[1,13]$, $(-1,15]$, $[3,11)$, $[0,14)$ and $\{7\}$.

19. Interval must be centered at $x = 1$, so: $(0,2)$, $(-5,7)$, $[0,2]$, $(-3,5]$, $(-9,11]$, $[0,2)$ and $\{1\}$.

21. $R = \frac{1}{2}(6-0) = 3$ 23. $R = \frac{1}{2}(8-2) = 3$

25. $\sum_{k=0}^{\infty} \frac{(x-3)^k}{3^k}$ is one possibility.

27. $\sum_{k=0}^{\infty} \frac{(5-x)^k}{k \cdot 3^k}$ is one possibility.

29. This is a geometric series with ratio $x-3$, so it converges precisely when:

$$|x-3| < 1 \Rightarrow -1 < x-3 < 1 \Rightarrow 2 < x < 4$$

On that interval:

$$\sum_{k=0}^{\infty} (x-3)^k = \frac{1}{1-(x-3)} = \frac{1}{4-x}$$

31. Geometric series with ratio $\frac{x-6}{5}$, converges if:

$$\left| \frac{x-6}{5} \right| < 1 \Rightarrow |x-6| < 5 \Rightarrow -5 < x-6 < 5 \Rightarrow 1 < x < 11$$

On that interval: $\sum_{k=0}^{\infty} (\frac{x-6}{5})^k = \frac{5}{11-x}$

33. This is a geometric series with ratio $\frac{1}{2}\sin(x)$. Because $\left|\frac{1}{2}\sin(x)\right| \leq \frac{1}{2} < 1$ for all values of x, the interval of convergence is $(-\infty, \infty)$ and:

$$\sum_{k=0}^{\infty} \left(\frac{1}{2}\sin(x)\right)^k = \frac{1}{1-\frac{1}{2}\sin(x)} = \frac{2}{2-\sin(x)}$$

35. This is a geometric series with ratio $x-a$, so it converges precisely when:

$$|x-a| < 1 \Rightarrow -1 < x-a < 1 \Rightarrow a-1 < x < a+1$$

37. Applying the Ratio Test:

$$\left| \frac{\frac{(x-a)^{k+1}}{k+1}}{\frac{(x-a)^k}{k}} \right| = \frac{k}{k+1} \cdot |x-a| \longrightarrow |x-a|$$

so the series converges when:

$$|x-a| < 1 \Rightarrow -1 < x-a < 1 \Rightarrow a-1 < x < a+1$$

and diverges when $x < a-1$ or $x > a+1$. At $x = a+1$ the series becomes the harmonic series, which diverges; at $x = a-1$, the series becomes the alternating harmonic series, which converges conditionally. The interval of convergence is therefore $[a-1, a+1)$.

39. This is a geometric series with ratio ax, so it converges precisely when:

$$|ax| < 1 \Rightarrow -1 < ax < 1 \Rightarrow -\frac{1}{a} < x < \frac{1}{a}$$

41. This is a geometric series with ratio $ax-b$, so it converges precisely when:

$$|ax-b| < 1 \Rightarrow -1 < ax-b < 1 \Rightarrow -\frac{b-1}{a} < x < \frac{b+1}{a}$$

Section 10.3

1. Starting with the geometric series:

$$\frac{1}{1-u} = \sum_{k=0}^{\infty} u^k = 1 + u + u^2 + u^3 + \cdots$$

and using the substitution $u = x^4$ yields:

$$\frac{1}{1-x^4} = \sum_{k=0}^{\infty} x^{4k} = 1 + x^4 + x^8 + x^{12} + \cdots$$

3. Substitute $u = -x^4$ in the geometric series:

$$\frac{1}{1+x^4} = \sum_{k=0}^{\infty} (-1)^k \cdot x^{4k} = 1 - x^4 + x^8 - x^{12} + \cdots$$

5. Rewrite the function as:

$$\frac{1}{5+x} = \frac{\frac{1}{5}}{1+\frac{x}{5}} = \frac{\frac{1}{5}}{1-(-\frac{x}{5})}$$

and put $u = -\frac{x}{5}$ in the geometric series:

$$\frac{1}{5+x} = \frac{1}{5}\sum_{k=0}^{\infty} (-\frac{x}{5})^k = \frac{1}{5} - \frac{x}{25} + \frac{x^2}{125} - \frac{x^3}{625} + \cdots$$

7. Substitute $u = -x^3$ in the geometric series:

$$\frac{1}{1+x^3} = \sum_{k=0}^{\infty} (-1)^k \cdot x^{3k} = 1 - x^3 + x^6 - x^9 + \cdots$$

and multiply the result by x^2:

$$\frac{x^2}{1+x^3} = \sum_{k=0}^{\infty} (-1)^k \cdot x^{3k+2} = x^2 - x^5 + x^8 - x^{11} + \cdots$$

9. Into the first result from Example 3:

$$\ln(1-u) = -\sum_{k=1}^{\infty} \frac{u^k}{k} = -u - \frac{1}{2}u^2 - \frac{1}{3}u^3 - \cdots$$

substitute $u = -x^2$ to get:

$$\ln(1+x^2) = \sum_{k=1}^{\infty} (-1)^{k+1} \cdot \frac{x^{2k}}{k} = x^2 - \frac{1}{2}x^4 + \frac{1}{3}x^6 - \cdots$$

11. Substitute $u = x^2$ to get into the second result from Example 3 and multiply by x:

$$\arctan(u) = \sum_{k=0}^{\infty} (-1)^k \cdot \frac{u^{2k+1}}{2k+1} = u - \frac{1}{3}u^3 + \frac{1}{5}u^5 - \frac{1}{7}u^7 + \cdots$$

$$\Rightarrow \quad \arctan\left(x^2\right) = \sum_{k=0}^{\infty} (-1)^k \cdot \frac{x^{4k+2}}{2k+1} = x^2 - \frac{1}{3}x^6 + \frac{1}{5}x^{10} - \frac{1}{7}x^{14} + \cdots$$

$$\Rightarrow x \cdot \arctan\left(x^2\right) = \sum_{k=0}^{\infty} (-1)^k \cdot \frac{x^{4k+3}}{2k+1} = x^3 - \frac{1}{3}x^7 + \frac{1}{5}x^{11} - \frac{1}{7}x^{15} + \cdots$$

13. Substitute $u = x^2$ into the result from Example 2:

$$\frac{1}{(1-u)^2} = \sum_{k=1}^{\infty} k \cdot u^{k-1} = 1 + 2u + 3u^2 + 4u^3 + 5u^4 + \cdots$$

$$\Rightarrow \frac{1}{(1-x^2)^2} = \sum_{k=1}^{\infty} k \cdot x^{2k-2} = 1 + 2x^2 + 3x^4 + 4x^6 + 5x^8 + \cdots$$

15. Differentiate the result from Example 2 and then divide by 2:

$$(1-x)^{-2} = \sum_{k=1}^{\infty} k \cdot x^{k-1} = 1 + 2x + 3x^2 + 4x^3 + 5x^4 + \cdots$$

$$\Rightarrow 2(1-x)^{-3} = \sum_{k=1}^{\infty} k(k-1) \cdot x^{k-2} = 2 + 6x + 12x^2 + 20x^3 + \cdots$$

$$\Rightarrow \frac{1}{(1-x)^3} = \sum_{k=1}^{\infty} \frac{k(k-1)}{2} \cdot x^{k-2} = 1 + 3x + 6x^2 + 10x^3 + \cdots$$

17. Replace x with $-x^2$ in the result from Problem 15:

$$\frac{1}{(1+x^2)^3} = \sum_{k=1}^{\infty} (-1)^k \frac{k(k-1)}{2} \cdot x^{2k-4} = 1 - 3x^2 + 6x^4 - 10x^6 + \cdots$$

19. Integrate the first result from Practice 1 between $x = 0$ and $x = \frac{1}{2}$:

$$\int_0^{\frac{1}{2}} \frac{1}{1-x^3}\, dx = \int_0^{\frac{1}{2}} \left[\sum_{k=0}^{\infty} x^{3k}\right] dx = \int_0^{\frac{1}{2}} \left[1 + x^3 + x^6 + \cdots\right] dx$$

$$= \sum_{k=0}^{\infty} \left[\frac{1}{3k+1} x^{3k+1}\right]_0^{\frac{1}{2}} = \left[x + \frac{1}{4}x^4 + \frac{1}{7}x^7 + \cdots\right]_0^{\frac{1}{2}} = \frac{1}{2} + \frac{1}{64} + \frac{1}{896} + \cdots \approx 0.5167$$

21. Integrate the result from Practice 2 between $x = 0$ and $x = \frac{3}{5}$:

$$\int_0^{\frac{3}{5}} \ln(1+x)\, dx = \int_0^{\frac{3}{5}} \left[\sum_{k=0}^{\infty} \frac{(-1)^k}{k+1} \cdot x^{k+1}\right] dx = \int_0^{\frac{3}{5}} \left[x - \frac{1}{2}x^2 + \frac{1}{3}x^3 - \cdots\right] dx$$

$$= \left[\sum_{k=0}^{\infty} \frac{(-1)^k}{(k+2)(k+1)} \cdot x^{k+2}\right]_0^{\frac{3}{5}} = \left[\frac{1}{2}x^2 - \frac{1}{6}x^3 + \frac{1}{12}x^4 - \cdots\right]_0^{\frac{3}{5}} = \frac{9}{50} - \frac{9}{250} + \frac{27}{2500} - \cdots$$

or about 0.1548 (adding up the first three terms of the numerical sum).

23. Multiply the second result from Example 3 by x^2 and integrate between $x = 0$ and $x = \frac{1}{2}$:

$$\int_0^{\frac{1}{2}} x^2 \cdot \arctan(x)\, dx = \int_0^{\frac{1}{2}} \left[\sum_{k=0}^{\infty} (-1)^k \cdot \frac{x^{2k+3}}{2k+1}\right] dx = \int_0^{\frac{1}{2}} \left[x^3 - \frac{1}{3}x^5 + \frac{1}{5}x^7 - \cdots\right] dx$$

$$= \left[\sum_{k=0}^{\infty} (-1)^k \cdot \frac{x^{2k+4}}{(2k+4)(2k+1)}\right]_0^{\frac{1}{2}} = \left[\frac{1}{4}x^4 - \frac{1}{18}x^6 + \frac{1}{40}x^8 - \cdots\right]_0^{\frac{1}{2}}$$

$$= \frac{1}{64} - \frac{1}{1152} + \frac{1}{10240} - \cdots \approx 0.01485$$

25. Integrate the result from Example 2 between $x = 0$ and $x = 0.3$:

$$\int_0^{0.3} \frac{1}{(1-x)^2}\, dx = \int_0^{0.3} \left[\sum_{k=1}^{\infty} k \cdot x^{k-1}\right] dx = \int_0^{0.3} \left[1 + 2x + 3x^2 + \cdots\right] dx$$

$$= \left[\sum_{k=1}^{\infty} \cdot x^k\right]_0^{0.3} = \left[x + x^2 + x^3 + \cdots\right]_0^{0.3} = 0.3 + 0.09 + 0.027 + \cdots \approx 0.417$$

27. If $x \neq 0$, divide the second result from Example 3 by x to get:

$$\frac{\arctan(x)}{x} = \sum_{k=0}^{\infty} (-1)^k \cdot \frac{x^{2k}}{2k+1} = 1 - \frac{1}{3}x^2 + \frac{1}{5}x^4 - \frac{1}{7}x^6 + \cdots$$

As $x \to 0$, the last expression approaches 1, so $\lim_{x\to 0} \frac{\arctan(x)}{x} = 1$.

29. If $x \neq 0$, divide the result from Practice 2 by $2x$:

$$\frac{\ln(1+x)}{2x} = \sum_{k=0}^{\infty} \frac{(-1)^k}{2k+2} \cdot x^k = \frac{1}{2} - \frac{1}{4}x + \frac{1}{6}x^2 - \frac{1}{8}x^3 + \cdots \longrightarrow \frac{1}{2} \text{ (as } x \to 0)$$

31. If $x \neq 0$, divide the power series for arctan (x) obtained in the solution to Example 4 by x^2 to get:

$$\frac{\arctan(x^2)}{x^2} = \sum_{k=0}^{\infty} \frac{(-1)^k}{2k+1} x^{4k} = 1 - \frac{1}{3}x^4 + \frac{1}{5}x^8 - \frac{1}{7}x^{12} + \cdots \longrightarrow 1 \text{ (as } x \to 0)$$

33. If $x \neq 0$, replace x with $-x^2$ in the power series for $\ln(1+x)$ and divide by $3x$ to get:

$$\frac{\ln(1-x^2)}{3x} = -\frac{1}{3x}\sum_{k=1}^{\infty} \frac{x^{2k}}{k} = -\sum_{k=1}^{\infty} \frac{x^{2k-1}}{3k} = -\frac{1}{3}x - \frac{1}{6}x^3 - \frac{1}{9}x^5 - \cdots \longrightarrow 0 \text{ (as } x \to 0)$$

(Check that you get the same result in Problems 27–34 from applying L'Hôpital's Rule.)

35. $\dfrac{1}{1+x} = \sum\limits_{k=0}^{\infty} (-1)^k \cdot x^k$ (a geometric series with ratio $-x$ precisely when $|-x| < 1 \Rightarrow |x| < 1 \Rightarrow -1 < x < 1$, so the interval of convergence is $(-1,1)$.

37. From Example 3, we know that:

$$\ln(1-x) = -x - \frac{1}{2}x^2 - \frac{1}{3}x^3 - \frac{1}{4}x^4 - \cdots = -\sum_{k=0}^{\infty} \frac{1}{k+1} \cdot x^{k+1}$$

Applying the Ratio Test to this series:

$$\left| \frac{-\frac{x^{k+1}}{k+1}}{-\frac{x^k}{k}} \right| = \frac{k}{k+1} \cdot |x| \longrightarrow |x|$$

as $k \to \infty$, so the series converge when $|x| < 1$ and diverges when $|x| > 1$. At $x = 1$, the series becomes a multiple of the harmonic series, which diverges; at $x = -1$, the series becomes a multiple of the alternating harmonic series, which converges conditionally. So the interval of convergence is $[-1,1)$.

39. From Example 3, we know that:

$$\arctan(x) = \sum_{k=0}^{\infty} \frac{(-1)^k}{2k+1} x^{2k+1} = x - \frac{1}{3}x^3 + \frac{1}{5}x^5 - \frac{1}{7}x^7 + \cdots$$

Applying the Ratio Test to this series:

$$\left| \frac{\frac{(-1)^{k+1}x^{2k+3}}{2k+3}}{\frac{(-1)^k x^{2k+1}}{2k+1}} \right| = \frac{2k+1}{2k+3} \cdot x^2 \longrightarrow 1$$

as $k \to \infty$, so the series convereges when $x^2 < 1 \Rightarrow |x| < 1$ and diverges when $x^2 > 1 \Rightarrow |x| > 1$. At $x = \pm 1$, the series converges conditionally (by the Alternating Series Test—check this) so the interval of convergence is $[-1,1]$.

41. From Example 2 we know that:

$$(1-x)^{-2} = \sum_{k=1}^{\infty} k \cdot x^{k-1} = 1 + 2x + 3x^2 + 4x^3 + 5x^4 + \cdots$$

Applying the Ratio Test to this series:

$$\left| \frac{(k+1) \cdot x^k}{k \cdot x^{k-1}} \right| = \frac{k+1}{k} \cdot |x| \longrightarrow 1$$

as $k \to \infty$, so the series converges when $|x| < 1$ and diverges when $|x| > 1$. At $x = 1$, the series becomes $\sum_{k=1}^{\infty} k$, which diverges; at $x = -1$, the series becomes $\sum_{k=1}^{\infty} k \cdot (-1)^{k-1}$, which also diverges. The interval of convergence is $(-1,1)$.

Section 10.4

1. With $f(x) = \ln(1+x)$, $f(0) = \ln(1) = 0$ and:

$$f'(x) = (1+x)^{-1} \Rightarrow f'(0) = 1$$
$$f''(x) = -(1+x)^{-2} \Rightarrow f''(0) = -1$$
$$f'''(x) = 2(1+x)^{-3} \Rightarrow f'''(0) = 2$$
$$f^{(4)}(x) = -6(1+x)^{-4} \Rightarrow f^{(4)}(0) = -6$$

and so on, so the first few terms of the MacLaurin series for $f(x)$ are:

$$0 + 1 \cdot x + \frac{-1}{2!}x^2 + \frac{2}{3!}x^3 + \frac{-6}{4!}x^4 + \cdots$$
$$= x - \frac{1}{2}x^2 + \frac{1}{3}x^3 - \frac{1}{4}x^4 + \cdots$$

3. $f(x) = \arctan(x) \Rightarrow f(0) = \arctan(0) = 0$ and:

$$f'(x) = \frac{1}{1+x^2} \Rightarrow f'(0) = 1$$
$$f''(x) = -\frac{2x}{(1+x^2)^2} \Rightarrow f''(0) = 0$$
$$f'''(x) = \frac{6x^2-2}{(1+x^2)^3} \Rightarrow f'''(0) = -2$$
$$f^{(4)}(x) = \frac{24\left(x - x^3\right)}{(1+x^2)^4} \Rightarrow f^{(4)}(0) = 0$$

and so on, so the first few terms of the MacLaurin series for $f(x)$ are:

$$0 + 1 \cdot x + \frac{0}{2!} \cdot x^2 + \frac{-2}{3!}x^3 + \frac{0}{4!} \cdot x^4 + \cdots$$
$$= x - \frac{1}{3}x^2 + \frac{1}{3}x^3 - \frac{1}{4}x^4 + \cdots$$

5. With $f(x) = \cos(x)$, $f(0) = \cos(0) = 1$ and:

$$f'(x) = -\sin(x) \Rightarrow f'(0) = 0$$
$$f''(x) = -\cos(x) \Rightarrow f''(0) = -1$$
$$f'''(x) = \sin(x) \Rightarrow f'''(0) = 0$$
$$f^{(4)}(x) = \cos(x) \Rightarrow f^{(4)}(0) = 1$$

From here the derivatives repeat the same pattern, so $f^{(5)}(0) = 0$, $f^{(6)}(0) = -1$ and the first few terms of the MacLaurin series for $f(x)$ are:

$$1 - \frac{1}{2!}x^2 + \frac{1}{4!}x^4 - \frac{1}{6!}x^6 + \cdots$$

7. With $f(x) = \sec(x)$, $f(0) = \sec(0) = 1$ and:

$$f'(x) = \sec(x)\tan(x) \Rightarrow f'(0) = 0$$
$$f''(x) = \sec^3(x) + \sec(x)\tan^2(x) \Rightarrow f''(0) = 1$$
$$f'''(x) = 5\sec^3(x)\tan(x) + \sec(x)\tan^3(x)$$

so that $f'''(0) = 0$, while $f^{(4)}(x) = 5\sec^5(x) + 18\sec^3(x)\tan^2(x) + \sec(x)\tan^4(x) \Rightarrow f^{(4)}(0) = 5$; the first terms of the MacLaurin series are:

$$1 - x^2 + \frac{5}{4!}x^4 + \cdots = 1 - x^2 + \frac{5}{24}x^4 + \cdots$$

9. With $f(x) = \ln(x)$, $f(1) = \ln(1) = 0$ and:

$$f'(x) = x^{-1} \Rightarrow f'(1) = 1$$
$$f''(x) = -x^{-2} \Rightarrow f''(1) = -1$$
$$f'''(x) = 2x^{-3} \Rightarrow f'''(1) = 2$$
$$f^{(4)}(x) = -6x^{-4} \Rightarrow f^{(4)}(1) = -6$$

and so on, so the first few terms of the Taylor series are:

$$(x-1) - \frac{1}{2}(x-1)^2 + \frac{1}{3}(x-1)^3 - \frac{1}{4}(x-1)^4 + \cdots$$

11. With $f(x) = \sin(x)$, $f\left(\frac{\pi}{2}\right) = \sin\left(\frac{\pi}{2}\right) = 1$ and:

$$f'(x) = \cos(x) \Rightarrow f'\left(\frac{\pi}{2}\right) = 0$$
$$f''(x) = -\sin(x) \Rightarrow f''\left(\frac{\pi}{2}\right) = -1$$
$$f'''(x) = \cos(x) \Rightarrow f'''\left(\frac{\pi}{2}\right) = 0$$
$$f^{(4)}(x) = \sin(x) \Rightarrow f^{(4)}\left(\frac{\pi}{2}\right) = 1$$

From here the derivatives repeat the same pattern, so $f^{(5)}\left(\frac{\pi}{2}\right) = 0$, $f^{(6)}\left(\frac{\pi}{2}\right) = -1$ and the first few terms of the Taylor series:

$$1 - \frac{1}{2!}\left(x - \frac{\pi}{2}\right)^2 + \frac{1}{4!}\left(x - \frac{\pi}{2}\right)^4 - \frac{1}{6!}\left(\frac{x-\pi}{2}\right)^6 + \cdots$$

13. With $f(x) = \sqrt{x}$, $f(9) = \sqrt{9} = 3$ and:

$$f'(x) = \frac{1}{2}x^{-\frac{1}{2}} \Rightarrow f'(9) = \frac{1}{6}$$
$$f''(x) = -\frac{1}{4}x^{-\frac{3}{2}} \Rightarrow f''(9) = -\frac{1}{108}$$
$$f'''(x) = \frac{3}{8}x^{-\frac{5}{2}} \Rightarrow f'''(9) = \frac{1}{648}$$

so the first few terms of the Taylor series are:

$$3+\frac{1}{6}(x-9)-\frac{1}{216}(x-9)^2+\frac{1}{3888}(x-9)^3-\cdots$$

15. Using $P_4(x)=1-\frac{1}{2}x^2+\frac{1}{24}x^4 \approx \cos(x)$:

x	$\cos(x)$	$P_4(x)$
0.1	0.995004165	0.995004167
0.2	0.98006657	0.98006666
0.5	0.87758	0.87604
1.0	0.54030	0.54167
2.0	−0.4161	−0.3333

17. Using $P_4(x)=x-\frac{1}{3}x^3+\frac{1}{5}x^5 \approx \arctan(x)$:

x	$\arctan(x)$	$P_4(x)$
0.1	0.09966865	0.09966867
0.2	0.197396	0.197397
0.5	0.4636	0.4646
1.0	0.7854	0.8667
2.0	1.1071	5.7333

19. With $\sin(u)=u-\frac{1}{6}u^3+\frac{1}{120}u^5-\cdots$, put $u=x^2$ so that $\sin\left(x^2\right)=x^2-\frac{1}{6}x^6+\frac{1}{120}x^{10}-\cdots$ and:

$$\int \sin\left(x^2\right)\,dx = C+\frac{1}{3}x^3-\frac{1}{42}x^7+\frac{1}{1320}x^{11}-\cdots$$

21. With $\sin(u)=u-\frac{1}{6}u^3+\frac{1}{120}u^5-\cdots$, put $u=x^3$ so that $\sin\left(x^3\right)=x^3-\frac{1}{6}x^9+\frac{1}{120}x^{15}-\cdots$ and:

$$\int \sin\left(x^3\right)\,dx = C+\frac{1}{4}x^4-\frac{1}{60}x^{10}+\frac{1}{1920}x^{16}-\cdots$$

23. With $e^u=1+u+\frac{1}{2}u^2+\cdots$, put $u=-x^2$ so that $e^{-x^2}=1-x^2+\frac{1}{2}x^4+\cdots$ and:

$$\int e^{-x^2}\,dx = C+x-\frac{1}{3}x^3+\frac{1}{10}x^5-\cdots$$

25. With $e^u=1+u+\frac{1}{2}u^2+\cdots$, put $u=-x^3$ so that $e^{-x^3}=1-x^3+\frac{1}{2}x^6+\cdots$ and:

$$\int e^{-x^3}\,dx = C+x-\frac{1}{4}x^4+\frac{1}{14}x^7-\cdots$$

27. Multiply $\sin(x)=x-\frac{1}{6}x^3+\frac{1}{120}x^5-\cdots$ by x to get $x\cdot\sin(x)=x^2-\frac{1}{6}x^4+\frac{1}{120}x^6-\cdots$ so:

$$\int x\cdot\sin(x)\,dx = C+\frac{1}{3}x^3-\frac{1}{30}x^5+\frac{1}{840}x^7-\cdots$$

29. Multiply $\sin(x)=x-\frac{1}{6}x^3+\frac{1}{120}x^5-\cdots$ by x^2 to get $x\cdot\sin(x)=x^3-\frac{1}{6}x^5+\frac{1}{120}x^7-\cdots$ so:

$$\int x^2\cdot\sin(x)\,dx = C+\frac{1}{4}x^4-\frac{1}{36}x^6+\frac{1}{960}x^8-\cdots$$

31. Subtract $\cos(x)=1-\frac{1}{2}x^2+\frac{1}{24}x^4-\cdots$ from 1 and divide by x^2 to get:

$$\frac{1-\cos(x)}{x^2}=\frac{\frac{1}{2}x^2-\frac{1}{24}x^4+\cdots}{x^2}=\frac{1}{2}-\frac{1}{24}x^2+\cdots$$

which has limit $\frac{1}{2}$ as $x\to 0$.

33. Subtract $e^x=1+x+\frac{1}{2}x^2+\cdots$ from 1 and divide by x to get:

$$\frac{1-e^x}{x}=\frac{x+\frac{1}{2}x^2+\cdots}{x}=1+\frac{1}{2}x+\cdots$$

which has limit 1 as $x\to 0$.

35. Dividing $\sin(x)=x-\frac{1}{6}x^3+\frac{1}{120}x^5-\cdots$ by x:

$$\frac{x-\frac{1}{6}x^3+\frac{1}{120}x^5-\cdots}{x}=1-\frac{1}{6}x^2+\frac{1}{120}x^4\cdots$$

yields a limit of 1 as $x\to 0$.

37. Subtract $\sin(x)=x-\frac{1}{6}x^3+\frac{1}{120}x^5-\frac{1}{5040}x^7\cdots$ from $x-\frac{1}{6}x^3$ and divide by x^5 to get:

$$\frac{-\frac{1}{120}x^5+\frac{1}{5040}x^7-\cdots}{x^5}=\frac{1}{120}-\frac{1}{5040}x^2+\cdots$$

which has a limit of $\frac{1}{120}$ as $x\to 0$.

39. Starting with $e^u = 1 + u + \frac{1}{2!}u^2 + \frac{1}{3!}u^3 + \frac{1}{4!}u^4 + \cdots$, put $u = x$, then $u = -x$ to get:

$$\sinh(x) = \frac{e^x - e^{-x}}{2} = \frac{1}{2}\left[1 + x + \frac{1}{2!}x^2 + \frac{1}{3!}x^3 + \frac{1}{4!}x^4 + \cdots\right] - \frac{1}{2}\left[1 - x + \frac{1}{2!}x^2 - \frac{1}{3!}x^3 - \frac{1}{4!}x^4 + \cdots\right]$$
$$= x + \frac{1}{3!}x^3 + \frac{1}{5!}x^5 + \frac{1}{7!}x^7 + \cdots$$

41. $\mathbf{D}(\sinh(x)) = \mathbf{D}\left(x + \frac{1}{3!}x^3 + \frac{1}{5!}x^5 + \frac{1}{7!}x^7 + \cdots\right) = 1 + \frac{1}{2!}x^2 + \frac{1}{4!}x^4 + \frac{1}{6!}x^6 + \cdots = \cosh(x)$

43. $e^{i\left(\frac{\pi}{2}\right)} = \cos\left(\frac{\pi}{2}\right) + i \cdot \sin\left(\frac{\pi}{2}\right) = 0 + i \cdot 1 = i$, while $e^{\pi i} = \cos(\pi) + i \cdot \sin(\pi) = -1 + i \cdot 0 = -1$.

45. $\binom{3}{0} = 1$ by definition, while $\binom{3}{1} = \frac{3}{1!} = 3$, $\binom{3}{2} = \frac{3 \cdot 2}{2!} = 3$ and $\binom{3}{3} = \frac{3 \cdot 2 \cdot 1}{3!} = 1$; these agree with the numbers 1, 3, 3, 1 from Pascal's triangle and with the coefficients of $(1+x)^3 = 1 + 3x + 3x^2 + x^3$.

47. The MacLaurin series for $(1+x)^{\frac{5}{2}}$ is:

$$1 + \frac{5}{2}x + \left(\frac{5}{2}\right)\left(\frac{3}{2}\right) \cdot \frac{1}{2!}x^2 + \left(\frac{5}{2}\right)\left(\frac{3}{2}\right)\left(\frac{1}{2}\right) \cdot \frac{1}{3!}x^3 + \left(\frac{5}{2}\right)\left(\frac{3}{2}\right)\left(\frac{1}{2}\right)\left(-\frac{1}{2}\right) \cdot \frac{1}{4!}x^4 + \cdots$$

49. Using the Binomial Series Theorem, the MacLaurin series for $\frac{1}{\sqrt{1+u}} = (1+u)^{-\frac{1}{2}}$ is:

$$1 - \frac{1}{2}u + \left(\frac{1}{2}\right)\left(\frac{3}{2}\right) \cdot \frac{1}{2!}u^2 - \left(\frac{1}{2}\right)\left(\frac{3}{2}\right)\left(\frac{5}{2}\right) \cdot \frac{1}{3!}u^3 + \left(\frac{1}{2}\right)\left(\frac{3}{2}\right)\left(\frac{5}{2}\right)\left(\frac{7}{2}\right) \cdot \frac{1}{4!}u^4 + \cdots$$

Putting $u = -x^2$ gives $\frac{1}{\sqrt{1-x^2}} = (1-x^2)^{-\frac{1}{2}} = 1 + \frac{1}{2}x^2 + \frac{3}{8}x^4 + \frac{5}{16}x^6 + \frac{35}{128}x^8 + \cdots$, so integrating term-by-term (and using the fact that $\arcsin(0) = 0$) yields:

$$\arcsin(x) = x + \frac{1}{6}x^3 + \frac{3}{40}x^5 + \frac{5}{112}x^7 + \frac{35}{1152}x^9 + \cdots$$

51. With $f(x) = (1+x)^m \Rightarrow f(0) = 1$, $f'(x) = m(1+x)^{m-1} \Rightarrow f'(0) = m$, $f''(x) = m(m-1)(1+x)^{m-2} \Rightarrow f''(0) = m(m-1)$, $f'''(x) = m(m-1)(m-2)(1+x)^{m-3} \Rightarrow f'''(0) = m(m-1)(m-2)$ and $f^{(4)}(x) = m(m-1)(m-2)(m-3)(1+x)^{m-4} \Rightarrow f^{(4)}(0) = m(m-1)(m-2)(m-3)$ so the MacLaurin series is:

$$(1+x)^m = 1 + \frac{m}{1!}x + \frac{m(m-1)}{2!}x^2 + \frac{m(m-1)(m-2)}{3!}x^3 + \frac{m(m-1)(m-2)(m-3)}{4!}x^4 + \cdots$$

Section 10.5

1. $P_0(x) = 0$, $P_1(x) = P_2(x) = x$, $P_3(x) = P_4(x) = x - \frac{1}{6}x^3$; use technology to create graphs.

3. $f(x) = \ln(x) \Rightarrow f'(x) = x^{-1} \Rightarrow f''(x) = -x^{-2} \Rightarrow f'''(x) = 2x^{-3} \Rightarrow f^{(4)}(x) = -6x^{-3}$, so $P_0(x) = 0$, $P_1(x) = x$, $P_2(x) = x - \frac{1}{2}x^2$, $P_3(x) = x - \frac{1}{2}x^2 + \frac{1}{3}x^3$ and $P_4(x) = x - \frac{1}{2}x^2 + \frac{1}{3}x^3 - \frac{1}{4}x^4$

5. $P_0(x) = 1$, $P_1(x) = 1 + (x-1) = P_2(x) = P_3(x) = P_4(x)$

7. $f(x) = (1+x)^{-\frac{1}{2}} \Rightarrow f'(x) = -\frac{1}{2}(1+x)^{-\frac{3}{2}} \Rightarrow f''(x) = \frac{3}{4}(1+x)^{-\frac{5}{2}} \Rightarrow f'''(x) = -\frac{15}{8}(1+x)^{-\frac{7}{2}} \Rightarrow f^{(4)}(x) = \frac{105}{16}(1+x)^{-\frac{9}{2}}$, so $P_0(x) = 1$, $P_1(x) = 1 - \frac{1}{2}x$, $P_2(x) = 1 - \frac{1}{2}x + \frac{3}{8}x^2$, $P_3(x) = 1 - \frac{1}{2}x + \frac{3}{8}x^2 - \frac{5}{16}x^3$, $P_4(x) = 1 - \frac{1}{2}x + \frac{3}{8}x^2 - \frac{5}{16}x^3 + \frac{35}{128}x^4$

9. $f(x) = \sin(x) \Rightarrow f'(x) = \cos(x) \Rightarrow f''(x) = -\sin(x) \Rightarrow f'''(x) = -\cos(x) \Rightarrow f^{(4)}(x) = \sin(x)$ so $P_0(x) = 1 = P_1(x)$, $P_2(x) = 1 - \frac{1}{2}\left(x - \frac{\pi}{2}\right)^2 = P_3(x)$ and $P_4(x) = 1 - \frac{1}{2}\left(x - \frac{\pi}{2}\right)^2 + \frac{1}{24}\left(x - \frac{\pi}{2}\right)^4$

11. $R_5(x) = \frac{\sin(z)}{6!}(x-0)^6$

$|R_5(x)| \leq \frac{1}{720}\left(\frac{\pi}{2}\right)^6 = \frac{\pi^6}{46080} \approx 0.021$

13. $R_5(x) = \frac{\cos(z)}{6!}(x-0)^6$

$|R_5(x)| \leq \frac{1}{720}(\pi)^6 = \frac{\pi^6}{720} \approx 1.335$

15. $R_{10}(x) = \frac{-\cos(z)}{10!}(x-0)^{10}$

$|R_{10}(x)| \leq \frac{1}{362880} \cdot 2^6 = \frac{1}{56700} \approx 0.0000176$

17. $R_6(x) = \frac{e^z}{6!}(x-0)^6$

$|R_6(x)| \leq \frac{e^2}{720} \cdot 2^6 \leq \frac{2.72^2 \cdot 64}{720} \approx 0.658$

19. $|R_n(x)| \leq \frac{1}{(n+1)!} < \frac{1}{1000} \Rightarrow n \geq 6$

21. $|R_n(x)| \leq \frac{1.6^{n+1}}{(n+1)!} < \frac{1}{100000} \Rightarrow n \geq 10$

23. $|R_n(x)| \leq \frac{2.72^2 \cdot 2^{n+1}}{(n+1)!} < \frac{1}{1000} \Rightarrow n \geq 10$

25. Any derivative of $f(x) = \cos(x)$ is either $\pm\sin(x)$ or $\pm\cos(x)$, so $\left|f^{(n+1)}(z)\right| \leq 1$ for any z. Hence:

$$|R_n(x)| = \frac{\left|f^{(n+1)}(z)\right|}{(n+1)!}|x-0|^{n+1} \leq \frac{|x|^{n+1}}{(n+1)!}$$

As noted in the solution to Example 5, this expression approaches 0 as $n \to \infty$ (for any x).

27. (a) Using the definition of the derivative:

$$f'(0) = \lim_{h\to 0} \frac{f(h)-f(0)}{h} = \lim_{h\to 0} \frac{e^{-h^{-2}}-0}{h}$$

(b) For $h > 0$, let $y = \frac{1}{h} \Rightarrow h = \frac{1}{y}$ so that:

$$f'(0) = \lim_{y\to\infty} \frac{e^{-y^2}}{\frac{1}{y}} = \lim_{y\to\infty} \frac{y}{e^{y^2}}$$

By L'Hôpital's Rule:

$$f'(0) = \lim_{y\to\infty} \frac{1}{2y \cdot e^{y^2}} = 0$$

The process for $h < 0$ is quite similar.

29. (a) $\sum_{k=1}^{5} \frac{(-1)^{k+1} \cdot 4}{2k-1} = \frac{1052}{315} \approx 3.33968$

(b) $\left|\frac{(-1)^{50+1} \cdot 4}{2 \cdot 50 - 1}\right| = \frac{4}{99} \approx 0.0404$

(c) $\left|\frac{(-1)^{k+1} \cdot 4}{2k-1}\right| < \frac{1}{10000} \Rightarrow k \geq 20001$

31. (a) $4\arctan\left(\frac{1}{5}\right) \approx \frac{4}{5} - \frac{4}{375} + \frac{4}{15625} \approx 0.789589$ and $\arctan\left(\frac{1}{239}\right) \approx 0.004184$ so:

$$\pi \approx 4\left[0.197397 - 0.004184\right] \approx 3.14162$$

(b) We are using smaller values of x in the arctan series, and the powers of these smaller values of x approach 0 more quickly than the values of x used in Methods I and II.

Section 12.1

5. $\|\mathbf{u}\| = \sqrt{1^2+4^2} = \sqrt{17}$, $\|\mathbf{v}\| = \sqrt{3^2+2^2} = \sqrt{13}$ directions: $\left\langle \frac{1}{\sqrt{17}}, \frac{4}{\sqrt{17}} \right\rangle$, $\left\langle \frac{3}{\sqrt{13}}, \frac{2}{\sqrt{13}} \right\rangle$; slopes: 4, $\frac{2}{3}$; angles: $\arctan(4) \approx 76.0°$, $\arctan\left(\frac{2}{3}\right) \approx 33.7°$
7. $\sqrt{29}$, $\sqrt{58}$; $\left\langle \frac{-2}{\sqrt{29}}, \frac{5}{\sqrt{29}} \right\rangle$, $\left\langle \frac{3}{\sqrt{58}}, \frac{-7}{\sqrt{58}} \right\rangle$; -2.5, $-\frac{7}{3}$; $\arctan(-2.5) \approx -68.2°$, $\arctan\left(-\frac{7}{3}\right) \approx -66.8°$
9. $\|\mathbf{u}\| = 5$, $\|\mathbf{v}\| = 5$; $\left\langle -\frac{4}{5}, -\frac{3}{5} \right\rangle$, $\left\langle \frac{3}{5}, -\frac{4}{5} \right\rangle$; 0.75, $-\frac{4}{3}$; $\arctan(0.75) \approx 36.9°$, $\arctan\left(-\frac{4}{3}\right) \approx -53.1°$
13. $\mathbf{u} = \langle -1, 3 \rangle$; $\mathbf{v} = \langle 3, 5 \rangle$
15. $\mathbf{v} = 3\langle 0.6, 0.8 \rangle = \langle 1.8, 2.4 \rangle$
17. $\mathbf{v} = \langle 5\cos(35°), 5\sin(35°) \rangle \approx \langle 4.10, 2.87 \rangle$
19. $\mathbf{v} = \left\langle \frac{7}{\sqrt{10}}, \frac{21}{\sqrt{10}} \right\rangle$ or $\left\langle -\frac{7}{\sqrt{10}}, -\frac{21}{\sqrt{10}} \right\rangle$
21. $f'(x) = 2x+3 \Rightarrow f'(1) = 5$ so direction is $\mathbf{v} = \left\langle \frac{1}{\sqrt{26}}, \frac{5}{\sqrt{26}} \right\rangle$ or $\left\langle -\frac{1}{\sqrt{26}}, -\frac{5}{\sqrt{26}} \right\rangle$
23. $\mathbf{v} = \langle 1, 0 \rangle$ or $\langle -1, 0 \rangle$ 25. $\mathbf{v} = \langle 1, 0 \rangle$ or $\langle -1, 0 \rangle$
31. x-axis: $1\mathbf{i} + 0\mathbf{j}$; y-axis: $0\mathbf{i} + 4\mathbf{j}$
33. $5\mathbf{i} + 0\mathbf{j}$; $0\mathbf{i} - 2\mathbf{j}$ 37. $\langle -4, -6 \rangle$
39. parallel: $\langle 60\cos(65°), 0 \rangle \approx \langle 25.36, 0 \rangle$ perpendicular: $\langle 0, 60\sin(65°) \rangle \approx \langle 0, 54.37 \rangle$
41. 119.5 lbs; $\arctan\left(\frac{50\sin(45°)+70\sin(35°)}{50\cos(45°)+70\cos(35°)}\right) \approx 39.2°$
43. 268.45 lbs; 23.9°
45. (a) The velocity is $\langle -3, 4 \rangle + \langle 0, 18 \rangle = \langle -3, 22 \rangle$, which corresponds to a speed of $\sqrt{493} \approx 22.2$ mph with heading 7.8° west of due north. (b) The captain should head 9.6° east of due north.
47. (a) The tension in each rope is 90.45 lbs. (b) The tension in the shorter part is 97.1lbs; the tension in the longer part is 59.2 lbs.

Section 12.2

13. $\text{dist}(A,B) = \sqrt{5}$, $\text{dist}(A,C) = \sqrt{11}$, $\text{dist}(A,D) = 5$, $\text{dist}(B,C) = \sqrt{6}$, $\text{dist}(B,D) = 2\sqrt{2}$ and $\text{dist}(C,D) = 3\sqrt{2}$; no three points are collinear.
15. $\text{dist}(A,B) = 6$; $\text{dist}(A,C) = 3$; $\text{dist}(A,D) = \sqrt{6}$; $\text{dist}(B,C) = 9$; $\text{dist}(B,D) = 5\sqrt{2}$; $\text{dist}(C,D) = \sqrt{11}$; $\text{dist}(B,A) + \text{dist}(A,C) = \text{dist}(B,C)$ so A, B and C are collinear.
17. $(1,4,1)$, $(4,4,1)$, $(1,2,3)$, $(4,2,3)$, $(4,4,3)$; 12
19. $(1,4,0)$, $(1,5,0)$, $(4,4,0)$, $(4,4,3)$, $(4,5,3)$; 9
25. $(x-4)^2 + (y-3)^2 + (z-5)^2 = 9$
27. $(x-5)^2 + (y-1)^2 + z^2 = 25$
29. center: $(3,-4,1)$; radius: 4
31. Completing the square:
$$(x-2)^2 + (y-3)^2 + (z-4)^2 = 100$$
center: $(2,3,4)$; radius: 10
33. empty set; point; line 35. empty set; point; circle
37. (a) circle (b) circle 39. cylinder
41. (a) & (b) sinusoidal curve (c) sinusoidal cylinder
43. (a) line (b) line (c) plane
45. (a) $\frac{16\pi}{3}$ (b) $\frac{8\pi}{3}$ (c) $\frac{4\pi}{3}$
47. yz-plane: $(0,2,3)$; xz: $(1,0,3)$; xy: $(1,2,0)$
49. yz-plane: $(0,b,c)$; xz: $(a,0,c)$; xy: $(a,b,0)$
51. yz-plane: $(0,2,4)$ and $(0,4,3)$; xz-plane: $(1,0,4)$ and $(1,0,3)$; xy-plane: $(1,2,0)$ and $(1,4,0)$; yes.
53. Line segment from $(0,0,0)$ to $(0,0,3)$; original triangle; line segment from $(0,0,0)$ to $(4,0,0)$. No.
55. In the yz-plane: triangle or line segment connection $(0,b,c)$, $(0,q,r)$ and $(0,y,z)$. (Other shadows are similar.) No.
57. 0; 12

Section 12.3

5. $\langle 6, -3, 18 \rangle$; $\|\mathbf{u}\| = 7$, $\|\mathbf{v}\| = 11$, $\|\mathbf{w}\| = 3\sqrt{41}$
7. $\langle 14, -3, 32 \rangle$; $\|\mathbf{u}\| = 15$, $\|\mathbf{v}\| = 9$, $\|\mathbf{w}\| = \sqrt{1229}$
9. $21\mathbf{i} + 18\mathbf{j} - 2\mathbf{k}$; $\|\mathbf{u}\| = 9$, $\|\mathbf{v}\| = 13$, $\|\mathbf{w}\| = \sqrt{769}$
11. $\langle 26, 25-2 \rangle$; $\|\mathbf{u}\| = 15$, $\|\mathbf{v}\| = 9$, $\|\mathbf{w}\| = \sqrt{1305}$
13. $\langle -8, -6, 6 \rangle$
15. $\langle -3-e, -9-\pi, -1 \rangle$
17. smallest: $\|\mathbf{c}\| = \sqrt{90}$; largest: $\|\mathbf{d}\| = \sqrt{171}$
19. shadows: $\langle 0,4,3 \rangle$, $\langle 2,0,3 \rangle$, $\langle 5,2,0 \rangle$
21. shadows: $\langle 0,2,3 \rangle$, $\langle 5,0,3 \rangle$, $\langle 2,4,0 \rangle$
23. $\langle 1,0,0 \rangle$, $\langle 0,3,0 \rangle$, $\langle 5,-4,0 \rangle$ are all perpendicular to $\mathbf{a}$, as are all non-zero vectors with z-coordinate equal to 0 (all vectors parallel to the xy-plane).

25. $\langle 0,0,2\rangle$, $\langle -2,1,0\rangle$, $\langle 2,-1,7\rangle$ are all perpendicular to $\mathbf{c}$; infinitely many.

31. $x(t) = 3+4t$, $y(t) = 5-t$, $z(t) = 1$

33. $x(t) = 2+3t$, $y(t) = 3$, $z(t) = 6-5t$

Section 12.4

1. $\mathbf{a}\cdot\mathbf{b} = 3$; $\mathbf{b}\cdot\mathbf{a} = 3$; $\mathbf{a}\cdot\mathbf{a} = 14$; $\mathbf{a}\cdot(\mathbf{b}+\mathbf{a}) = 17$; $(2\mathbf{a}+3\mathbf{b})\cdot(\mathbf{a}-2\mathbf{b}) = -101$

3. $\mathbf{u}\cdot\mathbf{v} = 2$; $\mathbf{u}\cdot\mathbf{u} = 41$; $\mathbf{u}\cdot\mathbf{i} = 6$; $\mathbf{u}\cdot\mathbf{j} = -1$; $\mathbf{u}\cdot\mathbf{k} = 2$; $(\mathbf{v}+\mathbf{k})\cdot\mathbf{u} = 8$

5. $\mathbf{s}\cdot\mathbf{t} = -3$; $\mathbf{t}\cdot\mathbf{u} = -4$; $\mathbf{t}\cdot\mathbf{t} = 35$; $(\mathbf{s}+\mathbf{t})\cdot(\mathbf{s}-\mathbf{t}) = -4$; $(\mathbf{s}\cdot\mathbf{t})\,\mathbf{u} = \langle -3,-9,-6\rangle$

7. angle between $\mathbf{a}$ and $\mathbf{b}$ is:

$$\arccos\left(\frac{\mathbf{a}\cdot\mathbf{b}}{\|\mathbf{a}\|\,\|\mathbf{b}\|}\right) = \arccos\left(\frac{3}{\sqrt{14}\sqrt{21}}\right) \approx 79.9^\circ$$

angle between $\mathbf{a}$ and x-axis is:

$$\arccos\left(\frac{\mathbf{a}\cdot\mathbf{i}}{\|\mathbf{a}\|\,\|\mathbf{i}\|}\right) = \arccos\left(\frac{1}{\sqrt{14}}\right) \approx 74.5^\circ$$

angle between $\mathbf{a}$ and y-axis: $\approx 57.7^\circ$
angle between $\mathbf{a}$ and z-axis: $\approx 36.7^\circ$

9. 86.7°; 20.4°; 99.0°; 71.8°

11. 96.3°; 64.1°; 150.5°; 77.4°

13. 90°; 71.1°; 119.1°; 35.8°

15. 148.7°; 21.8°; 111.8°; 90°

Made in the USA
Monee, IL
03 June 2021